CRUCIFERS
OF
GREAT BRITAIN
AND
IRELAND

B.S.B.I. Handbook No.6

T.C.G. RICH

Illustrated by

C. DALBY, P. DONOVAN, G.M.S. EASY,

T.G. EVANS, H.S. THOMPSON

AND H.A.D. YAGOIN

Botanical Society of the British Isles

London

1991

ISBN 0 901158 20 8

Published by the Botanical Society of the British Isles

c/o The Natural History Museum

Cromwell Road, London SW7 5BD

Printed by The Cambrian Printers, Aberystwyth

ACKNOWLEDGEMENTS

This Cruciferae Handbook is a BSBI team effort. I have lost count of the number of people who have tested keys, collected, showed or cultivated material for me, read manuscripts, translated papers, collated records, pointed out problems, advised, helped, tolerated my enthusiasm and supported me throughout its production - I am grateful to one and all. I would particularly like to thank Eric Clement whose invaluable initial guidance and encouragement was the foundation on which the book was built. Arthur Chater has been a similar source of wisdom, advice and common sense, a veritable oracle. Duggie Kent checked the momenclature, Ian Hedge critically appraised the manuscript, Eimear Nic Lughadha, Gerard Campbell and Alan Newton helped proof read the copy.

I am very grateful to Kery Dalby, Trevor Elkington and Bengt Jonsell for their accounts of *Cochlearia, Erophila* and *Rorippa* repectively. Their knowledge of these critical genera has added considerable depth to my superficial coverage.

I would like to thank the artists, who (with one exception) are amateurs and have provided their artistic services free. Without their co-operation and contributions this book would never have been started, let alone published. I would also like to thank Jane Croft and Chris Preston for their help producing the maps, and Yvonne Bloom for typing and producing the camera-ready copy.

My family and friends have been very supportive and tolerant during the 6 years it has taken me to write the book. Sincere apologies to Sarah.

This book is dedicated to Peggy Stemson, John Clare, Geoff Halliday and Donald Pigott who have taught me so much, and to Doris Ashby, Dave Earl, Len and Pat Livermore and Margret Baecker with whom I have had so much fun and seen so many plants.

T.C.G. Rich December, 1989

ABBREVIATIONS

agg.	-	aggregate
c.	-	circa, about
s.l.	-	sensu lato
s.s.	-	sensu stricto
sect.	-	section
sp.	-	species (singular)
spp.	-	species (plural)
subsp.	-	subspecies
T.S.	-	transverse section
var.	-	variety
±	-	more or less

CONTENTS

INTRODUCTION

This book is an aid to the identification of British and Irish Cruciferae. It covers the 138 species most likely to be found in the field, including all native species, aliens with more than 5 records since 1950, and a few miscellaneous extras as space permits.

Crucifers have a reputation for being difficult to identify for five main reasons:-

i) A large number of species occur, many of which are superficially similar.

ii) Both flowers and fruit are often required for the keys, and specimens are frequently found without one or the other. This remains a problem to a degree here, although the synoptic key should help with non-fruiting material. Fruit set may also be variable.

iii) Some of the characters previously used in keys apply to herbarium material (e.g. "veins on the valves") and are nearly impossible to see on fresh specimens. Some of the characters also required considerable experience to interpret.

iv) There are a few critical genera (e.g. *Lepidium, Erophila, Rorippa, Camelina,* etc) which require careful observation and often some experience for correct identification. The taxonomy of *Cochlearia* is still a matter for debate.

v) Some species are very variable. Some are genetically variable (e.g. *Sinapis arvensis*), and others are environmentally plastic (e.g. *Barbarea verna*).

The taxonomy is primarily based on *Flora Europaea*, modified with recent updates and personal opinion. The sequence of genera and species adopted here is completely artificial; as far as possible the most similar-looking plants, or those likely to be mistaken for one another, have been placed next to each other (the species of each genus are kept together). No evolutionary or taxonomic significance can be drawn from this sequence; for instance, *Crambe* and *Neslia* are placed next to each other but are not closely related. A taxonomic sequence better reflecting assumed relationships is given on page 3.

The descriptions and keys relate specifically to Britain and Ireland; more variation is present in Europe and elsewhere. They have been drawn up principally from material seen in the wild, augmented with herbarium and cultivated material, and occasionally from the literature. Material in the following herbaria has been examined in detail or part: ABRN, BEL, BM, BRIST, CGE, DBN, DGS, E, GLAM, K, LANC, LTR, NMW, OXF, RNG and TCD. Much material from BSBI members has also been seen, the collections of T.B. Ryves being of particular importance.

THE CRUCIFERAE

The Cruciferae (or Brassicaceae) contains about 3500 species in 350 genera. The species form a large, well-defined, natural family, easily distinguished by the 4 petals arranged in a cross (cruciform, hence Cruciferae) and 6 stamens (tetradynamous, i.e. a short outer pair and 2 longer inner pairs), and by the distinctive fruits. They are found virtually throughout the world in habitats ranging from high arctic tundra to deserts and to tropical forests. They are most abundant in the northern hemisphere, with particular centres of diversity around the Mediterranean and in C. and S.W. Asia.

The consistency in basic flower and fruit structure throughout the family suggests strongly that it is of monophyletic origin. It is related to the Capparaceae, some members of which show marked similarities in flower and fruit structure to the Cruciferae, and it is probable they have been derived from a common ancestral stock. Cruciferae and Capparaceae are usually placed in the order Capparales with Resedaceae and Moringaceae.

In Britain and Ireland about 300 species of Cruciferae have been recorded, of which only 50 are native. The characters of the family as it occurs in the British Isles can be described as follows:-

Annual to perennial herbs, rarely woody below, 1-250 cm tall. Leaves alternate, without stipules. Inflorescences usually racemose; bracteoles rarely present. Flowers usually hermaphrodite, actinomorphic, rarely zygomorphic, hypogynous. Sepals 4, free, in 2 decussate pairs. Petals 4, rarely absent, free, alternating with the sepals. Stamens 6 (rarely 4 or fewer), tetradynamous. Ovary of 2 carpels, superior, syncarpous with 2 parietal placentas, usually bilocular, sometimes transversely plurilocular. Style 1 or absent. Stigma entire to bilobed. Fruit usually a specialised capsule opening from below by 2 valves; sometimes indehiscent, breaking into 1-seeded portions or not; rarely transversely articulate with dehiscent and indehiscent segments, sometimes separating at maturity into 1-seeded portions. Seeds in 1 or 2 rows in each loculus.

There are two main tribal classifications, that of Schulz (1936) and its modification by Janchen (1942). Both are more satisfactory for some tribes than others, for instance the Brassiceae forms a fairly natural tribe whilst the Alysseae is highly artificial, but neither arrangement can be considered the final word. These classifications have been reviewed by Hedge (1976).

The tribal classification given below is modified from Janchen (1942), and is the arrangement adopted in *Flora Europaea* (note, the arrangement of species within this book is artificial).

SISYMBRIEAE: *Sisymbrium, Descurainia, Alliaria, Arabidopsis, Myagrum, Isatis, Bunias.*

HESPERIDEAE: Erysimum (including *Cheiranthus*), *Hesperis, Malcolmia, Matthiola.*

ARABIDEAE: Barbarea, Rorippa (including *Nasturtium*), *Armoracia, Cardamine, Arabis* (including *Cardaminopsis*), *Aubrieta.*

ALYSSEAE: Lunaria, Alyssum, Aurinia, Berteroa, Lobularia, Draba, Erophila.

LEPIDIEAE: Cochlearia, Camelina, Neslia, Capsella, Hornungia, Teesdalia, Pachyphragma, Thlaspi, Iberis, Lepidium (including *Cardaria*), *Coronopus, Subularia.*

BRASSICEAE: Conringia, Diplotaxis, Brassica, Sinapis, Eruca, Erucastrum, Coincya (including *Rynchosinapis*), *Hirschfeldia, Carrichtera, Cakile, Rapistrum, Crambe, Raphanus.*

The genera are usually distinct, natural groupings of species, but they are exasperatingly difficult to define (one German taxonomist, Krause, even resorted to putting all the species into one genus, *Crucifera*). A few genera are large containing over 100 species (e.g. *Lepidium, Erysimum*), some are small, and many are monotypic containing only one species. The extent to which the genera are lumped or spilt is very much a matter of personal opinion and alternative views are often equally valid.

The species on the other hand are relatively easy to delimit except in the critical genera. Many species are variable throughout their range, and sometimes this variation is recognised at subspecific rank. Little recognition is given in this Handbook to infraspecific taxa below subspecies. Many of the described varieties and forms are phenotypic variants and cultivation experiments are required to distinguish these from the genotypic variants. There is much scope for further work in this area.

BIOLOGY

Some aspects of the biology deserve a mention. Pollination mechanisms and flower structure are closely related. Many species are insect-pollinated and tend to have large, brightly coloured petals with a long claw, have erect, saccate sepals, and usually secrete nectar. Some of these species are obligate out-breeders (e.g. *Brassica oleracea, Eruca sativa*) and are pollinated chiefly by bees. Others (e.g. *Sinapis arvensis*) are self-compatible and may be visited by a variety of insects. The flowers of *Matthiola longipetala* open at night and are said to be pollinated by moths. Many of these species are also protogynous to promote out-breeding. The self-pollinated species tend to have small flowers with small, white petals (e.g. *Capsella, Arabidopsis*) but many of these will outbreed at low frequency. The flowers of *Subularia* are pollinated in bud (cleistogamous). There is complete intergradation between self- and insect-pollinated species. For further details of pollination, see Proctor and Yeo (1973).

In contrast to most flowering plant families, the Cruciferae generally lack a mycorrhizal association with the roots (Gerdemann 1968), though they are known in at least 8 species (Medve 1983). This is probably related to the weedy nature of the family.

Despite the variation in fruit structure there are few specialised seed dispersal mechanisms and seeds generally fall close to the parent. The fruits of *Cardamine* species dehisce explosively and may scatter seeds a metre or more. The corky fruits of *Cakile, Crambe* and *Raphanus* are dispersed by the tides. Seeds of some species are dispersed by birds (e.g. *Brassica oleracea, Rorippa islandica s.s.*), but the bulk of seeds are small and light and are dispersed naturally by wind and water.

Some species (e.g. *Armoracia, Lepidium draba*) seem to be effectively dispersed as root fragments in soil - the former especially so as it rarely sets seed. Whatever the dispersal mechanism, man is probably one of the most effective agents; numerous species have been distributed both accidentally and deliberately throughout the world in association with his activities.

There have been numerous cytological studies in the Cruciferae. The first major survey (Manton 1932) is still the basic source of reference. The chromosomes are usually small with a few exceptions. Base numbers range continuously from 4 to 13, with about one third of the species based on 8. Aneuploidy and polyploidy are frequent, and supernumerary fragments or accessory chromosomes occur occasionally. Individual counts range from n=4 to n=128 (Al-Shehbaz 1985). Cytological studies have often clarified taxonomic problems (e.g. in *Erophila, Brassica, Cochlearia*). The work of Howard & Manton (1946) on watercress is one of the classic applications of this technique.

The chemical constituents of many species have been analysed mainly because of their economic importance (Vaughan *et al.* 1976). The most characteristic chemicals are the glycosinolates which are probably present in all species and are responsible for the smell and taste of crucifers. Seed fatty acids have also been extensively investigated as they are of economic importance. Some species of *Alyssum, Cochlearia* and *Thlaspi* accumulate heavy metals.

ECOLOGY

If there is one ecological generalisation about crucifers, it is that most are weeds. These are usually annual opportunists with a broad, unspecialised ecology. They are generally intolerant of competition, ephemeral and rarely tolerated by man. They may be introduced by a variety of means; many are contaminants of grain and other crop imports, others occur in wool shoddy (Lousley 1961), bird seed (Hanson & Mason 1985), docks (e.g. Sandwith 1933), railways (e.g. Messenger 1968), brewery waste, ballast, etc. (Lousley

1953). Some are relicts of cultivation or escapes from gardens. Many are rare, unpredictable in occurrence, great fun to seek out and exciting to identify.

There are, however, native species with very much more precise ecological requirements. *Crambe maritima* and *Cakile* are very characteristic of shingle and sand, *Cochlearia* spp. and *Lepidium latifolium* of saltmarshes. *Draba* and *Arabis* spp. occur at high altitudes on mountains, and *Subularia* is a specialised aquatic. *Teesdalia* is typical of sandy heaths and *Thlaspi caerulescens* of heavy metal sites. There are many fascinating problems to answer, such as why is *Thlaspi perfoliatum* so characteristic of oolitic limestone and *Iberis amara* of chalk.

There are also many anomalies in distribution - why is *Arabis alpina* restricted to Skye, and is *Cardamine bulbifera* a recent arrival or a relict? Pigott & Walters (1954) give fascinating discussion of discontinuous distributions. The maps here (pages 282-312) are largely of introduced species whose occurrence is related to the activities of man.

HYBRIDS

Interspecific hybrids are generally rare in Britain and Ireland, though they are frequent in *Cochlearia* and *Rorippa* and occasional in *Cardamine* and *Raphanus*. Intergeneric hybrids have been synthesised but not reported from the wild. Details of most of the hybrids are given in Stace (1975).

Recognition of hybrids in the field is not easy. It is important to know the species and their variation before determining hybrids (especially in *Cochlearia* and *Rorippa*), and local parent material should always, if possible, be collected with putative hybrids to account for local variability. Hybrids must be assessed on a combination of characters, the sepal and petal sizes often being very useful in this respect. Introgression can produce a bewildering array of intermediates in some groups (e.g. *Rorippa amphibia* x *R. sylvestris*) and in these cases it is most practical to define parent species fairly narrowly and the hybrids more broadly.

Although many hybrids have reduced fertility, sterility or poor fruit set may also result from self-incompatibility and lack of cross pollination, or environmental factors such as flooding and cold weather. Some individuals, however, may be genetically completely sterile and will not set fruit even if cross-pollinated.

Hybridization has probably been of importance in the evolution of species in a number of genera where hybridization followed by doubling of chromosome number has resulted in the origin of new species (e.g. *Diplotaxis muralis*, *Cardamine flexuosa*). In *Brassica*, for instance, *B. napus* is believed to be an allopolyploid derived from *B. oleracea* x *B. rapa*, *B. juncea* from *B. nigra* x *B. rapa*, and *B. carinata* from *B. nigra* x *B. oleracea*.

CRUCIFERS AND MAN

Crucifers have been widely utilized for many centuries. They are probably one of the 10 economically most important families to man.

Crucifers are extensively used for food. They may be eaten as vegetables (e.g. Cabbages, Swedes), salads (e.g. Watercress, "Mustard and Cress") or as condiments (e.g. Horseradish, Mustard). The range and variety bred into some species is staggering - *Brassica oleracea* derivatives include Cabbage, Brussels Sprouts, Broccoli, Kohlrabi, Purple-sprouting, Kale, Savoy Cabbage, etc. Further details of crop plants can be found in Simmonds (1976) and Tsunoda *et al.* (1980).

Crucifers may be grown for oil-seed (e.g. Oil-seed rape) and forage and animal feed (e.g. Turnips). However, due to the glycosinolate content, crucifers and their products often have to be used in moderation and some species may be toxic if consumed in quantity.

Some species are grown in gardens for ornament. Wallflowers, Stocks, Alyssums, Candytufts, Aubretias and Arabis are particularly conspicuous in early summer. Woad is grown for a different type of ornament - the fermented leaves produce the blue dye of great historic importance.

Other species in gardens and fields are not grown for food or flowers - these are the weeds. There can be few more familiar garden weeds than Shepherd's Purse or Bittercress, and Charlock and Hoary Pepperwort are noxious weeds of cultivated land. Crucifers are notorious as serious weeds throughout the world.

FURTHER READING

There is an enormous volume of literature published on the Cruciferae. The following are recommended for starters; Al-Shehbaz (1985) and subsequent papers, Tsunoda *et al.* (1980) and Vaughan *et al.* (1976). For those preferring a more light-hearted approach, Kington (1983) is highly recommended.

THE CHARACTERS

Consistent assessment of the characters is crucial for accurate identification, and this chapter sets out how they are used in the keys and descriptions. The characters are given here and in the descriptions in the following order: Plant, roots, stems, leaves, inflorescences, flowers, pedicels, fruits, seeds, chromosome number, flowering time. Some terms are explained in the glossary.

HABIT

Crucifers may be annual, biennial or perennial, but this is not easy to use as a discriminatory character in the field even with experience.

Annuals tend to be small and easily uprooted with poorly developed roots (e.g. *Cochlearia danica*) but equally may be tall and robust (e.g. *Brassica nigra*). Biennials tend to be more robust with a stouter rootstock, and a proportion of vegetative, first year plants is often present in the population. Perennials usually have well-developed rootstocks (e.g. *Armoracia rusticana*), non-flowering rosettes (e.g. *Arabis caucasica*), spreading rhizomes (e.g. *Rorippa sylvestris*) or are woody at the base (e.g. *Erysimum cheiri*). Sometimes the remains of the previous year's fruits can be found to show that the plant has flowered more than once. However, these indicators may all break down in practice and the only true test is to cultivate or mark individual plants. Even then habit may vary depending on the environment and some species (e.g. *Cardamine flexuosa*) may be annual, biennial or perennial. Note that overwintering plants are described as annual if they complete their life cycle within one year of germination.

HEIGHT

Typical height measurements are given after the habit. Heights are approximate and refer to plants in flower and/or fruit. Height may be greatly influenced by environmental conditions - plants of *Sinapis arvensis* in poor, dry soils may be only 5 cm tall (the same size as *Erophila*) or over 2 m tall in rich, damp soils (nearly as tall as *Brassica nigra*).

HAIRS

Hairs, if present, are of considerable value for identification. The general descriptions of pubescence relate here to the stems and leaves only. Where hairs on the flowers or fruits are of diagnostic value they are noted separately. Five basic hair types are used here for identification (Fig. 1); in most cases it should be possible to see their structure with a x10 lens, though a x20 lens is better. Plants may have one or more types of hair, or lack hairs altogether. Hairs should be looked at on the lower stem and leaves, in silhouette and from above. The presence of at least some forked or stellate hairs is the most significant feature.

1) Simple. These hairs are unbranched, long or short, coarse or soft, appressed or spreading, bulbous-based or linear. They are the commonest hair type (Fig. 1 a-d).
2) Forked. These hairs are branched once near their apex, and stand out clearly in silhouette (Fig. 1 e,f).
3) Stellate (including stellate-dendroid). These hairs have 3 or more rays attached at a central point, and the rays may be spreading or appressed. They are often best seen from above, although when sparse can be seen in silhouette. In some species (e.g. *Matthiola incana*), they form a dense, soft, white mat, and in others (e.g. *Erysimum cheiranthoides*) a rough, sparse cover (Fig. 1 g-k).
4) Medifixed. These are appressed hairs which are attached at their middle - but even with a good lens they may appear simple. Good examples can be seen (or not seen as the case may be) in *Lobularia* and *Erysimum cheiri*. These hairs can be misleading and consequently plants with only this hair type are keyed out twice (Fig. 1 l,m). There are two simple ways to identify medifixed hairs (A.O. Chater, pers. comm.):
    i) Tear the leaf in two; if hairs protrude from both torn ends of the leaf they are more or less certain to be medifixed (appressed simple hairs would project from only one end).
    ii) Under a microscope, push one end of a hair with a needle; if the other end moves it is medifixed.
5) Glandular. These are simple hairs with a little, swollen head (gland) which, when fresh, glisten if held up to the light. They are best examined in silhouette, especially on pedicels (e.g. *Hesperis*), and may, when abundant, make the plant sticky. In dried material they are often difficult to see (Fig. 1 n,o).

COLOUR OF PLANT

A general indication of plant colour is given, but this is quite variable and difficult to describe clearly. Colour may be a useful "jizz" character, for instance, in picking out *Barbarea stricta* from *B. vulgaris* prior to closer investigation.

Plants in woodland, or in shady or damp places are often lighter green and thinner in texture than plants from exposed, dry sites which tend to be darker green and purplish. A glaucous, waxy bloom is present on some species (e.g. *Brassica oleracea*), and in others a dense covering of hairs may turn the plants whitish.

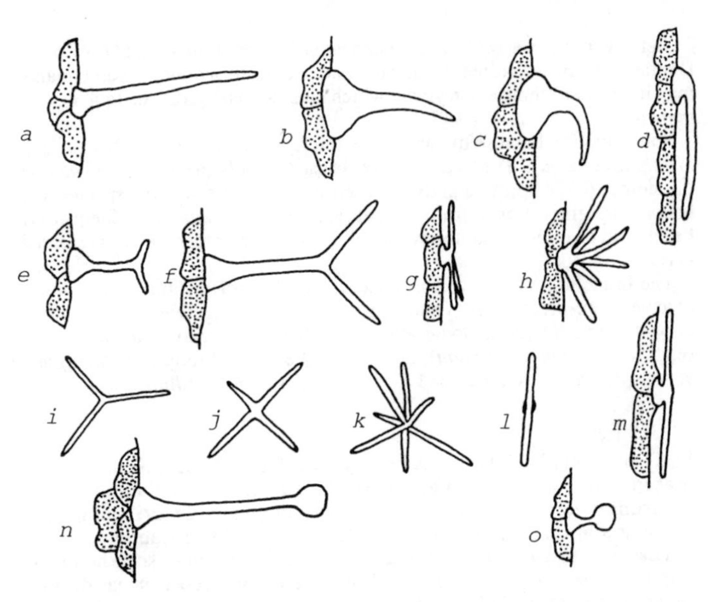

Figure 1. Schematic diagram of hair types. a-d Simple hairs (side view). e,f Forked hairs (side view). g,h Stellate hairs (side view). i-k Stellate hairs (view from above). l Medifixed hair (view from above). m Medifixed hair (side view). n,o Glandular hair (side view). Not to scale.

ROOTS

The underground roots and rhizomes are generally not described here as they usually require destructive examination. In some cases they are useful - for instance, in swede and radish which have swollen storage tap-roots, or in *Rorippa* where *R. sylvestris* and *R. austriaca* have spreading rhizomes.

Some species characteristically root at the nodes (e.g. *Cardamine amara, Rorippa amphibia*) and some other species may also root at the nodes if the stem comes into contact with the soil.

STEMS

Undamaged central or main stems should be examined. Side shoots are often more variable.

Stems vary from erect (e.g. *Brassica napus*) to prostrate (e.g. *Coronopus*). Damaged plants are often decumbent and should be avoided. Some plants may have erect stems when young which later become prostrate (e.g. *Draba muralis*).

Some species have stems which are thickened and "woody" below (e.g. *Matthiola incana*). The term "woody" is used loosely in Cruciferae because the stems are not lignified in its strict sense. The stems of many species will become toughened once fruits are set (e.g. *Matthiola sinuata*). Stems may be smooth or ridged, and hollow or solid; these characters are not presented here.

The branching of the stem is a useful "jizz" character, but branches can change from erect when young to widely spreading when mature (e.g. *Sisymbrium officinale*). Some species are characteristically branched above (e.g. *Sisymbrium altissimum*), others branched below (e.g. *Lobularia*) and some species have unbranched stems (e.g. *Cardamine bulbifera*).

LEAVES

Leaf characters have been extensively used for identification as they are present on both flowering and fruiting plants. As there is much environmental and genetic variation, they are described briefly concentrating on the most important characters for identification.

Leaves from damaged plants should be avoided, but small-scale herbivory generally does not greatly affect leaf characters. Small rosettes may develop in the upper leaf axils late in the season - these often resemble the lower leaves and should be ignored as they are atypical. First-year rosette leaves or biennials may be different from second-year rosettes (e.g. *Brassica*; see Wiggington & Graham 1981), and both sets of leaves are described although only the second-year leaves will be present at flowering.

The positioning of the leaves is described as follows (Fig. 2):

1)Rosette or basal. Arising from a central point at ground level (not on a stem). This includes leaves arising directly from a rhizome (e.g. in *Cardamine*).

2)Lower. Leaves arising on the stem from the bottom to half way up the leafy portion of the stem.

3)Middle. Leaves arising on the middle third of the leafy part of the stem. These are not described in all cases, and note that the definition overlaps with lower and upper leaves.

4)Upper. Leaves arising in the upper half of the leafy part of the stem (excluding bracteoles, but including inflorescence bracts).

5)Uppermost. The top 2-4 leaves on a stem (excluding bracteoles but including inflorescence bracts).

6)Bracteoles. To avert potential confusion, the word bract is avoided here as it is broadly defined as a leaf-like structure which subtends an inflorescence or a flower. Most crucifers have stem leaves which develop inflorescences in their axils - these "inflorescence bracts" are described as leaves for the purpose of identification.

Leaf length is measured from where the leaf joins the stem or the centre of the basal rosette to the leaf apex (i.e. including the petiole). Sizes given are approximate due to environmental plasticity of leaf development. Relative petiole and blade (or lamina) lengths are sometimes also given.

Petioles may be partly winged, and then it is difficult to decide where the petiole ends and the lamina begins. Leaves lacking a petiole are sessile, and those with a very short or obscure petiole are sub-sessile. If the leaf has a petiole, then the auricles should not clasp the stem.

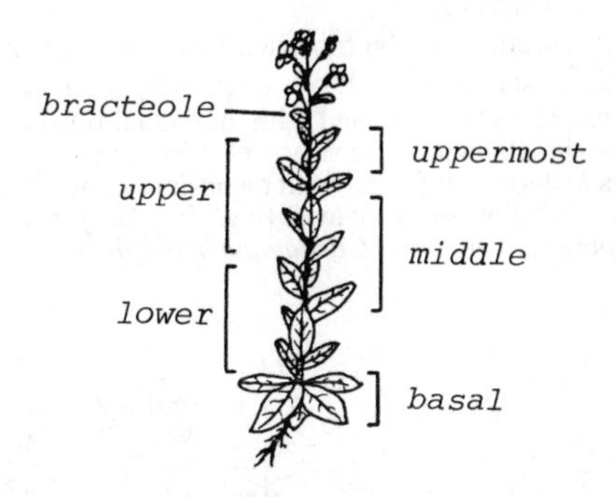

Figure 2. Diagram showing terms used to described positioning of leaves on stems.

The general leaf shape is described as an outline if the leaves are divided. Leaves differ in their degree of division, varying from simple (undivided) to lobed, pinnatifid, pinnatisect, pinnate, bipinnate, tripinnate or even quadripinnate (Fig. 3; the terms are defined in the glossary). (The term "lyrate-pinnatifid" has been avoided as it has been too variably applied). If the leaves are deeply divided into distinct lobes, the terminal lobe is first described and then the lateral lobes. The number of lateral lobes is given as "pairs" - this is more to imply lobes are on opposite sides of the petiole rather than that they are opposite each other (they rarely are opposite at the base of the leaf). Note that the leaf margins are described separately.

The change in division of leaves from the bottom to the top of the stem can be worth noting. Some species have the lower leaves more divided than the upper leaves (e.g. *Brassica rapa*) or less divided (e.g. *Erucastrum gallicum*). In others, all the leaves may be similar (e.g. *Erysimum*). Sessile leaves sometimes have auricles at their base (small ear-like projections of leaf tissue) which clasp the stem - the leaves may in fact completely encircle the stem (perfoliate). These are very useful, easily seen features for identification (Fig. 3).

The leaf base and apex vary between species and are described unless they are implied by the leaf shape (i.e. an oblanceolate leaf has a cuneate base). "Cordate" is used here only to describe the base of the leaf or fruit and not the general shape.

The leaf margin may be entire, sinuate or toothed (anything more than toothing is described under leaf division, cf. Fig. 3). The type of toothing can be very distinctive but is variable and has not been described in detail here. The margins of the leaf may be undulate, up and down in the plane of the leaf (e.g. *Brassica oleracea*).

Leaf shape and dissection can be modified by environmental conditions such as shade and fluctuating water levels (especially in *Rorippa*).

A few species have bracts associated with individual flowers (in addition to the inflorescence bracts) which are very useful for identification; these are termed bracteoles to avoid confusion with the inflorescence bracts.

Some species only have rosette leaves (e.g. *Erophila*), some only stem leaves (e.g. *Sinapis alba*) and one, *Cardamine heptaphylla*, only has upper stem leaves at maturity.

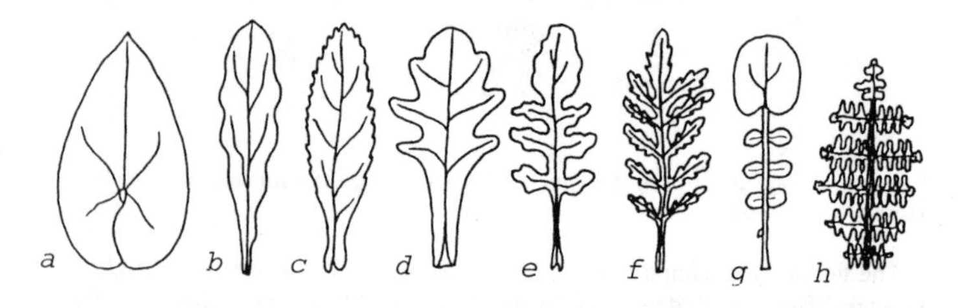

Figure 3. Schematic diagram of leaf dissection. a Leaf simple, sessile and clasping stem; margin entire. b Leaf simple, broadly petiolate; margin sinuate. c Leaf simple, sessile, with auricles; margins toothed. d Leaf lobed, sessile; margin entire. e Leaf pinnatifid, sessile, with inconspicuous auricles; margins sinuate. f Leaf pinnatisect, petiolate; margins irregularly toothed. g Leaf pinnate, petiolate; margins entire. h Leaf bipinnate, petiolate, but note the bottom left lobes may appear to clasp the stem; the margins are difficult to describe once the lobes become subdivided. Not to scale.

INFLORESCENCES

There are two main types of inflorescence. The commonest type is a raceme which elongates as the flowers and/or fruit mature. *Crambe* and *Iberis*, however, have corymbose inflorescences which are densely contracted with all the flowers and fruit at about the same level. In *Coronopus*, an inflorescence may be produced at ground level in the axils of the basal leaves, and a flowering stem may not be produced at all.

In addition to the main terminal inflorescence, most species have lateral inflorescences which arise in the axils of stem leaves. In *Coronopus* and *Carrichtera*, however, each lateral inflorescence arises on the opposite side of the stem to a leaf. This character is best looked at on the 3rd or 4th lateral inflorescence away from the terminal inflorescence.

Generally, the individual flowers of an inflorescence do not have bracteoles. In some species, bracteoles may be present on the lowest 1-3 flowers, and in a very few species (e.g. *Erucastrum gallicum*) the whole inflorescence (or at least the lower third) may be bracteolate. Lateral inflorescences may lack bracteoles even if the main inflorescence has them.

The relative positions of the buds, flowers and/or fruit is a useful character in some groups (e.g. *Brassica*) but requires more care with examination than is at first apparent. Before assessing the relative positions of the buds, flowers and fruits it is crucial to check that the inflorescence is developing normally and that a range of bud sizes is present. If the new apical buds abort (often due to adverse weather conditions), those buds past a certain stage of development can go on to flower, leaving a tight cluster of small, dead, brownish buds at the apex of the raceme. Thus an inflorescence which usually has buds overtopping flowers or fruits may as a result appear to have flowers or fruits overtopping buds (see illustration in Rich 1987a). 70% of material of *Brassica napus* in one herbarium showed this character which might lead to mis-identification as *B. rapa*. Sometimes it is difficult to judge whether the buds overtop the flowers or *vice versa*, and the character should then not be used.

Flowers may be arranged in dense or lax clusters at the ends of the racemes. The rate at which racemes elongate varies enormously, giving a characteristic look to the inflorescence of many species.

FLOWERS

The flowers are crucial for identification and provide many useful characters. Ideally, they should be selected from the main inflorescence as here they are most consistent, but side shoots can also be examined. Flowers damaged by pollen beetles or other insects, or from damaged plants or shoots (especially secondary shoots), should be avoided. Flowers produced late in the season after the plant has fruited are also atypical.

Iberis and *Teesdalia* have asymmetrical flowers in which the outer two petals and sepals are larger than the inner two. Inner and outer petals are described separately for these species.

Mutations are occasionally found in wild plants (e.g. "double" flowers in *Cardamine pratensis*) or horticultural material (e.g. Stocks or Wallflowers). Double-width petals are also a feature of cultivated plants.

SEPALS

Sepals reach a maximum size at about anthesis and can be measured on open flowers of all ages (avoid buds). They shrink only slightly when dried (in contrast to petals) and the measurements given here can be applied to both fresh and herbarium material. Length is maximum length and includes the saccate pouch or the awn if they are present.

Sepals are quite variable (Fig. 4) and have been under-utilized here for identification. They can vary from linear to broadly ovate or triangular. In *Sinapis*, they are inrolled and appear linear (Fig. 4d). Two of the sepals may be larger and more developed than the other two (Fig. 4a,b) in insect-pollinated species.

The base of the sepal may be saccate (or pouched) (Fig. 4a) in insect-pollinated species - this is where nectar collects. The apex may be slightly hooded or even awned (Fig. 4e) in some species; this is most clearly seen in bud. The margins of the sepals may be membranous or whitish. The sepals may be the same colour as the petals or different, but can be quite variable (e.g. the tips are often purple or reddish).

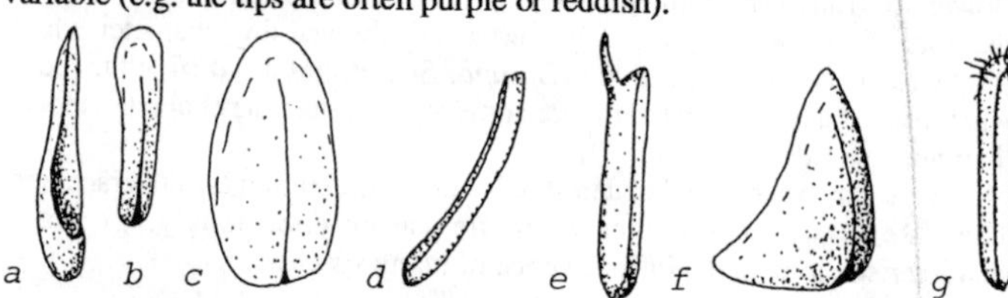

Figure 4. Variation in sepals. a,b Large, saccate, outer sepal and smaller, inner sepal of a typical insect-pollinated flower (*Lunaria annua*). c Ovate, with broad margin (*Armoracia rusticana*). d Inrolled, linear (*Sinapis arvensis*). e Strongly awned (*Sisymbrium strictissimum*). f Triangular (*Teesdalia nudicaulis*). g Hairy (*Raphanus maritimus*). Not to scale.

The angle at which the sepals are held relative to the ovary is a useful character (Fig. 5) but requires experience to use. It is best seen on fresh flowers and is often obscured in pressed material. In *Lepidium draba* and some other species, the angle can change from erect to patent and back again as the flowers age. The variation in angle has been examined carefully here but is only used for identification in a few instances. The angle of sepals is used as a major dichotomy early in the Cruciferae key in *Flora Europaea* but is very difficult to use and prone to error.

Occasionally, the sepals are persistent in fruit but they are usually shed with the petals or sometimes before (e.g. *Alliaria*).

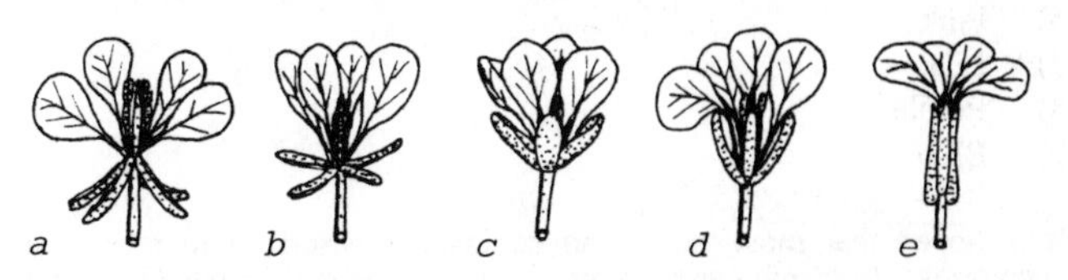

Figure 5. Angles at which the sepals are held. a Reflexed (*Sinapis arvensis*). b Patent (*Sinapis arvensis*). c Inclined (*Brassica nigra*). d Ascending (*Brassica juncea*). e Erect (*Raphanus maritimus*). Not to scale.

PETALS

Petals vary enormously between species, principally related to the different pollination mechanisms. Occasionally petals are reduced and rudimentary (less than 1 mm long) or absent altogether.

Petals tend to increase in size from first opening of the flowers to their maturity, and the measurements given cover this variation (except in *Rorippa* where mature flowers only must be examined). It is best to select at least 5 petals, differing in age, and calculate the average length and width.

Petal measurements refer to fresh material only. Dried petals shrink inconsistently and are unreliable. Petals are best measured by dissecting a flower and sticking the petals (and sepals!) onto Sellotape, and then sticking the Sellotape onto paper. They can then be measured with ease to the nearest 0.5 mm. Length is maximum length and includes the claw, width is maximum width and is measured at the widest point.

Flower diameter has been avoided (with the exception of *Cochlearia*) because it is ambiguous to measure. It is partly a function of petal size, and this latter character is thus preferred because it can be measured precisely. Shape is described for the petal as a whole when the claw is indistinct, or the limb and claw are described separately (Fig. 6). The limb may be ovate to oblong or obovate, elliptic or rarely bifid (e.g. *Erophila*). The apex may be rounded (e.g. *Sinapis*), obtuse, truncate or emarginate (e.g. *Aurinia*).

The petal colour is a very important character. Due to variation in flower colour, and widespread differences in definitions and interpretation, colour is here divided only into the following general categories:

1) White
2) Cream (definitely not white or yellow)
3) Yellow { pale yellow (including very pale yellow)
{ yellow
{ deep yellow

4) Orange
5) Red
6) Pink
7) Lilac
8) Purple
9) Blue

It is hoped that most petals can be instantly ascribed to one of these categories. In a few cases where there may be difficulties (e.g. *Arabis*), plants have been keyed out twice to cover difficulty in interpretation. Petal colour fades in herbarium material and cannot be relied on without colour notes. In *Berteroa*, the white petals may turn yellowish when dried. In *Alyssum*, the petals may fade from yellow to white whilst still on the plant. Petals may be multicoloured (e.g. *Erysimum*), in which case the general background impression should be taken, or variegated (e.g. *Matthiola*) in which case the colour, and not the white, should be used for the purposes of keying. Venation may be quite marked on the petals (e.g. *Eruca, Raphanus*) but this can be ignored when determining the colour.

The claw may be indistinct or absent (Fig. 6 c,d,g,h), or distinctly demarcated from the limb (Fig. 6 a,b,e,f). It may be short or about as long as the limb. It is usually narrowly triangular but can be linear, and in some *Arabis* species may have 2 small teeth (Fig. 6 h). The colour is often paler or greener than that of the limb.

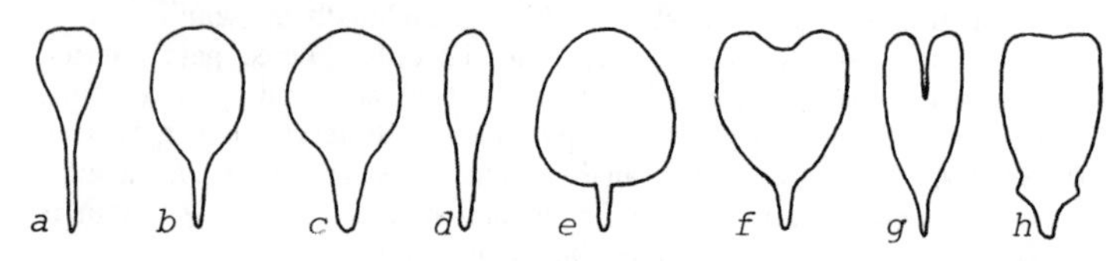

Figure 6. Variation in petal shape. a Limb obovate, truncate; claw distinct, long (*Raphanus*). b Limb obovate; claw distinct, short (*Sinapis*). c Limb elliptic; claw indistinct (*Brassica napus*). d Petal narrowly oblanceolate; claw indistinct (*Alyssum*). e Limb ovate; claw distinct, short (*Lobularia*). f Limb obovate, emarginate; claw distinct, short (*Aurinia*). g Petal bifid, unclawed (*Berteroa*). h Petal obovate, more or less truncate; claw indistinct with small teeth at each side (*Arabis petraea*). Not to scale.

PETAL TO SEPAL LENGTH RATIO

The size of petals relative to sepals is an instantly visible character. The ratio, however, increases as the petals mature, and is quite variable in any case. The most useful distinction is between those species with petals about as long as sepals and those with petals (1.4-)1.5 or more times as long as sepals. Many of the ratios given by Clapham *et al.* (1952-1987) are over-generalizations and should be treated with caution. The individual measurements of sepals and petals are more reliable and often diagnostic.

STAMENS

There are usually 6 stamens but this may be reduced to 4 or 2 in some of the self-pollinated species. The two outer stamens may be quite reduced, and flowers must therefore be dissected carefully.

The filaments are usually simple, but those of the inner stamens may be toothed or with a distinct appendage (e.g. *Aubrieta*) or have a small scale at the base (e.g. *Teesdalia*). In some species (e.g. *Coronopus*) some stamens may be sterile.

Anther colour may be a useful character but must be examined before the anthers open as the pollen is usually yellow, irrespective of the anther colour.

OVARY

Ovaries are small and difficult to work with, and are thus generally best avoided for identification purposes. However, shape, pubescence, and number of ovules are useful identification features in some species. In non-fruiting material, it may be possible to establish fruit shape and structure from the ovary, but this requires experience.

Ovules can be counted in one of three ways though all require practice and patience. The advantages of counting ovules rather than seeds is that the number of ovules is usually genetically determined whilst the number of ovules which develop into seeds depends on pollination which may be variable. In all cases at least 5 ovaries should be examined. Ovules are most easily observed in ovaries from mature flowers. In young flowers, the ovules are too under-developed. In older flowers, once the ovaries begin to develop into fruits, the ovary wall thickens and the number of ovules (rather than developing seeds) is difficult to discern. In bruised or dried ovaries, the bottom of the ovary (which appears as a line) can be mistaken for an ovule.

i) Slice the ovary longitudinally with a razor and dissect out the ovaries. This is an accurate but fiddly method.

ii) Stick a few ovaries onto Sellotape, then stick the Sellotape onto paper. Leave then to dry out slowly (it may take a week), and the ovules will stand out as dark green blobs.

iii) Gently bruise the ovary between your fingernails and the ovules bruise the ovary wall and appear a darker green. Start bruising gently until the ovules can be seen - too much pressure and they will disintegrate. They can then be counted with a x10 lens.

STIGMA AND STYLE

All the stylar characters used relate to fruits (see below).

The main stigma character is the shape (Fig. 7), and this is most useful for species which have a bilobed stigma (e.g. *Erysimum cheiri*). The lobes may be appressed and difficult to discern in the flower (Fig. 7f) but can usually be seen more clearly in fruits.

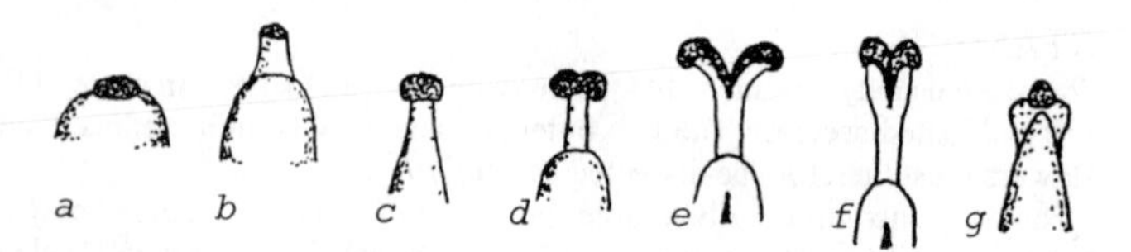

Figure 7. Types of stigma. a Sessile (*Crambe maritima*). b Entire (*Cardamine hirsuta*). c Capitate, entire (*Barbarea vulgaris*). d Capitate, emarginate (*Diplotaxis tenuifolia*). e Bilobed (*Erysimum cheiri*). f Bilobed, but lobes erect and stigma thus appearing emarginate (*Erysimum cheiri*). g Stigma with 2 lateral processes (*Matthiola incana*). Not to scale.

PEDICELS

Pedicel characters refer to pedicels of ripe fruits unless otherwise stated. Pedicel length is measured from the junction with the stem to the receptacle. Width measurements are only given in *Sisymbrium*, otherwise these are referred to using relative terms such as 'stout' or 'slender'.

The angle at which pedicels are held relative to the stem is another useful character. Note the angle of the pedicel is given irrespective of how the fruits are orientated.

FRUITS

Fruits come in all shapes and sizes (Fig. 8). This variation is one of the most useful features for identification of crucifers, and many genera and species have instantly diagnostic fruits. Fruit characters are extensively used in the keys and it is important to understand the basic structure and its variations. Fruit should be selected from the lower half of the main inflorescence as these are the most typical. Fruits towards the top of the inflorescence are often smaller and poorly developed. All the characters described here refer to fresh, mature fruit unless otherwise stated.

Ideally, the fruit should have ripe (usually brown, not green) seeds but in practice this is rarely possible early in the season. Fruits with green seeds can be examined provided the seeds look fully formed (i.e. all the same shape and size and hard, not soft). Fruits will also generally have reached maturity when most of the lower fruits have reached the same shape and size. Note that fruits will continue to fatten after they have stopped elongating. If ripe fruits are not available, pick the largest available and use the keys and descriptions with caution.

Fruit set can be variable due to climate, soils, lack of cross-pollination, insect or physical damage, hybrid origin or inherent genetic disorders. Atypical fruits are often produced late in the season following a second

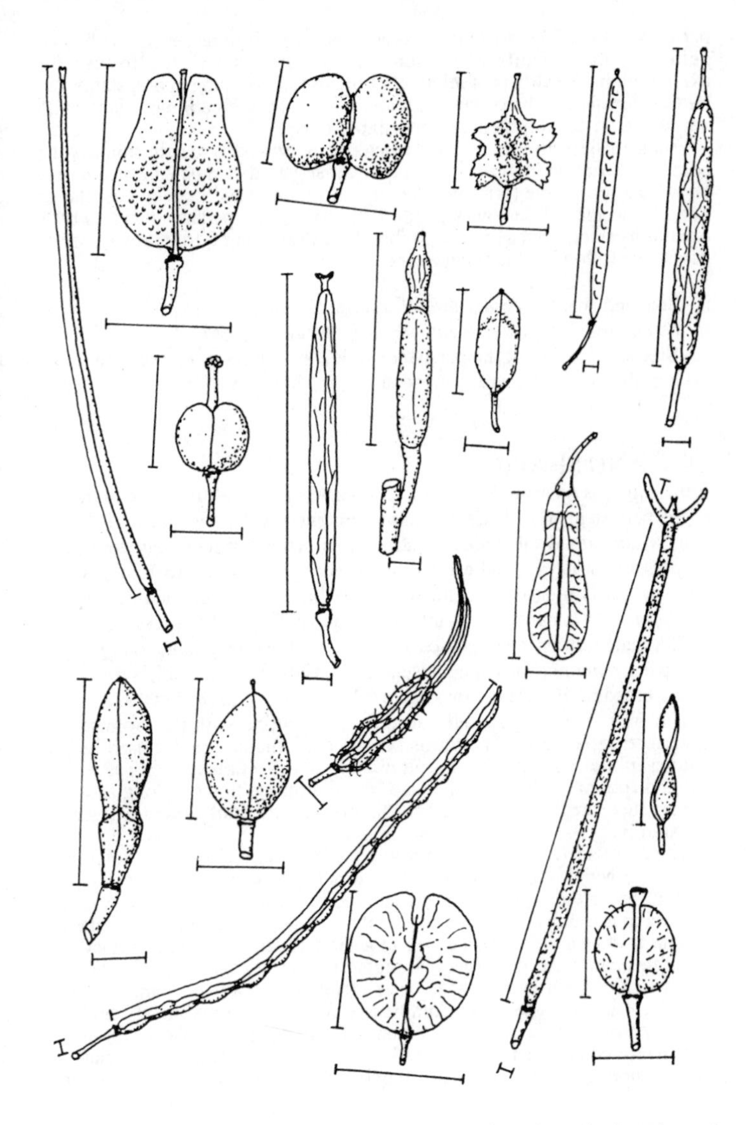

Figure 8. Variation in crucifer fruits. The bars show how the lengths and widths are measured.

period of flowering, so if there is a mixture of inflorescences with both dehisced, dried fruits and young, fresh ones, select the former for examination. Sterility is of relatively little use in assessing hybrid status, but partial sterility can be of much greater value. Identification of plants with variable fruit development requires experience.

Fruits are described here mainly in terms of two parts to help simplify the complex variation in structure. The lower segment is situated immediately above the receptacle (ignoring the stipe if present) and may be valvular or not. At the apex of the lower segment there may be a persistent style, a beak or an upper, terminal segment. When first examining a fruit, always try to assess the following four characters:-

i) Size and shape (average size of at least 5 fruits).
ii) Structure of the lower segment - does it have valves?
iii) Size and fertility of the persistent style, beak or terminal segment.
iv) Number and arrangement of seeds.

SIZE AND SHAPE

Fruit length is measured from the receptacle or base of the valve to the top of the persistent style, beak, terminal segment or valve, whichever is longer (Fig. 8) and always includes the stipe. Fruit length is relatively unaffected by pressing or drying and can be measured on fresh or herbarium material. Fruit width is measured at the widest point (except for identification of *Rorippa* sect. *Nasturtium*). Fruit width may be distorted by pressing and can shrink markedly on drying (Rich 1989). Fruit breadth (or thickness) is only given in *Camelina*, but in angustiseptate fruits the septum width is equivalent to the breadth. Breadth is also distorted by pressing or drying.

The fruit shape is generally described in outline with the cross-section described separately. The terms siliqua and silicula are not used here (a siliqua (plural = siliquae) is a fruit more than three times as long as wide, a silicula (plural = siliculae) is a fruit less than three times as long as wide). Where there are two distinct parts or segments to the fruit, these are described separately.

The cross-section of the fruit provides a few useful characters, notably the degree to which the fruits are flattened and the positioning of the septum within the valvular portion (if present) of the fruit (Fig. 9). The degree of flattening is usually described as ± terete, 4-angled, compressed or flattened. Terete fruits can be rolled between the fingers with ease and distinct sides or angles cannot be discerned. 4-angled fruits are approximately square in cross-section with 4 equal sides. Fresh fruits which are terete may shrink to become 4-angled when dry. Compressed fruits are approximately elliptic or oval in cross-section and have 2 distinctly flatter sides - these can usually also be rolled between the fingers. Flattened fruits are usually very flattened and cannot be rolled between the fingers. There is a complete gradation from terete to compressed to flattened and the terms applied are relative. Other types of cross-section are occasionally found and described separately. Cross-sections are best examined by sectioning in the middle of the fruit with a razor blade.

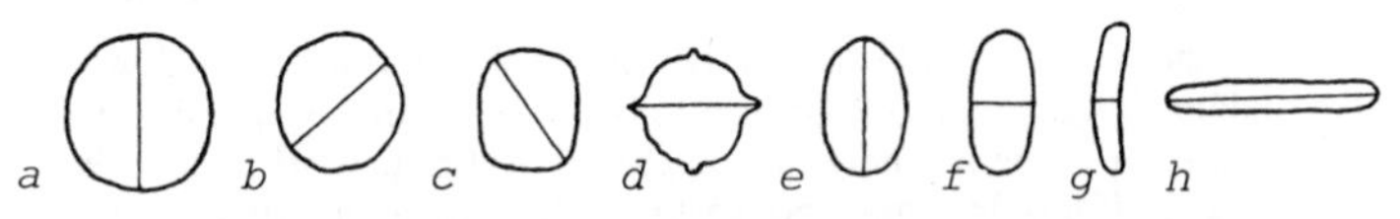

Figure 9. Cross-section of fruits. a, b ± Terete. c, d 4-angled. e Compressed, latiseptate. f Compressed, angustiseptate. g Flattened, angustiseptate. h Flattened, latiseptate. Not to scale.

The positioning of the septum relative to the cross-section is a very useful character because it can even be determined on immature fruit. The septum is a membranous partition across the valvular part of the fruit which separates the two loculi (cavities the seeds develop in). The classic example of a septum is in Honesty (*Lunaria annua*) where the septa are used in dried flower arrangements. When the septum is across the broadest width of the cross-section, the fruit is said to be latiseptate (*lati-* = broad). When the septum is across the narrowest width it is angustiseptate (*angusti-* = narrow). Terete fruits are included with latiseptate fruits but could be described as mediseptate. Measurements of septum widths, measured at the widest point, are given for angustiseptate fruits, and can be assumed to be equivalent to fruit widths in the case of latiseptate or mediseptate fruits. Cross-section of beaks or terminal segments are described separately.

A stipe is a distinct stalk (Fig. 10) between the receptacle and the bottom of the valve, a useful characteristic of a few species (e.g. *Brassica elongata*). Many species have a small (less than 0.5 mm) stalk. The stipe can be seen in immature fruit.

The angle the fruits are held at relative to the stem is often distinctive, and is given irrespective of the angle of the pedicels.

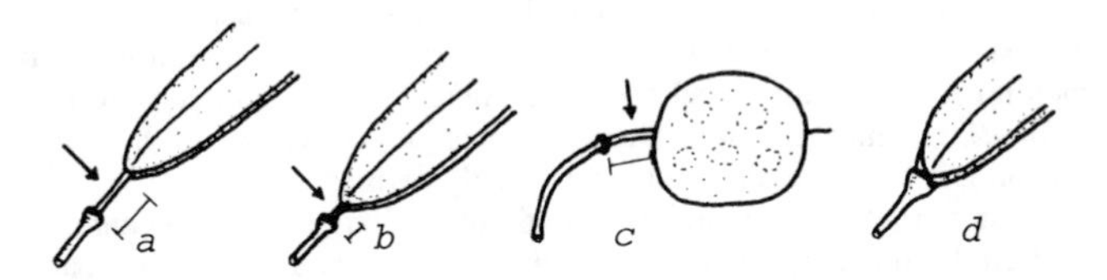

Figure 10. Location and measurement of stipe (a distinct stalk between the receptacle and base of the valve). a-c Stipe distinct. d Stipe absent. Not to scale.

LOWER SEGMENT

The most important feature to look for on the lower segment is the presence or absence of valves (Fig. 11). A valve is a (usually) deciduous covering over the seed or seeds. In general, the riper the fruit the easier it will be to examine valves.

One of the best ways to look for valves is to hold the fruit upside down at the middle, and then pull the pedicel to one side. If in the right direction (perpendicular to the septum) the valves will peel or pop off at their base (Fig. 11a,b). If this does not work first time, try pulling the pedicel at right angles to the original direction, or round and round. This method works best with the ripe fruit. Alternatively examine the base of the fruit immediately above the receptacle with a lens (Fig. 11c-d) where it should be possible to see the base of the valves (if present).

If there are no valves, the lower segment is usually indehiscent and contains 1 seed (rarely 2 seeds) or is sterile. Some species with highly specialised fruits (e.g. *Crambe*) have very reduced, sterile lower segments. In some species, it is difficult to decide whether there are valves or not; *Cakile* does not have valves but may appear to, whilst *Lepidium draba* has valves but they are usually indehiscent. These are keyed out twice.

Length of the valves is measured from the base to the apex. The base is usually clearly visible but the apex may be indistinct (e.g. *Sisymbrium*). Valves can be measured *in situ* on the fruit, or, better still, pulled off and measured separately.

The number and relative strengths of the veins on the valves is an important taxonomic character but is of considerably less practical use for identification. Indeed, it is the use of this character in keys by Clapham *et al*. (1952-1987) and the use of the "calyx open or closed" character in *Flora Europaea* that have caused the majority of difficulties in identification of crucifers.

The veins on the valves are most clearly seen on dried specimens, and are virtually impossible to see on fresh material. The veins are strips of vascular tissue which stand out as the softer tissue shrinks around them. Veins may thus be made clearer by leaving fresh fruits to dry in a warm, dry place for a week or so. Immature fruits may shrink and distort to give untypical venation patterns. When veins are present, it is the large, longitudinal veins which must be examined and the fine, reticulate veins ignored (except in *Cochlearia*). The strength of lateral veins is judged relative to the central vein, but it is an impression and is not strictly defined.

Because of the problems involved with interpreting these vein characters they have been avoided as far as possible, but may be required to separate some *Brassica* species from *Sinapis*.

The valves may be winged (e.g. *Thlaspi*, *Iberis*).

Some valves have little bulges corresponding to the seeds underneath (e.g. *Sisymbrium irio*) and are described as torulose (beaded). Torulose fruits can be quite distinctive when fresh. This character is unreliable in pressed fruit.

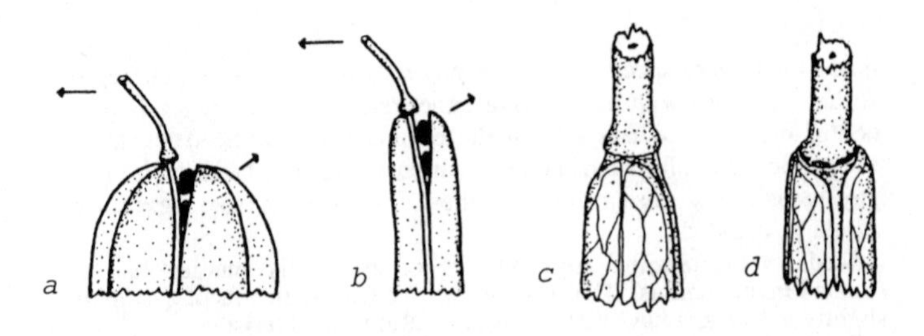

Figure 11. Examination of valves (see text for details). a,b Valves peeling off at their base. c-d Base of valves enlarged.

PERSISTENT STYLE, BEAK OR TERMINAL SEGMENT

The second major feature to look for on a fruit is the persistent style, beak or terminal segment (i.e. portion between the lower segment and the stigma). The difference between these is essentially one of size and they intergrade.

If this upper portion is fertile then it is a terminal segment. In some taxa (e.g. *Sinapis*), where the upper segment is fertile or sterile depending on the particular specimen, it is called a terminal segment for consistency. If the upper portion is essentially linear, or less than 3.5 x 1 mm, then it is a persistent style. A beak is a larger (more than 4 mm long), sterile, upper portion. Whilst these definitions are relatively unimportant, the size and fertility of the upper fruit portion is very important.

The persistent style, beak or terminal segment is measured from the tip of the stigma (which persists in ripe fruit) to the top of the lower segment (this is usually also the top of the valves). Width is measured at the widest point. Some styles are easily lost when fruits mature (e.g. *Neslia*) and unripe fruits will have smaller upper segments. In species which have winged valves (e.g. *Iberis*), the persistent style is measured from where it joins the valves. Shape and structure are described in a similar way to those of the lower segments.

As the terminal segments are indehiscent, establishing their fertility can be difficult. It is easiest to section a terminal segment lengthwise with a razor blade, but they can often be dissected with a fingernail (easiest with ripe fruits).

The stigma at the tip of the style/beak/terminal segment is often more distinct in fruit than in flower and is described again in this condition. In *Matthiola*, the apex of the fruit may be developed into long horns.

SEEDS

Species with very specialised fruits may only have one seed per fruit (e.g. *Neslia*). Species with valves have either one, two or more seeds in each loculus and may also have seeds in the terminal segment. Seed development is dependent on pollination and fertilization, which may be variable. Insects may parasitize seeds and prevent their development, so maggoty fruits should be avoided.

Seeds in the terminal segments are often slightly smaller than those situated in the loculus. The seeds in the two fruit segments of *Cakile* differ slightly in size and have different germination characteristics.

Seeds themselves provide some useful characters. Seed length is maximum length, including the wing if present. At least 10 seeds should be measured and the average calculated. The shape can differ from round to square or cylindrical, and if the fruits are flattened then the seeds are usually flattened too. A wing, if present, is a membranous margin to the seed. Seed sculpturing is of crucial importance in *Rorippa* and requires a microscope to observe properly. This must be observed on ripe, brown seeds as unripe, green seeds can be very misleading. Seed colour also refers to ripe seeds.

The arrangement of seeds inside each loculus can be of diagnostic value. A few taxa with yellow flowers (notably *Diplotaxis*) and many with white flowers have seeds arranged in 2 rows in each loculus (biseriate). This feature must be examined in the middle of the fruit as the seeds may be in one row at the ends (Fig. 12). The majority of yellow-flowered species have the seeds arranged in 1 row (uniseriate).

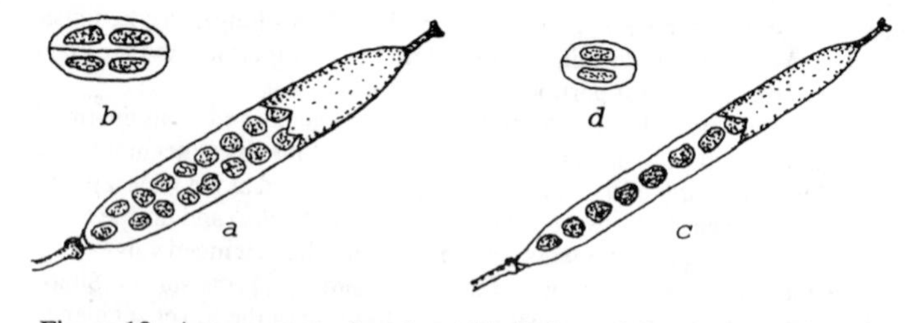

Figure 12. Arrangement of seeds within fruits. a,b Seeds in 2 rows in each loculus (biseriate). a, Fruit with lower portion of valve removed; b, Fruit T.S. Note the seeds may appear to be in 1 row at the base of the fruit, hence this character must be examined in the middle of the fruit. c,d Seeds in 1 row in each loculus (uniseriate). c, Fruit with lower portion of valve removed; d, fruit T.S. Not to scale.

CHROMOSOME NUMBER

Chromosome numbers have been compiled from the literature and from the chromosome index held by Dr. R.J. Gornall, Leicester University. Counts confirmed from wild British or Irish material are indicated with an asterisk. Unconfirmed, very atypical, queried or definitely non-British counts are given in brackets.

FLOWERING TIMES

An indication of typical flowering times is given. This varies with year, habitat, location and the genotype concerned, hence the months cited are approximations. The bulk of species flower in late spring or summer, though some may flower all year in mild winters. Less typical months in which flowering has been observed are given in brackets.

COLLECTION OF MATERIAL

Specimens should only be collected for identification if conservation considerations allow. *Arabis alpina, A. scabra, Coincya wrightii* and *Alyssum alyssoides* are protected in Britain under the Wildlife and Countryside Act 1981, *Arabis petraea, Cardamine impatiens* and *Rorippa islandica* s.s. are protected in the Republic of Ireland under the Wildlife Act 1976, and *Teesdalia nudicaulis* is protected in N. Ireland under the Wildlife (NI) Order 1985; these species must not be collected under any circumstances. There is also no point in collecting unless it is done properly. In most cases identification can be made or confirmed from parts selected from the plant and there is no need to collect whole plants. Unless otherwise stated, select fruits, flowers, upper leaves and basal leaves (in that order of priority) if available from the main or central stem (side shoots may be atypical). Petal size and sepal angle should be noted in the field, and flower colour noted before pressing.

Full details of location, habitat, date, collector and preferably grid reference should be kept for each specimen collected.

THE KEYS

There are two types of key, the traditional indented dichotomous key, and an alternative "synoptic" key. The characters used, and how they are interpreted, are described above and must be understood before the keys are used.

The main indented dichotomous key to groups and genera will give the greatest accuracy but usually requires ripe fruits, flowers and leaves. Keys to the species usually include both flower and fruit characters. The synoptic key is principally intended for non-fruiting material and is inevitably less accurate. If both these keys fail, the list of rare characters and species which show them may help.

Bear in mind the following golden rules when using the keys:-

i) Either lead can be followed where intermediate plants occur, or where characters are liable to be misinterpreted in some species (eg medifixed hairs in *Lobularia*).

ii) The keys are based on fresh material and are designed to be used in the field with a x10 lens and a ruler. If material is collected to examine later, note petal colour and size, and the angle of the sepals in the field.

iii) Measurements need only be made to the nearest 0.5 mm. Population averages are much more reliable than single measurements.

iv) Damaged plants, or plants flowering out of season, are often highly atypical.

v) Unless otherwise stated, the characters are of equal importance and reliability.

vi) When you have keyed a species out, check the fruit and petal sizes against the description; if either does not match reasonably well, start again!

KEYS TO GROUPS AND GENERA

1 Plant not setting ripe fruit — **Group A** (p.27)
1 At least some ripe fruit present
  2 At least some lateral inflorescences arising opposite leaves, or flowering stem absent and inflorescence in axil of basal rosette leaves — **Group B** (p.28)
  2 Inflorescences not leaf-opposed
    3 Fruits without valves — **Group C** (p.28)
    3 Fruits with valves

4 Most mature fruits with a terminal segment/ beak/style 4 mm or more long  **Group D** (p.29)

4 Fruits with a terminal segment/beak/style less than 3.5(-4) mm long

5 Fresh petals yellow (rarely with reddish flushes or darker veins)

6 Upper stem leaves with auricles clasping stem 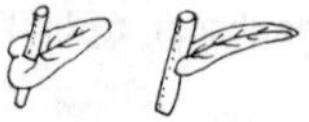 **Group E** (p.30)

6 Upper stem leaves sessile or petiolate, without auricles, or stem leaves absent  **Group F** (p.31)

5 Fresh petals white, purple, pink, red, orange or cream, or petals absent

7 Most valves more than 4 times as long as fruit width 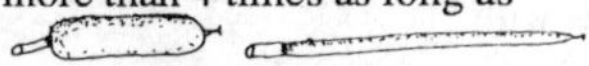 **Group G** (p.32)

7 Valves less than 3.5(-4) times as long as fruit width  **Group H** (p.33)

GROUP A

Plants not setting ripe fruit.

1 Basal leaves 30-100 cm, long-stalked, the blade ± oblong, crenate (these look like large Dock leaves); flowers often imperfect; petals 5-8 mm, white; pedicels ± appressed, with aborting, elliptic ovaries rarely reaching 4 mm 117 **Armoracia rusticana**

1 Not as above.

Plants may be self-sterile individuals, hybrids (which are rare with one or two exceptions) or aliens which dislike our climate. Try the synoptic key, matching against the illustrations, or Kington (1983). Taxa which are not hybrids and which may regularly fail to set fruit include:

Cardamine amara, C. bulbifera, C. pratensis, C. raphanifolia
Eruca vesicaria
Lepidium draba
Raphanus maritimus, R. raphanistrum, R. sativus
Rorippa amphibia, R. austriaca, R. sylvestris
Sisymbrium orientale, S. volgense

GROUP B

At least some lateral inflorescences arising opposite leaves, or flowering stem absent and infloresence in axils of basal rosette leaves.

1 Petals 6-8 mm; fruits 6-8 mm — 135 **Carrichtera annua**
1 Petals 0.5-2 mm or absent; fruits 1.3-3.5 mm — **Coronopus** (p.40)

GROUP C

Fruits without valves, the seeds enclosed in an indehiscent fruit casing.

1 Upper stem leaves with auricles clasping stem 
  2 Fruits 9-21 mm, pendent 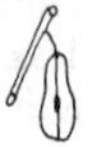 — 116 **Isatis tinctoria**
  2 Fruits 2-8 mm, erect to patent (angle to stem 0-100 degrees) 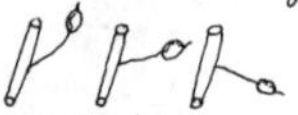
    3 Petals white; rhizomatous perennial — **Lepidium** (p.43)
    3 Petals yellow; rare annuals
      4 Fruits 2-3.5 mm, ± spherical  — **Neslia** (p.45)
      4 Fruits 5-8 mm, ± obovate  — 137 **Myagrum perfoliatum**
1 Upper stem leaves without auricles, sessile or petiolate 
  5 Petals yellow, often with dark veins
    6 Fruits (10-)15-90 mm; petals 11-25 mm — **Raphanus** (p.46)
    6 Fruits (3-)5-12 mm; petals 4-11 mm
      7 Plants without large glands; fruits with 1-2 segments  — **Rapistrum** (p.47)
      7 Plant with large glands, at least on pedicels; fruit with 1 segment  — **Bunias** (p.37)
  5 Petals white, pink or purple (rarely creamish when young)
    8 Terminal segment of fruit less than twice as long as wide, inner stamens with toothed filaments  — **Crambe** (p.40)
    8 Terminal segment of fruit more than twice as long as wide; inner filaments without a tooth 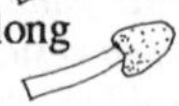
      9 Leaves fleshy; petals 6-13 mm; fruits with 1-2(-3) seeds — 12 **Cakile maritima**
      9 Leaves not fleshy; petals 11-25 mm; fruits with 1-10 seeds — **Raphanus** (p.46)

GROUP D

Fruit with valves; terminal segment, beak or persistent style 4 mm or more long.

1 Petals yellow, sometimes with darker veins
  2 Stems and leaves with medifixed or stellate hairs (x10-x20 lens) — **Erysimum** (p.42)
  2 Stems and leaves with simple hairs, or glabrous
    3 Seeds in 2 rows in each loculus, at least in middle of fruit — 26 **Eruca vesicaria**
    3 Seeds in 1 row in each loculus
      4 Fruits 7-16 mm, appressed (angle with stem 0-15(-30) degrees)
        5 Beak with (0-)1(-2) seeds, swollen; petals 5-10 mm; fruits 7-16 x 1-1.8(-2) mm — 15 **Hirschfeldia incana**
        5 Beak sterile, linear to narrowly conical; petals (7-)9-13 mm; fruits 8-25(-33) x (1.5-)2-4.5 mm — 16 **Brassica nigra**
      4 At least the larger fruits 16-100 mm, appressed to patent (angle with stem 0-105 degrees)
        6 Upper stem leaves with auricles usually clasping stem — **Brassica** (p.36)
        6 Upper stem leaves sessile or petiolate, not clasping stem
          7 Sepals ascending to reflexed (angle with ovary 15-150 degrees)
            8 At least some flowers with sepals patent to reflexed (angle with ovary 80-150 degrees); dried valves with 3-7 ± equally strong veins — **Sinapis** (p.49)
            8 Sepals erect to inclined (rarely patent) (angle with ovary 0-75(-90) degrees); dried valves with 1 strong central vein and with or without weaker lateral veins — **Brassica** (p.36)
          7 Sepals erect (angle with ovary 0-15 degrees)
            9 Terminal segment/beak sterile — **Brassica** (p.36)
            9 Terminal segment/beak with (0-)1-4 seeds
              10 Petals 4-7 mm — 21 **Brassica tournefortii**
              10 Petals (11-)13-26 mm — **Coincya** (p.40)

1 Petals white, pink or purple, or purple and yellowish-brown on the back only, with or without darker veins
  11 Stem and leaves with at least some forked or stellate hairs (x10 lens)
    12 Fruits with 2-3 horns projecting at apex (like a miniature pitchfork) — 80 **Matthiola longipetala**
    12 Fruits with a simple, linear style

13 Mat-forming perennial; fruits 2.5-4 mm wide; valves elliptic — **86 Aubrieta deltoidea**
13 Erect annual; fruits 1-2 mm wide; valves linear — **81 Malcolmia maritima**
11 Stems and leaves with simple hairs, or glabrous
14 Fruits 15-35 mm wide — **Lunaria** (p.44)
14 Fruits 2-9 mm wide
15 Petals 6-13 mm; seeds 1-2(-3) per fruit — **12 Cakile maritima**
15 Petals (11-)13-26 mm; seeds numerous
16 Seeds in 1 row in each loculus — **Cardamine** (p.38)
16 Seeds in 2 rows in each loculus, at least in middle of fruit — **26 Eruca vesicaria**

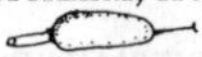

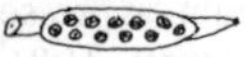

GROUP E

Petals yellow; fruit with valves and a short, persistent style or beak; upper stem leaves with clasping auricles.

1 Fruits pendent; seeds 1(-2) — **116 Isatis tinctoria**
1 Fruits erect to reflexed; seeds 2-many
2 Seeds in 2 rows in each loculus, at least in middle of fruit 
3 Fruits 30-70 mm — **45 Arabis glabra**
3 Fruits 3-26 mm
4 Fruits obpyriform, (2-)3-8 mm wide — **Camelina** (p.37)
4 Fruits linear, elliptic, oblong or globose, 0.8-3 mm wide  — **Rorippa** (p.47)
2 Seeds solitary or in 1 row in each loculus  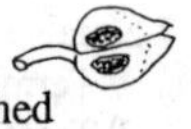
5 All leaves simple, margins entire or toothed
6 Stems and leaves glabrous — **73 Conringia orientalis**
6 Stems and leaves hairy — **46 Arabis turrita**
5 At least the lower leaves deeply lobed, pinnate or finely divided
7 Petals 1-2 mm
8 Upper stem leaves entire — **103 Lepidium perfoliatum**
8 Upper stem leaves finely divided — **41 Descurainia sophia**
7 Petals 4-30 mm
9 Petals 16-30 mm; leaves glaucous — **17 Brassica oleracea**
9 Petals 4-10 mm; leaves green, shining — **Barbarea** (p.35)

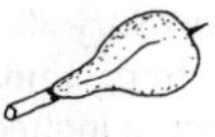

GROUP F

Petals yellow; fruit with valves and a short terminal segment/beak/persistent style; upper stem leaves without auricles.

1 Stems and leaves with forked, medifixed or stellate hairs, at least below (x10-x20 lens)
 2 Fruits 10-100 mm, linear
  3 Leaves simple — **Erysimum** (p.42)
  3 Leaves finely divided — 41 **Descurainia sophia**
 2 Fruits 2-8(-12) mm, ± orbicular
  4 Sepals persistent at least whilst the fruit is green; petals 2-3(-4) mm — 89 **Alyssum alyssoides**
  4 Sepals deciduous, not persistent in fruit; most petals 3-8 mm — 90 **Aurinia saxatilis**
1 Stems and leaves with simple hairs, or glabrous
 5 Seeds in 2 rows in each loculus, at least in middle of fruit
  6 At least larger fruits more than 23 mm; petals (4-)5.8-13 mm — **Diplotaxis** (p.41)
  6 Fruits 5-23 mm; petals 1-6.2 mm
   7 Leaves entire, margins ciliate — 118 **Draba aizoides**
   7 At least some leaves toothed, lobed or divided — **Rorippa** (p.47)
 5 Seeds in 1 row in each loculus
  8 Flowers of at least the lower third of the main infloresence with pinnatifid to bipinnatifid bracteoles — 29 **Erucastrum gallicum**
  8 Bracteoles simple, with entire or toothed margins, or bracteoles absent
   9 Terminal segment of fruit with (0-)1(-2) seeds; petals 5-10 mm; fruits 7-16 x 1-1.8(-2) mm — 15 **Hirschfeldia incana**
   9 Terminal segment/beak/style sterile
    10 Petals 6-17 mm (intermediates key out in both leads)
     11 Seeds globose; fruits 1.5-9 mm wide; dried valves with a strong central vein and weak lateral veins — **Brassica** (p.36)
     11 Seeds oblong to cylindrical; fruits 0.5-2 mm wide; dried valves with weak veins only — **Sisymbrium** (p.50)
    10 Petals 1-6 mm
     12 Perennial with non-flowering rosettes, forming patches with creeping rhizomes — 64 **Rorippa sylvestris**
     12 Annuals without non-flowering rosettes and creeping rhizomes — **Sisymbrium** (p.50)

GROUP G

Petals white or coloured but not yellow; fruit with valves at least 4 times as long as fruit width; persistent style short or absent.

1 Stem leaves pinnate, pinnatisect, trifoliate or digitate 
  2 Fruits ± terete; lateral lobes of leaves usually sessile  — **Rorippa** (p.47)
  2 Fruits ± flattened; lateral lobes of at least some leaves stalked (rarely all sessile) — **Cardamine** (p.38)
1 Stem leaves simple, toothed or lobed to 1/2 way to midrib, or absent 
  3 Stem leaves absent; petals bifid for 1/4-3/4 of their length — **Erophila** (p.42)
  3 Stem leaves present; petals entire to emarginate 
    4 Stem leaves with auricles at least half clasping the stem 
    5 Plant glabrous — 73 **Conringia orientalis**
    5 Stems and leaves hairy at least below — **Arabis** (p.34)
    4 Upper stem leaves without auricles, or auricles small and only slightly clasping stem (if in doubt take this lead) 
    6 Petals 2.5-10 x 0.6-4 mm
      7 Most fruits more than 18 mm (intermediates key out in both leads)
        8 Upper stem leaves petiolate; fresh plant smelling of garlic when crushed — 83 **Alliaria petiolata**
        8 Upper stem leaves sessile; fresh plant not smelling of garlic — **Arabis** (p.34)
      7 Fruits 5-16(-18) mm
        9 Fruit lanceolate, usually twisted — 121 **Draba incana**
        9 Fruit linear, usually straight 
          10 Petals 2.5-4.5 mm; slender annual — 42 **Arabidopsis thaliana**
          10 Petals 4-10 mm; biennials or perennials — **Arabis** (p.34)
    6 At least larger petals 10-33 x (3-)4-20 mm
      11 Fruit with 2-3 distinct horns at apex (like a miniature pitch-fork) — 80 **Matthiola longipetala**
      11 Fruit with a linear to 2-lobed persistent style 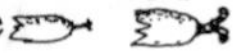
        12 Seeds irregularly cylindrical 
          13 Petals (9-)10-18 mm; slender annual 15-50(-100) cm tall — 81 **Malcolmia maritima**

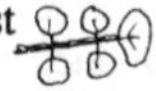

13 Petals 17-30 mm; stout biennial to perennial
30-100(-150) cm tall 82 **Hesperis matronalis**
12 Seeds flattened, square to oblong 
14 Flowers orange **Erysimum** (p.42)
14 Flowers white, pink, red or purple
15 Stem and leaves woolly with dense, stellate hairs **Matthiola** (p.44)
15 Stem and leaves green with medifixed hairs (see p.9) **Erysimum** (p.42)

GROUP H

Petals white or coloured but not yellow; fruit with valves usually less than 3.5 times as long as fruit width; style short.

1 Fruits compressed or markedly flattened with septum across the narrowest diameter (angustiseptate) (intermediates key out in both leads) 
2 Outer petals larger than inner petals
3 Outer petals less than 2 mm; persistent style less than 0.3 mm, ± included in apical notch of fruit 98 **Teesdalia nudicaulis**
3 Outer petals 3-16 mm; persistent style 1-4.5 mm, ± equalling or exceeding notch  **Iberis** (p.43)
2 Petals of ± equal size, or absent
4 Seeds 1 per loculus  **Lepidium** (p.43)
4 Seeds 2-many per loculus, or if seeds 1-2 then plant smelling of garlic when crushed  
5 Fruit winged
6 Stem leaves petiolate 136 **Pachyphragma macrophyllum**
6 Upper stem leaves sessile with auricles clasping stem **Thlaspi** (p.51)
5 Fruit not winged
7 Stem leaves regularly pinnate; fruit 1.8-3 mm 99 **Hornungia petraea**
7 Stem leaves entire to irregularly lobed; fruits (3-)4-15 mm
8 Fruits obtriangular to obovate; leaves not fleshy; usually hairy 115 **Capsella bursa-pastoris**
8 Fruits ovate, elliptic or globose; leaves fleshy; usually glabrous   **Cochlearia** (p.39)

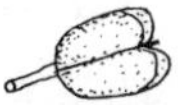

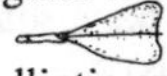

1 Fruits ± terete, or compressed or flattened with the septum across the broadest diameter (latiseptate) 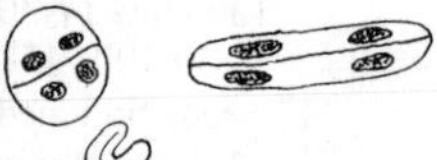
  9 Petals bifid for 1/4-3/4 of their length
    10 Stem leaves absent — **Erophila** (p.42)
    10 Stem leaves numerous — **88 Berteroa incana**
  9 Petals entire or emarginate 
    11 Fruits 15-35 mm wide — **Lunaria** (p.44)
    11 Fruits 1-10 mm wide
      12 Stems and leaves glabrous
        13 All leaves basal, linear  — **125 Subularia aquatica**
        13 Stem leaves and basal leaves with broad expanded blade  — **Cochlearia** (p.39)
      12 Stem and leaves hairy, at least below (x10 lens)
        14 Fruits with 1-2 seeds in each loculus
          15 Fruits with 1 seed per loculus — **87 Lobularia maritima**
          15 Fruits with (1-)2 seeds per loculus — **89 Alyssum alyssoides**
        14 Fruits many-seeded
          16 Fruits elliptic to lanceolate, flat; persistent style 0-0.5 mm  — **Draba** (p.41)
          16 Fruits obpyriform, inflated; persistent style 1.3-3 mm  — **Camelina** (p.37)

KEYS TO THE SPECIES

ARABIS L.

Arabis species have at least some stellate hairs, white, pink, purple or yellow petals, and linear, usually flattened, fruits. The combination of stellate hairs and linear fruits also occurs in a number of other genera, but *Arabis* species are only likely to be confused with *Arabidopsis* (which is a slender annual with petals less than 4.5 mm).

Arabis here includes the genus *Cardaminopsis* (C.A. Meyer) Hayek, as the differences between the genera are both unconvincing and unworkable.

1 Stem leaves cuneate to rounded at base, sometimes just clasping stem (intermediates key out in both leads)
  2 Fruits erect; all leaves with entire to toothed margins — **43 A. hirsuta**
  2 Fruits ascending to patent; leaves entire to pinnatisect
    3 Petals cream; claw without teeth  — **44 A. scabra**
    3 Petals white to purple; claw with a pair of teeth  — **50 A. petraea**
1 Stem leaves with auricles at least half clasping stem

4 Petals pale yellow to yellow; fruits (30-)43-120 mm
5 Upper leaves glabrous; seeds biseriate **45 A. glabra**
5 Upper leaves hairy; seeds uniseriate **46 A. turrita**
4 Petals white, pink or purple; fruits (9-)18-70(-90) mm
6 Stems erect; fruits erect
7 Petals 4-6.2 x 0.9-1.6 mm, white **43 A. hirsuta**
7 Petals 6-10 x 2-4 mm, pink to purple (rarely white) **48 A. collina agg.**
6 Stems decumbent; fruits ascending to patent
8 Petals 9.5-18 mm **47 A. caucasica**
8 Petals 5-8 mm **49 A. alpina**

BARBAREA R.Br.

Barbarea species have small, yellow flowers, clasping upper stem leaves, fruits with dehiscent valves and a short persistent style. They are usually glabrous or sparsely hairy below and typically have dark green, shiny, pinnate lower leaves. *Erysimum* species are roughly hairy with appressed medifixed and stellate hairs, and entire or toothed leaves. Fruits of *Brassica* species usually have longer beaks. Yellow-flowered *Rorippa* species usually have seeds in two rows in each loculus. For a full account of the genus see Rich (1987d).

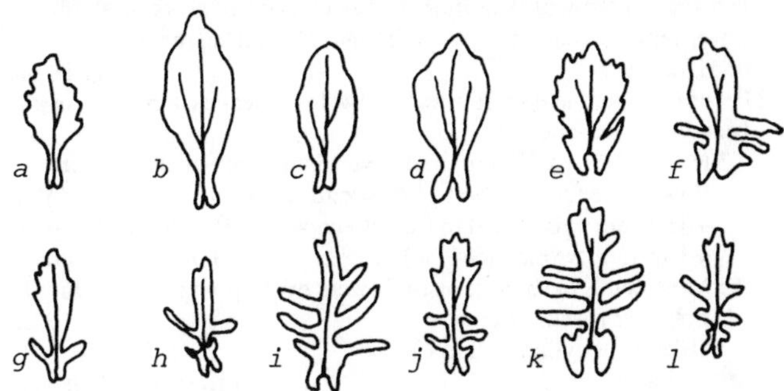

Figure 13. Uppermost stem leaves of *Barbarea* species. a-c *B. stricta*; d-f *B. vulgaris*; g-i *B. intermedia*; j-l *B. verna*.

1 Uppermost stem leaves simple, toothed or shallowly lobed with 1(-2) pair(s) of linear, lateral lobes; terminal lobe broad, obovate to ovate (Fig. 13 a-f); mean seed size *c.* 1.6 mm (measure 10 seeds)
2 Flower buds at least sparsely hairy*; persistent style 0.5-1.8(-2.3) mm, stout **69 B. stricta**
2 Flower buds glabrous; persistent style (1.7-)2-3.5(-4) mm, slender **70 B. vulgaris**

1 Uppermost stem leaves deeply lobed with (1-)2-5 pairs of lateral lobes; terminal lobe narrow, oblong to oblanceolate (Fig. 13 g-1); mean seed size *c*. 2.1 mm
3 Fruits 15-35(-40) mm; petals 4-6.3 mm — 71 **B. intermedia**
3 Most fruits more than 40 mm; most petals more than 6 mm — 72 **B. verna**

* Best seen in silhouette. Hairs rub off with age and in pressed material. Exceptionally, *B. vulgaris* may also have one or two hairs.

BRASSICA L.

Brassica is a large genus which is difficult to define. The species in Britain all have yellow petals, are glabrous or have simple hairs, have fruits with a distinct beak which is usually sterile or 1(-3)-seeded, valves with a strong central vein and usually weaker lateral veins (this is most obvious on dried fruits but requires experience to use and knowledge of other genera such as *Sinapis*), and globose seeds. A number of other *Brassica* species have been recorded as very rare casuals.

For a simpler, illustrated key see Rich (1988d).

1 Upper stem leaves sessile, with auricles usually clasping stem
2 Petals (16-)18-30 mm; stems woody, covered with many leaf scars often with 1 or more distinct whorls of scars on stem; lower leaves usually fleshy, glabrous; terminal segment with (0-)1(-2) seeds — 17 **B. oleracea**
2 Petals 6-18 mm; leaf scars on stem few; lower leaves thin, hairy or glabrous; beak usually sterile
3 Most petals 13-18 mm; buds overtopping or equalling open flowers (see p. 13); first year rosettes glaucous — (18 **B. napus**)
4 Tap-root cylindrical to Carrot-shaped — 18a **B. napus subsp. oleifera**
4 Tap-root swollen (Swede) — 18b **B. napus subsp. rapifera**
3 Most petals 6-13 mm; open flowers overtopping or equalling buds; first year rosettes green — (19 **B. rapa**)
5 Tap-root cylindrical to Carrot-shaped — 19a **B. rapa subsp. sylvestris**
5 Tap-root swollen (Turnip) — 19b **B. rapa subsp. rapa**
1 Upper stem leaves petiolate, not clasping stem
6 Fruits and pedicels appressed (angle to stem 0-20 degrees) — 16 **B. nigra**
6 Fruits and pedicels ascending to patent (angle to stem 30-90 degrees)
7 Fruit with a distinct stipe (0.8-)1.5-5 mm — 20 **B. elongata**
7 Fruit with stipe less than 1 mm, or stipe absent
8 Petals 4-7 mm; lower stem leaves with 4-10 pairs of lateral lobes, acutely toothed; beak 10-20 mm, (0-)1(-3) seeded — 21 **B. tournefortii**
8 Petals 9-17 mm; lower stem leaves with 0-3 pairs of lateral lobes, sinuate or shallowly to acutely toothed; beak 2.5-12 mm, sterile

9 Petals 9-14 mm; beak (4-)5-9(-12) mm **22 B. juncea**
9 Petals 13-17 mm; beak 2.5-6(-7) mm **23 B. carinata**

BUNIAS L.

A useful, more or less diagnostic character of *Bunias* is the presence of large, yellow glands, most conspicuous on the pedicels and inflorescence branches (x10 lens). Glands like these also occur in *Matthiola*, though many other species have small glandular hairs.

1 Fruit irregularly ovoid, with low, warty protuberances; perennial **6 B. orientalis**
1 Fruit with 4 conspicuous, crested wings; annual **7 B. erucago**

CAMELINA Crantz

The key and accounts are based on Mirek (1981), and supersede previous accounts in British Floras.

Examination of herbarium material shows that *C. alyssum* and *C. microcarpa* have been widely overlooked, probably due to the inadequacy of previously available accounts and due to the belief that *C. sativa* was the common species. Historical records, unless supported by critically determined herbarium specimens, should be referred to *C. sativa* sensu lato (i.e. including *C. alyssum*, *C. microcarpa* and *C. rumelica*). The synonomy is complex.

Specimens need to be determined on a combination of characters. Mature fruits with ripe seeds should be examined from the lower part of the main inflorescence. Collection of voucher specimens is desirable.

1 Ripe seeds (1.5-)1.6-2.9 mm; ripe fruits 3.1-6.5 mm thick*; stems and leaves glabrous or with branched hairs, rarely with additional simple hairs.
  2 Ripe seeds (1.5-)1.6-2(-2.1) mm; ripe fruits 3.5-5.5 mm wide; petals 3.7-4.6(-5) mm **91 C. sativa sensu stricto**
  2 Ripe seeds (1.8-)2.1-2.9 mm; ripe fruits (4.6-)5.1-7 mm wide; petals 4.6-5.8(-6.3) mm **92 C. alyssum**
1 Ripe seeds (0.9-)1-1.4(-1.5) mm; ripe fruits 1.7-3.1 mm thick*; stems and leaves with many simple hairs and with or without branched hairs (x 10 lens)
  3 Petals 5-8(-9) mm; simple hairs dense, *c.* 1.5-3.5 mm long, branched hairs few or absent; basal rosette distinct, usually persistent until flowers open or later **94 C. rumelica**
  3 Petals (2.2-)2.6-4(-4.2) mm; simple hairs less than 2 mm long; basal rosette usually not persisting to when the flowers open **93 C. microcarpa**

* Fruit thickness or breadth is measured as follows (it is often distorted in pressed fruits):-

T.S. Fruit

CARDAMINE L.

Cardamine species have pinnate, digitate or trifoliate leaves which typically have clearly stalked lateral lobes. This latter character is a very useful "jizz" character but not all material may show it and sometimes more or less stalked leaflets occur in other genera. The petals are white, pink or purple or are sometimes absent altogether. The fruits are long and flattened and the valves often spring off forcibly when ripe to disperse the seeds.

Hybrids also occur rarely; these are not keyed out.

1 Petals 1.5-4.8(-5.2) x 0.2-2.2 mm, white, or petals absent.
  2 Stem leaves with conspicuous, acute, clasping auricles 53 **C. impatiens**
  2 Stem leaves with small, rounded auricles, or auricles absent
    3* Most flowers with 4 stamens; main stem with (0-)1-4(-5) leaves; lower part of stems usually (*c*.90% of all plants) glabrous 51 **C. hirsuta**
    3 Most flowers with 6 stamens; main stem with (3-)4-10 leaves; lower part of stems usually (*c*. 90% of all plants) hairy 52 **C. flexuosa**
1 Petals 4.5-20 x 2.8-10 mm, white to purple or pink
  4 Axils of upper leaves with small, purple-brown bulbils 58 **C. bulbifera**
  4 Axils of upper leaves without bulbils
    5 Basal leaves trifoliate; stem leaves trifoliate, simple or absent 57 **C. trifolia**
    5 At least some leaves pinnate with 2 or more pairs of leaflets
      6 Leaves 3-5, all on upper part of stem; terminal lobe of uppermost leaf 60 mm or more long 138 **C. heptaphylla**
      6 Upper and lower stem leaves, and often also basal rosette leaves present; terminal lobe of uppermost stem leaf less than 50(-65) mm
        7 Undehisced anthers purple; petals usually pure white (rarely pinkish-purple) 54 **C. amara**
        7 Undehisced anthers yellow; petals usually pink or purple (rarely pure white)
          8 Terminal lobe of upper stem leaves linear to oblong; persistent style of fruit 0.3-2 mm 55 **C. pratensis**
          8 Terminal lobe of upper stem leaves ovate to elliptic; persistent style of fruit 2-3.5 mm 56 **C. raphanifolia**

* Characters in lead 3 are given in order of importance. Examine at least 5 flowers if possible and note that the two outer stamens of *C. flexuosa* may be small and inconspicuous. Petioles of both species are usually hairy irrespective of whether the stem is hairy or not.

COCHLEARIA L. *

Plants should be collected with ripe fruits, basal leaves, lower and upper stem leaves and flowers - in order of priority - if at all possible. Fruit valves must be examined for patterning only when they have lost their green pigments naturally and have become straw-coloured, as they are always smooth initially, and premature drying produces unnatural shrinkage patterns. Vegetative rosettes or young plants have broader leaves than flowering material. These can be misleading if applied to flowering/fruiting material which lack rosette leaves.

Intermediates between the species appear to be common; they may well often be hybrids. Hybrids should only be recorded after careful examination of local populations. It is best to get to know the parent species well before invoking hybridization as an explanation for variation.

1 Rosette and lowest stem leaves cuneate to rounded at base (not truncate or cordate); mature fruit distinctly flattened with the septum across the narrowest diameter (angustiseptate) 132 **C. anglica**

1 Rosette and lowest stem leaves mostly cordate, truncate or rounded at base; mature fruit not flattened, often with septum across the broadest width

2 Annual with poorly-developed tap-root; lower stem leaves usually petiolate and palmately lobed; upper stem leaves (of main stem) stalked or sessile but without clasping auricles; petals (1.8-)2.5-4.5(-5.5) mm, often pale lilac 126 **C. danica**

2 Biennials or perennials with well-developed tap-root; lower stem leaves sometimes petiolate but not palmately lobed; upper stem leaves sessile, often clasping stem; petals 3.5-9.5 mm, often white 127-131 **C. officinalis aggregate**

3 Rosette and lowest stem leaves mostly truncate or shallowly cordate at the base, the blade less than 1.5 cm long; flowers up to 10 mm in diameter

4 Rosette and lowest stem leaves dark green and matt or slightly shiny, rather tough 128 **C. atlantica**

4 Rosette and lowest stem leaves often light- or yellow-green, fleshy and often very shiny 130 **C. scotica**

3 Rosette and lowest stem leaves mostly cordate, only very rarely truncate at the base, varied in size; flowers up to 15 mm in diameter

5 Rosette and lowest stem leaves with blades often more than 2 cm long; mature fruits globose to ovoid, with strongly veined valves (habitat generally near the coast or on road verges) 127 **C. officinalis sensu stricto**

5 Rosette and lowest stem leaves with blades less than 2 cm long (usually not more than 1 cm); mature fruits ovoid to ellipsoid with weakly veined valves (habitat generally distant from the sea except in northern Scotland)

* by D. H. Dalby.

6 Mature fruit valves mostly symmetrical with delicate reticulate patterning; rosette and lower stem leaves mid green and slightly shiny **129 C. pyrenaica**
6 Mature fruit valves often asymmetrical, almost or wholly lacking reticulate patterning; rosette and lower stem leaves dark green and very shiny **131 C. micacea**

COINCYA Rouy (Rhynchosinapis Hayek)

The two subspecies of *C. monensis* are sometimes difficult to distinguish and a combination of characters must be used. Voucher specimens are desirable provided conservation considerations allow. For full details, see Leadlay & Heywood (1990).

1 Young fruits and ovaries hairy (fruits glabrescent); perennial **13 C. wrightii**
1 Young fruits and ovaries glabrous; annual-biennial
  2 Stem and leaves glabrous to sparsely hairy; stems prostrate to ascending; leaf surfaces glabrous; seeds 1.3-2 mm; **14a C. monensis subsp. monensis**
  2 Stem and leaves sparsely to densely hairy (rarely glabrescent); stems erect (rarely prostrate); leaf surfaces hairy (especially on lower side); seeds 1.1-1.6 mm **14b C. monensis subsp. recurvata**

CORONOPUS Haller

Coronopus is easily distinguished by the lateral inflorescences which arise all or mostly opposite leaves and not in their axils (*Carrichtera* also has this character, but has petals 6-8 mm and fruits 6-8 mm).

1 Fruits notched at apex; petals 0.5 mm or absent **133 C. didymus**
1 Fruits apiculate at apex; petals 1-2 mm **134 C. squamatus**

CRAMBE L.

The white petals and ± globose, indehiscent, 1-2 seeded fruits are distinctive. If in doubt, the inner stamens have a tooth (see also *Aubrieta*).

1 Low bush 30-50 cm; leaves fleshy; flowers in dense corymbs **10 C. maritima**
1 Tall plant 100-200 cm; leaves not fleshy; flowers in long, lax racemes **11 C. cordifolia**

DIPLOTAXIS (L.) DC.

With practice, the petal shape, colour, size of plant and number of stem leaves (ignoring secondary rosettes in the leaf axils) are useful characters and the plants can be separated at a glance.

1 Petals 8-15 mm; fruits with stipe (0.3-)0.5-6.5 mm (Fig.14 a-b) — 27 **D. tenuifolia**
1 Petals 4-8(-8.5) mm; fruits without stipe (Fig. 14 c) — 28 **D. muralis**

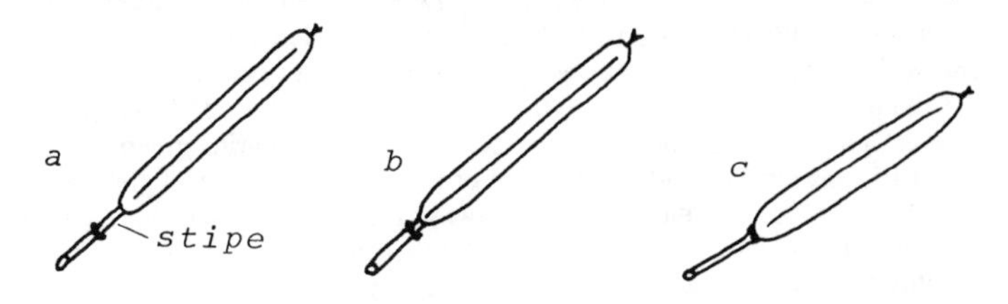

Figure 14. Fruits of *Diplotaxis* species to show stipe. a-b *D. tenuifolia*; c *D. muralis*. Not to scale.

DRABA L.

Draba species have lanceolate to elliptic, flattened, latiseptate, many-seeded fruits, small white or yellow petals, and usually stellate hairs (or simple hairs only in *D. aizoides*). They may be confused with *Erophila* which has bifid (rather than entire to emarginate) petals, or *Alyssum* which only has 2 seeds per loculus.

The only problem likely to arise is the distinction of dwarf *D. incana* from *D. norvegica*. The fruits of *D. incana* are usually twisted but this may be visible only in young or only in mature fruit, and rarely some plants in a population may have completely untwisted fruits.

1 Petals yellow — 118 **D. aizoides**
1 Petals white
  2 Stem leaves broadly ovate; pedicels and fruits inclined to patent — 119 **D. muralis**
  2 Stem leaves narrowly ovate to oblong-lanceolate; pedicels and fruits erect to ascending
    3 Fruits not twisted; stem leaves 0-1(-2); leaf margins entire — 120 **D. norvegica**
    3 At least some fruits twisted; stem leaves (0-)5 or more; leaf margins with acute teeth (rarely entire) — 121 **D. incana**

EROPHILA DC. *

Erophila is easily distinguished by the combination of bifid petals and leafless stems.

The density of the pubescence changes during development. The full leaf pubescence is developed before the leaves have reached their mature size, so that young leaves appear more pubescent than mature leaves. Also, hairs from both leaves and flowering stems tend to be lost with age, accentuating this effect.

1 Leaves densely pubescent, often appearing greyish; petioles not more than 1/2 as long as blade; lower parts of flowering stems densely pubescent, and with at least scattered hairs as far up as the lowest pedicel; seeds 0.3-0.5 mm — 122 **E. majuscula**
1 Leaves at most with a moderate pubescence, always appearing green; petioles at least 1/2 as long as blade; lower part of flowering stem almost glabrous to moderately pubescent, the lower parts always glabrous; seeds 0.5 mm or more
  2 Petals bifid to at least 1/2 and up to 3/4 their length; flowering stems always with at least scattered hairs on the lower parts — 123 **E. verna sensu stricto**
  2 Petals bifid to a maximum of 1/2 their length; flowering stems either with very scattered hairs on the lower parts or glabrous — 124 **E. glabrescens**

ERYSIMUM L.

Erysimum is particularly characterised by the yellow to orange flowers and appressed, forked or stellate hairs (x10-x20 lens). The forked hairs are medifixed but appear at first glance as simple. The genus *Cheiranthus* is here included in *Erysimum* (Snogerup 1967a).

Worldwide, a very difficult genus. In addition to the 4 species keyed out below, at least 10 other species have been recorded as casuals or garden escapes. Specimens not agreeing with the descriptions should be sent for identification.

1 Petals 2.9-10 mm; fruits 1-2 mm wide
  2 Petals 2.9-5 mm; fruits 12-27 mm — 74 **E. cheiranthoides**
  2 Petals 6-10 mm; fruits 45-100 mm — 75 **E. repandum**
1 Petals 14-33 mm; fruits *c.* 2-4 mm wide
  3 Fruits strongly 4-angled, *c.* 2 mm wide; petals clear, bright orange (rarely yellow) — 76 **E. allionii**
  3 Fruits flattened, 2-4 mm wide; petals yellow, orange or red — 77 **E. cheiri**

* By S.A. Filfilan & T.T. Elkington

IBERIS L.

Iberis is distinct in having large, asymmetrical petals.

Large-flowered and -fruited forms of *I. amara*, and hybrids with *I. umbellata*, are apparently cultivated and may escape.

1 Perennial with non-flowering rosettes; stems decumbent to ascending — 97 I. sempervirens
1 Annuals without non-flowering rosettes; stems erect
  2 Inflorescences elongating in fruit; valves 4-6 mm (larger in horticultural variants); lower stem and leaves usually hairy — 95 **I. amara**
  2 Inflorescences remaining densely contracted in fruit; valves (6-)7-10 mm; stem and leaves usually glabrous, rarely glandular — 96 **I. umbellata**

LEPIDIUM L.

Lepidium species have angustiseptate fruits with 1 seed per loculus (many-seeded in *Thlaspi*) and small, white to purplish or rarely yellow petals, or petals absent.

Lepidium is a difficult genus, and a world-wide revision is long overdue. Ripe fruits, selected from the lower half of the central inflorescence, are required for identification. About 40-50 species have been recorded in the British Isles but most are non-persistent casuals. The following key covers most taxa likely to be found; Ryves (1977) gives a good key and notes on most of the other taxa, and is recommended along with Hitchcock (1936, 1946), Mulligan (1961), Jonsell (1975a), Hewson (1981) and *Flora Europaea* I (ed. 2).

The main difficulties will be given by *L. campestre/L. heterophyllum* and the *L. virginium/L. densiflorum/L. ruderale* group (the exact species limits in the latter group are unclear). Collection of voucher specimens, especially of the latter group and of material not fitting the descriptions, is advisable.

1 Upper stem leaves with auricles clasping stem, or perfoliate
  2 Petals *c.* 1-1.5 mm, pale yellow; rosette and lower leaves finely divided with linear lobes — 103 **L. perfoliatum**
  2 Petals 2-4.5 mm, white; rosette and lower stem leaves with broad lobes or undivided
    3 Valves of fruit broadly winged at apex
      4 Undehisced anthers yellow; persistent style 0.1-0.7 mm, included in or just exceeding notch; annual — 109 **L. campestre**

4 Undehisced anthers red or purple, at least on sides; persistent style (0.4-)0.5-1.2 mm, projecting well beyond notch; biennial to perennial **110 L. heterophyllum**

3 Valves of fruit unwinged or only margined

5 Mature fruits emarginate to truncate at base **106 L. draba**

5 Mature fruits cuneate to rounded at base **107 L. chalepense**

1 Upper stem leaves sessile or petiolate, without clasping auricles

6 Fruits (4.5-)5-7 x 3-5.8 mm, broadly winged at apex **108 L. sativum**

6 Fruits 1.5-4 x 1.3-3.2 mm, the wing narrow or absent

7 Petals white, conspicuous, *c.* 1.5 times as long as sepals

8 Fruit notched at apex with style included in notch; annual **102 L. virginicum**

8 Fruit not notched at apex, the style prominent; perennial

9 Upper stem leaves ± linear **104 L. graminifolium**

9 Upper stem leaves ovate to lanceolate **105 L. latifolium**

7 Petals absent, reduced or shorter than sepals

10 Fruits 1.5-2(-2.3) mm wide, elliptic; seeds unwinged; plant foetid when fresh **100 L. ruderale**

10 Fruits 1.7-3.2(-4) mm wide, ± circular to obovate; seeds usually at least partly winged; plant not foetid

11 Fruits 2-3.5 mm, obovate **101 L. densiflorum**

11 Fruits (2.5-)3-3.7(-4) mm, ± circular **102 L. virginicum**

LUNARIA L.

The large (25-80 x 15-35 mm), flat fruits of *Lunaria* are very distinctive. *Thlaspi arvense* also has large, flat fruits but they are smaller (to 20 x 13 mm) and angustiseptate.

1 Valves of fruit truncate to rounded at ends; upper stem leaves cordate to ovate, with petioles 0-7(-10) mm **84 L. annua**

1 Valves of fruit cuneate to subacute at ends; upper stem leaves lanceolate, with petioles (5-)10-30 mm **85 L. rediviva**

MATTHIOLA R.Br.

The dense, stellate hairs (which give the plants a grey-green appearance) and large, pink-purple flowers will distinguish *Matthiola* species from most other crucifers. *Malcolmia* is sometimes confused with *Matthiola* but has a linear persistent style 2-6 mm.

M. incana is very variable with many cultivars which are not distinguished here. *M. tricuspidata* and *M. fruticulosa* have also been recorded (see page 187).

1 Fruits with 2-3 pronounced lateral horns at apex (like a miniature pitch-fork; Fig. 15a-b); pedicels of flowers or fruits 1-3 mm — 80 **M. longipetala**

1 Fruit without distinct lateral horns (Fig. 15 c-d); pedicels of flowers or fruits 4-25 mm

2 All leaves ± entire; large glands absent (naked eye or lens) but minute glandular hairs present (microscope) — 78 **M. incana**

2 At least lower leaves lobed; plant with many large yellow-black glands above (most conspicuous when fresh) — 79 **M. sinuata**

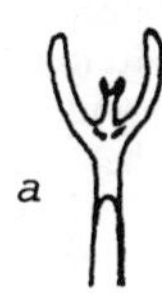

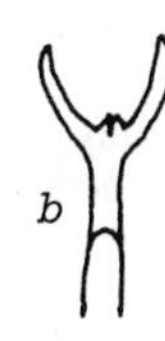

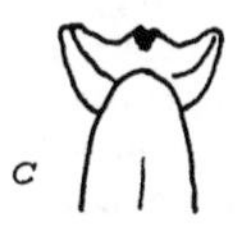

Figure 15. Fruit apices of *Matthiola* species. a-b *M. longipetala*; c *M. incana*; d *M. sinuata*. Not to scale.

NESLIA Desv.

Examine a range of fruits on the inflorescence. Intermediate plants are reported from Europe (Fig. 16c).

1 Mature fruits truncate at base and apex (Fig. 16a); longitudinal ribs 2 — 8 **N. paniculata s.s.**

1 Mature fruits apiculate at base and apex (Fig. 16b); longitudinal ribs 4 — 9 **N. apiculata**

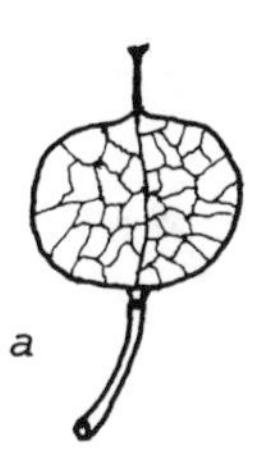

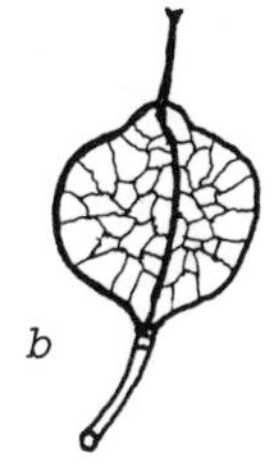

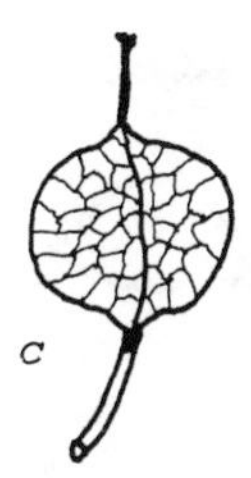

Figure 16. Mature fruits of *Neslia* species. a *N. paniculata*; b *N. apiculata;* c intermediate.

RAPHANUS L.

Fruit and seed set is often variable due to self-incompatibility and unpredictable pollination. Fresh fruits shrink by 0-2.5 mm (average 1 mm) on drying and it helps to measure the fresh, fleshy fruits separately from dried fruits (air-dried) for identification (see Rich 1989).

1 Fresh fruits 2.5-5.5(-6) mm wide (mean* width 3.5-4.5 mm) or dried fruits (1.5-)2-5 mm wide (mean* width 2.5-4 mm) — 1 **R. raphanistrum**
1 Fresh fruits 5-12 mm wide (mean* width 6-9 mm) or dried fruits 4-10 mm wide (mean* width 5-8 mm)
  2 Petals yellow (rarely white); fruits with 1-5(-6) seeds, with deep to shallow constrictions between the seeds — 2 **R. maritimus**
  2 Petals white to pink or purple (exceptionally yellow); fruits with 1-12 seeds, inflated or with shallow constrictions between the seeds; tap root swollen or not — 3 **R. sativus**

* average maximum width of at least 5 mature fruit.

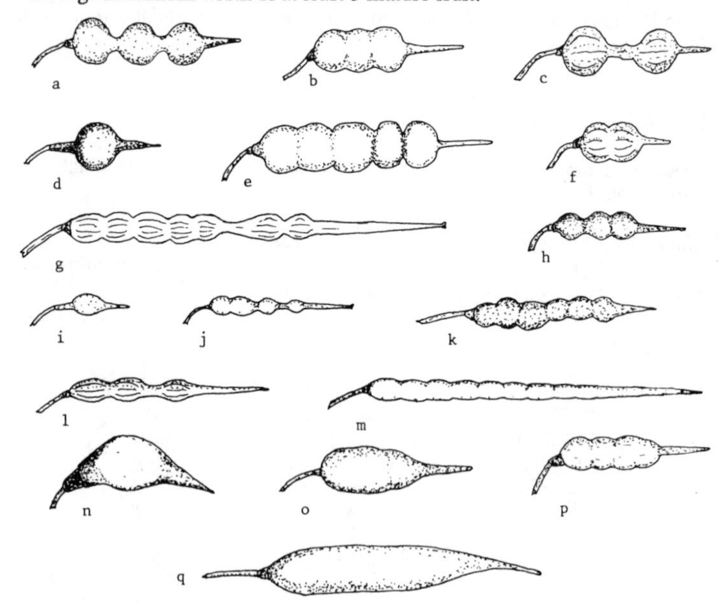

Figure 17. Variation in fresh fruits of *Raphanus* species. a-f *R. maritimus;* g-m *R. raphanistrum*; n-q *R. sativus.*

In addition, the seeds of *R. raphanistrum* and *R. maritimus* are more or less impossible to extract from the fruit casing (endocarp) and the fruits will break up into 1-seeded segments (more easily in the former than the latter). In *R. sativus*, although the fruits do not break into segments, the seeds are easily thrashed out or extracted from the fruit casing (Q.O.N. Kay, pers. comm.).

RAPISTRUM Crantz

1 Upper segment of fruit contracted into a distinct linear style (0.8-)1-5 mm (Figs. 18a-e); annual **4 R. rugosum sensu lato**

1 Upper segment of fruit tapering to an indistinct conical beak 0.5-1.2 mm (Fig. 18f); perennial **5 R. perenne**

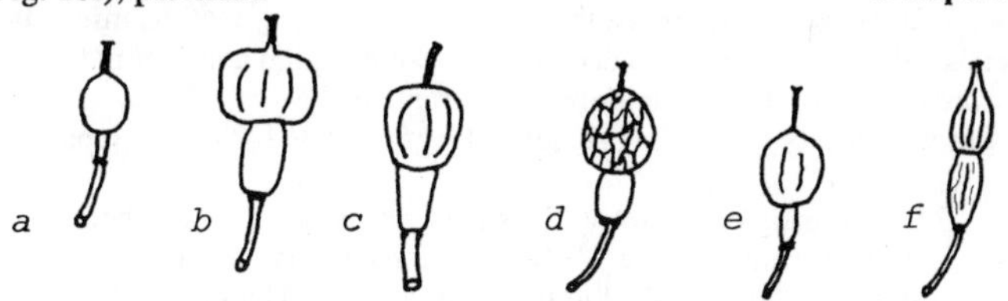

Figure 18. Fruits of *Rapistrum* species. a-e *R. rugosum* sensu lato; f *R. perenne*. Not to scale.

RORIPPA Scop. *

Rorippa species are variably fertile which often causes problems even identifying them to genus. They have yellow or white, 1-6(-6.2) mm petals, orbicular, elliptic or linear, 2.5-26 mm fruits with or without a persistent style, and numerous seeds in 1 or 2 rows in each loculus. They are usually glabrous but may be minutely hairy. *Barbarea* and *Brassica* species have larger flowers and/or fruits.

The genus *Nasturtium* R.Br. is traditionally recognised by British botanists but is more logically included in *Rorippa* on a world scale.

The variable fertility means that sterility or poor fruit set is of little value in assessing hybridity. The two commonest hybrids are included in the key, but another 3 also occur rarely. They are all ± intermediate between the parents. All taxa are variable morphologically and should be checked against the descriptions. Voucher specimens are required together with material of the parents from the same area to help account for local variability.

Descriptions of leaves apply to leaves on the main flowering stem(s); side shoots or rosette leaves are more variable. Petals and sepals lengths refer to fresh, mature (not newly opened) flowers. Length of fruit includes the style.

For full details of *Rorippa* s.str., see Jonsell (1968, 1975b) and for details of Section *Nasturtium*, see Rich (1987b).

* by B. Jonsell & T. C. G. Rich.

1 Petals white or purplish; stems decumbent at base, rooting at the nodes; leaves regularly pinnate (rarely simple) (*1)
2 Fruits 5-10(-11) mm, aborting or deformed; well-formed seeds 0-3(-4) per loculus; seeds with 10-14 depressions across width **61 R. x sterilis**
2 Fruits (9-)11-23(-24) mm, well-formed; well-formed seeds numerous; seeds with 6-20 depressions across width
3 Mature fruit (9-)11-19(-24) x (1.6-)1.9-2.7(-3) mm, up to 9.5 times as long as wide; seeds with (6-)7-12 depressions across width **59 R. nasturtium-aquaticum**
3 Mature fruit (15-)16-23(-24) x (1-)1.3-1.8(-2) mm, 10 or more times as long as wide; seeds with (11-)12-18(-20) depressions across width **60 R. microphylla**
1 Petals yellow; stems erect to decumbent, rooting at the nodes or not; leaves simple, or irregularly or regularly pinnate
4 Mature petals ± equalling sepals in length (petals 1-2.8 mm, sepals 1.1-3 mm)
5 Seeds finely colliculate (*2); stems ± prostrate to decumbent at base; fruits usually 2-3 times as long as pedicels (fruits 6-12 mm, pedicels 2-4 mm); petals 1-1.5(-1.7) mm; sepals 1.1-1.5(-2.4) mm **62 R. islandica sensu stricto**
5 Seeds more coarsely colliculate; stems ± erect from base; fruit usually 1-2 times as long as pedicels (fruit (4-)5-10 (-12) mm, pedicels 3-10 mm); petals (1.4-)1.7-2.7(-2.8) mm; sepals 1.6-3 mm **63 R. palustris**
4 Mature petals 1.4 or more times as long as sepals (petals (2.2-)2.8-6(-6.2) mm, sepals 1.8-4.3 mm)
6 Upper stem leaves pinnatisect to pinnatifid, lobed more than 1/2 way to midrib; auricles absent or small and inconspicuous (less than 1/2 width of stem)
7 Fruits 9-22 x 1-1.2 mm but variably fertile; persistent style 0.5-1(-1.2) mm; fruiting pedicels ascending to spreading; middle stem leaves equally deeply lobed at base and apex **64 R. sylvestris**
7 Fruits 3-10 x 1.2-2.5 mm but variably fertile; persistent style (0.8-)1.2-2.5(-3) mm; fruiting pedicels usually deflexed; middle stem leaves deeply lobed at base, more shallowly lobed at apex **66 R. x anceps**
6 Upper stem leaves simple, with margins entire, toothed or lobed to 1/3 of way to midrib (*3); auricles conspicuous or small (rarely absent)

8 Immature ovaries and valves of fruit elliptic to oblong; fruits (2.5-)3-6 mm; fruiting pedicels inclined to deflexed **65 R. amphibia**

8 Immature ovaries and valves of fruit orbicular; fruits *c*. 3 mm; fruiting pedicels usually ascending to inclined **68 R. austriaca**

*1 The taxa in Sect. *Nasturtium* can only be reliably distinguished in fruit (Rich 1987b). Select at least 5 fruits with ripe seeds from the lower half of the main inflorescence, measure length and width (width here is measured in the MIDDLE of the fruit, not necessarily the widest part) and average. Immature fruits continue to fatten after they have stopped elongating, hence fruit consistent in length and width must be selected for measuring.

The seed sculpting is the best character, most easily seen under a microscope. Mature, brown seeds are required (green seeds may be misleading); count the depressions across the width of at least 5 seeds and average. With familiarity, it is possible to distinguish seed sculpturing with a x10 (preferably x20) lens in the field.

Completely sterile plants could be either hybrids or sterile forms of either parent.

*2 Seed sculpturing is the best character (colliculate refers to low rounded swellings on the seed surface), but really requires microscopic examination with comparative material. The other characters given are reasonably diagnostic but confirmation from seed is still required. Voucher specimens for *R. islandica* s.s. are required, which need only consist of a few fruits with ripe seeds.

*3 Lower leaves of *R. amphibia* may be pinnatifid, especially if submerged, but these differ markedly from the upper leaves and usually do not persist.

SINAPIS L.

Both *Sinapis* species have yellow petals with a linear claw, and linear (inrolled) sepals, which are widely spreading to reflexed at maturity. Reflexed sepals are a ± diagnostic character of *Sinapis*, but are not always clearly shown. The fruits have a long beak/terminal segment which either has one seed or is sterile. The valves of the fruit have 3-7 strong veins when dried but these are usually poorly shown when fresh. A secondary character which may help to distinguish them from *Brassica* is that they are often coarsely hairy (*Brassica* species with stalked leaves are usually glabrous) but this is not diagnostic. *Sinapis* and *Brassica* are closely related and are sometimes treated as one genus. *Coincya* species have erect sepals.

Sinapis arvensis and *S. alba* are both very variable but are quite distinct and should not be confused.

1 Uppermost stem leaves simple or with 2 shallow, acute lobes at base (toothing very variable); terminal segment/beak conical, circular (orbicular) to elliptic in cross section **24 S. arvensis**
1 Uppermost stem leaves pinnatisect to pinnately lobed; terminal segment/beak flat (except at base if seed is present) **25 S. alba**

SISYMBRIUM L.

Sisymbrium is a difficult genus to define - perhaps the most distinctive feature is the very long, thin, ± terete fruits. The valves have 1-3 veins but these may be difficult to see. The species have long, simple hairs (or are sometimes glabrous) and small, usually obovate, yellow petals.

1 Flowers bracteolate
  2 Petals 1.5-2.5 mm; flowers in clusters of (1-)2-3(-5) in the axil of each bracteole **39 S. polyceratium**
  2 Petals 2.5-3.5 mm; flowers usually solitary in the axil of each bracteole **40 S. runcinatum**
1 Flowers without bracteoles (sometimes lowest 1-3 flowers with bracteoles)
  3 Fruits 10-18(-20) mm, appressed **34 S. officinale**
  3 Fruits (18-)20-120 mm, erect to recurved
    4 Pedicels about as wide as fruits, (0.3-)0.4-1.6 mm wide in middle
      5 Fruits 18-34 mm; petals 2-3.5 mm **33 S. erysimoides**
      5 Fruits (30-)40-120 mm; petals 4-11 mm
        6 Uppermost stem leaves entire, hastate or with 1-2, linear-oblong lateral lobes and a broader terminal lobe; young fruits hairy **35 S. orientale**
        6 Uppermost stem leaves with 2-5 filiform-linear lateral lobes and a similar linear terminal lobe; young fruits usually glabrous **36 S. altissimum**
    4 Pedicels *c.* 1/2 as wide as fruits, 0.15-0.5 mm wide in middle
      7 Annuals, easily uprooted; petals 2.5-7 mm; at least some leaves lobed
        8 Petals 2.5-4(-6) mm; fruits (20-)25-47 mm; young fruits overtopping flowers **31 S. irio**
        8 Petals 4.5-7 mm; fruits (7-)11-31 mm; some flowers overtopping fruits **32 S. loeselii**
      7 Perennials forming clumps or patches; petals 6-10 mm; all leaves simple or some lobed
        9 Lower leaves ovate to hastate; fruits 15-45(-60) mm, rarely developing properly **37 S. volgense**
        9 All leaves lanceolate; fruits (40-)50-75 mm **38 S. strictissimum**

THLASPI L.

Thlaspi species have white petals, clasping upper stem leaves, flattened, angustiseptate fruits which are winged (at least at apex) and have 2-many seeds in each loculus. *Lepidium* species are often confused with *Thlaspi* but have only one seed per loculus.

The revision of *Thlaspi* into 12 genera by Meyer (1973) is not accepted here.

1 Fruits (6-)9-20 mm, broadly winged all round (Fig. 19a) **113 T. arvense**

1 Fruits 3-10 mm, broadly to narrowly winged above, narrowly winged below (Figs. 19b-f)

  2 Persistent style of fruit (0.3-)0.5-1.5 mm, equalling or exceeding apical notch (Figs. 19b-d) **111 T. caerulescens**

  2 Persistent style of fruit 0-0.3 mm, shorter than or equalling apical notch (Figs. 19e-f)

    3 Stem leaves (1-)2-4(-6), the upper ovate to lanceolate; fruits 3-5.5(-7.5) mm; fresh plant not smelling of garlic when crushed **112 T. perfoliatum**

    3 Stem leaves 4-11, narrowly elliptic to oblong; fruits 5-10 mm; fresh plant smelling faintly of garlic when crushed **114 T. alliaceum**

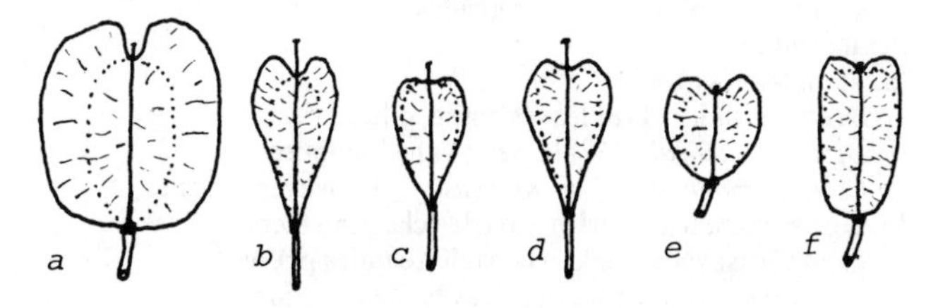

Figure 19. Fruits of *Thlaspi* species. a *T. arvense*; b-d *T. caerulescens*; e *T. perfoliatum*; f *T. alliaceum*. Not to scale.

SYNOPTIC KEY

This synoptic key is to help identify plants without fruit, but may also provide a quick (if less reliable) alternative to the main key.

Plants are keyed into groups using 3 basic characters (petals, stem leaves and hairs) and the taxa listed. Plants can then be identified by eliminating known plants, matching against illustrations and descriptions, or by working through the characters. The number in brackets either gives the species number or refers to the relevant page for the species keys.

The taxa in each group must be worked through strictly in order, as a diagnostic character used near the end of the list may also apply to the genus or species already eliminated earlier in the list using another character. For example in group 2b, *Coronopus didymus* is keyed out first using an inflorescence character, then *Cardamine* using linear fruits leaving *Lepidium* with 1 seed in each loculus; *Coronopus didymus* also has 1 seed in each loculus hence could be mis-identified as *Lepidium* if the characters are assumed to be diagnostic.

In some cases where it is not possible to identify individual genera quickly and concisely without fruit, either the genera are not separated or fruit characters are given. It should be possible to put a plant into a group with 95% reliability. Within the groups, the characters used are more generalised and less precise than in the main key and identification using the diagnostic characters may only be 80% reliable (less in the larger groups).

This key is partly experimental, being a compromise between multi-access and dichotomous keys, but it is anticipated that it will nonetheless be useful.

1a. Petals absent

- **2a. Upper stem leaves with auricles clasping stem**
  - *Conringia orientalis* (**73**) leaves entire
  - *Cardamine* (p.38) leaves pinnate
- **2b. Upper stem leaves sessile or petiolate**
  - *Coronopus didymus* (**133**) lateral inflorescences opposite leaves
  - *Cardamine* (p.38) fruits linear
  - *Lepidium* (p.43) seeds 1 per loculus

1b. Petals yellow

- **3a. Stem leaves absent**
  - *Draba aizoides* (**118**) leaves entire, ciliate
  - *Diplotaxis muralis* (**28**) leaves toothed or lobed
  - *Coincya monensis* subsp. *monensis* (**14a**) leaves pinnate
- **3b. Upper stem leaves with auricles clasping stem**
  - **4a. At least some forked or stellate hairs present**
    - *Descurainia sophia* (**41**) leaves finely divided
    - *Neslia* (p.45) fruits 2-3.5 mm, ± orbicular
    - *Camelina* (p.37) fruits 6-13 mm, obpyriform
    - *Arabis* (p.34) fruits 30-120 mm, linear
  - **4b. Glabrous or with simple hairs only**
    - **5a. Petals 1-6 mm**
      - *Isatis tinctoria* (**116**) fruit pendulous
      - *Lepidium* (p.43) upper leaves entire, lower leaves finely divided
      - *Myagrum perfoliatum* (**137**) fruit obovate
      - *Rorippa* (p.47) fruits 2-12 mm
      - *Barbarea* (p.35) fruits 11-71 mm

5b. Petals 6-30 mm

Rorippa (p.47) fruits 2-12 mm
Conringia orientalis (**73**) all leaves entire
Barbarea (p.35) beak of the fruit 0.5-3.5 mm, seeds irregularly oblong
Brassica (p.36) beak of the fruit 0.5-22 mm, seeds orbicular

3c. Upper stem leaves sessile or stalked

6a. At least some forked or stellate hairs present

Descurainia sophia (**41**) leaves finely divided
Alyssum alyssoides (**89**) petals 2-3 mm
Aurinia saxatilis (**90**) fruits ± orbicular
Erysimum (p.42) fruits linear

6b. Glabrous or with simple hairs only

Bunias (p.37) large glands present, especially on pedicels
Erucastrum gallicum (**29**) pinnatifid bracteoles present
Sisymbrium (p.50) toothed bracteoles present

Diplotaxis (p.41)	seeds in 2 rows in each loculus	beak 1-3.5 mm
Eruca vesicaria (**26**)		beak 4-11 mm
Raphanus (p.46)	valves absent, fruit indehiscent	petals 11-25 mm
Rapistrum (p.47)		petals 4-11 mm
Rorippa (p.47)	persistent style 0-2(-2.5) mm	petals 1-6 mm
Sisymbrium (p.50)		petals 2-11 mm

Sinapis (p.49) at least some flowers with reflexed sepals

Brassica
Coincya
Hirschfeldia
} try Group D, lead 4 onwards of the main key (p.29)

1c. Petals cream

7a. Upper stem leaves with auricles clasping stem

Arabis glabra (**45**) fruits linear
Camelina (p.37) fruits obpyriform

7b. Upper stem leaves sessile or petiolate

Arabis scabra (**44**) fruits flattened
Sisymbrium (p.50) fruits round in cross section

1d. Petals white

8a. Stem leaves absent

Erophila (p.42) petals bifid
Cardamine (p.38) lateral leaflets stalked
Subularia aquatica (**125**) leaves linear
Teesdalia nudicaulis (**98**) fruit notched at apex
Draba norvegica (**120**) high alpine rocks

8b. Upper stem leaves with auricles clasping stem

9a. At least some forked or stellate hairs present

Arabis (p.34) fruits linear, 17-90 mm
Camelina (p.37) persistent style 1-4 mm
Capsella bursa-pastoris (**115**) fruit notched at apex
Draba (p.41) fruit elliptic to lanceolate

9b. Glabrous or with simple hairs only
Cardamine (p.38) lateral leaflets stalked
Conringia orientalis **(73)** fruits 45-150 mm
Lepidium (p.43) 1 seed per loculus
Thlaspi (p.51) fruit winged
Capsella bursa-pastoris **(115)** fruit notched at apex
Cochlearia (p.39) plant generally fleshy

8c. Upper stem leaves sessile or petiolate

10a. At least some branched or stellate hairs present

11a. Petals 1-5 mm
Arabidopsis thaliana **(42)** } fruits linear { annual
Arabis (p.34) } { biennial-perennial
Alyssum alyssoides **(89)** sepals persistent in fruit
Lobularia maritima **(87)** fruit ± circular, 1 seeded
Draba (p.41) fruit elliptic to lanceolate, many seeded

11b. Petals 5-33 mm
Berteroa incana **(88)** petals bifid
Aubrieta deltoidea **(86)** inner filaments forked, valves elliptic
Arabis (p.34) petals 5-9 mm
Matthiola (p.44) seeds flat, winged
Hesperis matronalis **(82)** petals 17-30 mm
Malcolmia maritima **(81)** petals 9-18 mm

10b. Glabrous or with simple hairs only

12a. Petals 0.1-2 mm
Coronopus squamatus **(134)** lateral inflorescences opposite leaves
Hornungia petraea **(99)** stem leaves regularly pinnate
Teesdalia nudicaulis **(98)** outer petals larger than inner
Lepidium (p.43) 20-130 cm tall

12b. Petals 2-5 mm
Iberis (p.43) outer petals larger than inner
Lepidium (p.43) fruit winged
Lobularia maritima **(87)** valves ± circular
Alliaria petiolata **(83)** fresh plant smelling of garlic when crushed
Rorippa (p.47) fruits ± terete
Cardamine (p.38) lateral leaflets stalked
Arabidopsis thaliana **(42)** } fruits ± flattened { annual
Arabis (p.34) } { biennial-perennial

12c. Petals 5-11 mm
Iberis (p.43) outer petals larger than inner
Alliaria petiolata **(83)** } fresh plant smelling of
Pachyphragma macrophylla **(136)** } garlic when crushed
Carrichtera annua **(135)** lateral inflorescence opposite leaves

Crambe (p.40) inner filaments toothed
Armoracia rusticana **(117)** basal leaves 40-120 cm
Cakile maritima **(12)** plant fleshy

Arabis (p.34)	fruits ± flattened	lateral lobes sessile
Cardamine (p.38)		lateral leaflets stalked

Rorippa (p.47) fruits 5-26 mm
Sisymbrium (p.50) fruits 40-110 mm

12d. Petals 11-30 mm

Iberis (p.43) outer petals larger than inner
Lunaria (p.44) fruits 15-35 mm wide

Cardamine (p.38)	valves present	seeds in 1 row in each loculus
Eruca vesicaria **(26)**		seeds in 2 rows in each loculus
Raphanus (p.46)	valves absent	leaves green, thin
Cakile maritima **(12)**		leaves glaucous, fleshy

1e. Petals blue, purple, lilac, pink, red or orange

13a. Upper stem leaves with auricles clasping stem

Arabis (p.34) fruits linear
Thlaspi caerulescens **(111)** fruit winged
Capsella bursa-pastoris **(115)** fruit notched at apex
Cochlearia (p.39) plant fleshy

13b. Upper stem leaves sessile or stalked, without auricles

14a. At least some forked or stellate hairs present

Lobularia maritima **(87)** petals 2-4.5 mm
Aubrieta deltoidea **(86)** inner filaments toothed, valves of fruit elliptic
Arabis (p.34) petals 4-9 mm

Malcolmia maritima **(81)**	seeds ± cylindrical	petals 9-18 mm
Hesperis matronalis **(82)**		petals 17-30 mm

Erysimum (p.42) stigma 2-lobed
Matthiola (p.44) stigma with or without lateral horns

14b. Glabrous or with simple hairs only

Iberis (p.43) outer petals larger than inner
Lunaria (p.44) fruits 15-35 mm wide
Lepidium sativum **(108)** fruit winged
Lobularia maritima **(87)** fruits 2-4.5 mm and plant hairy

Cakile maritima **(12)**	fruit indehiscent	plant fleshy
Raphanus (p.46)		plant not fleshy

Cardamine (p.38) lateral leaflets stalked
Arabis (p.34) fruits linear
Cochlearia (p.39) plant fleshy, glabrous
Erysimum (p.42) plant not fleshy, hairy

LIST OF RARE CHARACTERS

The following list of rare or uncommon characters may help to identify some taxa.

Garlic-scented when fresh: *Alliaria, Thlaspi alliaceum, Pachyphragma*
Large glands present (pedicels): *Bunias, Matthiola sinuata*
Leaves thick and fleshy: *Brassica, Cakile, Crambe, Cochlearia*
Lateral leaflets stalked: *Cardamine*
All leaves linear, basal: *Subularia*
Stem leaves absent: *Cardamine, Coincya, Draba, Erophila, Subularia, Teesdalia*
Stem leaves with axillary bulbils: *Cardamine bulbifera*
Lateral inflorescences arising opposite leaves: *Coronopus, Carrichtera*
Inflorescences densely contracted in fruit: *Crambe, Iberis*
Most flowers bracteolate: *Erucastrum gallicum, Arabis turrita, Sisymbrium, Brassica nigra*
Sepals reflexed: *Sinapis*
Sepals persistent in fruit: *Alyssum, Coronopus, Lepidium*
Petals absent: *Cardamine, Conringia, Coronopus, Lepidium*
Petals blue-purple: *Aubrieta*
Petals orange: *Erysimum*
Outer petals larger than inner: *Iberis, Teesdalia*
Petals bifid: *Erophila, Berteroa*
Petals yellow, ± as long as sepals: *Descurainia, Rorippa, Sisymbrium irio*
Stamens mostly 4: *Arabidopsis, Cardamine hirsuta, Draba muralis, Lepidium*
Stamens mostly 2: *Coronopus, Lepidium*
Inner filaments toothed or lobed: *Aubrieta, Crambe, Teesdalia*
Fruits pendent: *Isatis*
Fruits flat, more than 15 mm wide: *Lunaria*
Fruits ± globose: *Crambe, Neslia, Rapistrum, Rorippa*
Fruits with stipe 1 mm or more long: *Brassica elongata, Diplotaxis tenuifolia, Sisymbrium strictissimum, Lunaria*
Fruits with valves and fertile beak: *Brassica, Coincya, Hirschfeldia, Sinapis*
Fruits with large, flat beak: *Sinapis, Eruca, Carrichtera*
Fruits linear, more than 80 mm long: *Arabis, Brassica, Sisymbrium*
Fruits winged (beak absent): *Bunias, Iberis, Lepidium, Pachyphragma, Thlaspi*
Fruits usually 1(-2) seeded: *Bunias, Crambe, Isatis, Neslia, Rapistrum*

DESCRIPTIONS AND FIGURES

The descriptions give the main characters required for identification and describe the variation present in Britain and Ireland only; there is usually more variation in material from Europe and elsewhere. Population samples should fall within the ranges cited, though individual plants may show more variation. Figures or characters in brackets either represent atypical extremes, unconfirmed literature data or in some cases variation seen in European material. Doubtful character states are indicated with a question mark.

All measurements refer to *fresh* material unless otherwise stated (dried material often shrinks and is not easy to measure accurately). Measurements under 10 mm are given to to nearest 0.1 mm (though measurements to the nearest 0.5 mm will mostly be adequate for identification). Measurements over 10 mm are given to the nearest mm or cm as stated. Measurements are length, or length x width unless otherwise stated.

The descriptions are largely consistent between accounts. Some characters are only given in specific cases - these are either rare character states (e.g. toothed filaments) or useful only in certain genera (e.g. hairy buds in *Barbarea*). Brief diagnoses only are given for some very rare or critical taxa. Note that "ovate to lanceolate" means that the shape varies from ovate to lanceolate, whilst "ovate-lanceolate" means intermediate between ovate and lanceolate.

Most of the illustrations have been drawn to a standard format so that the leaves appear on the left side of the page, the plant in the middle and the flowers and fruits on the right side; thus species can be identified from fruits simply by flicking through the book quickly. The petals/flowers can be coloured their appropriate colour so taxa can be picked out on colour quickly.

Scales are largely consistent within a genus, and are noted on the drawings. The parts illustrated are as follows:

A: whole plant

B: basal/rosette leaves

C: stem leaves

D: flower

E: sepal

F: petal

G: stamens and gynoecium

H: detail of stigma

I: fruit

J: fruit - side view

K: fruit T.S.

L: seed

1. Raphanus raphanistrum L.

Wild Radish, Runch, White Charlock

Annual (? biennial) 25-100 cm, with simple hairs mainly below. Stems erect, branched. Rosette leaves to 38 cm, petiolate, pinnate with an ovate, obtuse terminal lobe and 0-7 pairs of smaller, elliptic, usually not overlapping lateral lobes; margins sinuate to coarsely toothed. Lower stem leaves similar, but with a lanceolate, acute, terminal lobe and fewer lateral lobes. Upper stem leaves lanceolate to linear-lanceolate (rarely linear), acute; margins acutely toothed. Inflorescence lax. Sepals 6-11.5 mm, oblong, saccate, green to reddish, erect. Petals 11-24 x 4-9.5 mm; limb obovate, rounded to truncate at apex, white to yellow (rarely purple), often with darker veins; claw longer than limb, linear, whitish. Petals about twice as long as sepals. Stamens 6; anthers yellow. Ovules 4-14. Stigma ± entire. Pedicels in fruit 10-25 mm, slender, ascending to inclined. Fruits (13-)25-85 x 2.5-5.5(-6) mm ((1.5-)2-5 mm wide when dry), ± linear in outline, erect to ascending. Lower segment very reduced (*c.* 1-2 mm), sterile (rarely *c.* 3 mm with one seed), indehiscent. Terminal segment with 1-10 fertile segments, ± terete, constricted (sometimes deeply) or not between the segments, unribbed to deeply ribbed, the portion beyond the last fertile segment 6-24 mm, narrowly conical, with a sessile stigma. Seeds 1 per fertile segment, 2-2.9 mm, ± globose or compressed, brown. 2n=18*. Flowering June to October.

A weed or casual plant of fields, flower beds, cultivated land, docks, waste ground, roadsides, etc. Seeds germinate mainly in the spring but are often strongly dormant (Roberts & Boddrell 1983). Once a noxious weed but now controlled by selective herbicides. Common in England, frequent in Ireland, Wales and Scotland (map in Perring & Walters 1962). Probably native in Europe, N. Africa and the Near East, introduced and often a serious weed in Asia, the Far East, N. and S. America and Australasia.

A variable plant. Petals may be white, various shades of yellow or sometimes purple. Yellow-flowered plants are widely distributed, white-flowered plants rare in the far north and west (Perring 1968), and purple-flowered plants occur mainly in England. The pollination biology of the different morphs has been much studied (e.g. Kay 1976). The fruits are also variable in shape and size, and there is a clinal increase in width and decrease in constriction between segments northwards (Q.O.N. Kay, pers. comm.).

R. raphanistrum is sometimes treated as an aggregate which includes *R. maritimus* and other taxa as subspecies.

The large, indehiscent fruits of *Raphanus* are distinctive. Without fruit, *R. raphanistrum* is sometimes confused with *Sinapis arvensis* which has spreading sepals and darker yellow petals. For distinction from *R. maritimus* and *R. sativus*, see page 46 or Rich (1989).

Raphanus raphanistrum 1

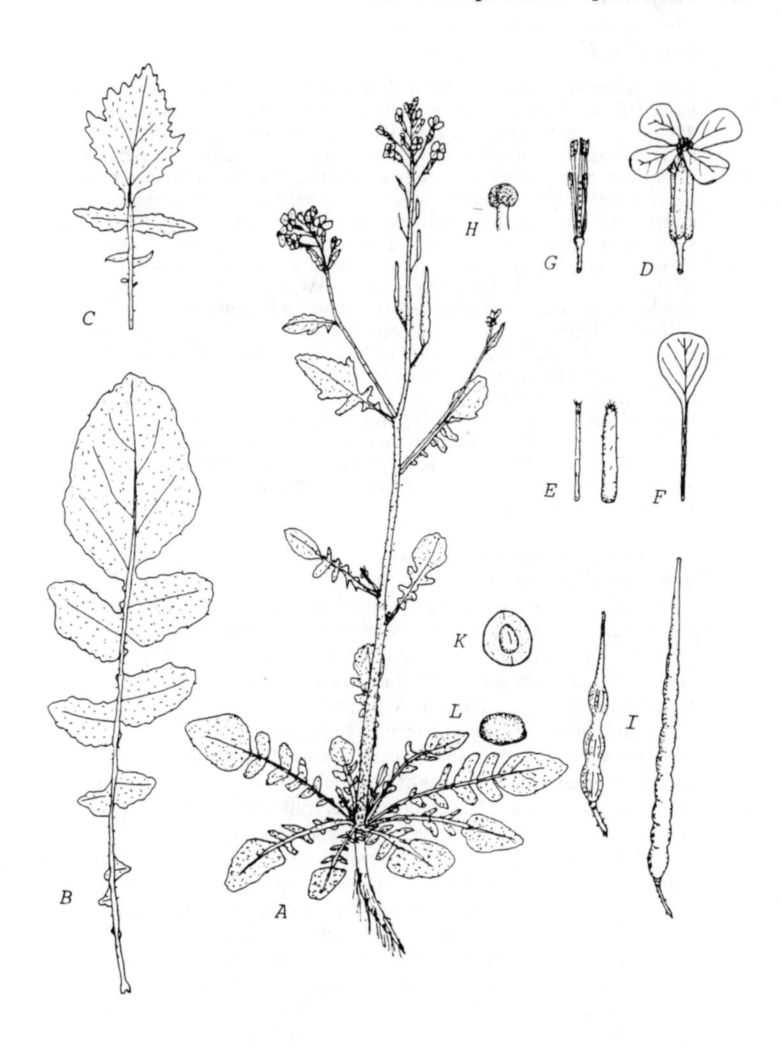

Ax0.2;B,Cx0.4;D-Gx1;Hx10;I,Kx1;Lx4.

2. **Raphanus maritimus** Sm.

Sea Radish

Biennial or perennial 30-130 cm, dark green to purple, with coarse, simple hairs below. Stems erect, branched mainly above. Rosette leaves to 40(-60) cm, petiolate, pinnate with an ovate, obtuse terminal lobe and (4-)5-10 pairs of smaller, elliptic, distant or overlapping lateral lobes; margins coarsely toothed. Lower stem leaves similar, with fewer pairs of lateral lobes. Upper stem leaves lanceolate to linear, acute at apex; margins entire to sparsely toothed. Inflorescence lax. Sepals (7-)8-10 mm, oblong, saccate, green to reddish, erect. Petals (14-)15-22(-25) x 4-7 mm; limb obovate, rounded to truncate at apex, deep yellow to pale yellow (rarely white), often with darker veins; claw ± longer than limb, linear, whitish. Petals about twice as long as sepals. Stamens 6; anthers yellow. Ovules 1-5(-6). Stigma ± entire. Pedicels in fruit (9-)13-22(-24) mm, slender, ascending. Fruits (10-)15-55(-57) x 5-10 mm ((4-)4.5-8.5 mm wide when dry), erect to ascending, indehiscent. Lower segment very reduced (*c.* 1-2 mm), sterile. Terminal segment with 1-5(-6) fertile, fleshy segments, ± terete, deeply (rarely shallowly) constricted between the segments, with shallow or medium ribs, the portion beyond the last fertile segment *c.* 5-25 mm, ± linear to narrowly conical, tapering to a sessile stigma. Seeds 1 per segment, 2.5-4.7 mm, ± globose to compressed, brown. 2n=18*. Flowering mainly May to July.

A locally abundant plant of sand dunes, shingle, cliffs, open coastal grassland, disturbed ground near the sea and docks. Fruits are dispersed by the tide. Scattered around the south and west coasts of Britain, much rarer in the east and far north possibly due to frost sensitivity. Occasional around the coast of Ireland (map in Perring & Walters 1962). In Europe from the Black Sea and the Mediterranean around the Atlantic coasts to the Netherlands. Introduced to Australasia.

Variable in leaf shape and fruit development. White-flowered plants have been recorded from the Channel Islands, Cornwall and the Clyde.

Often treated as a subspecies of *R. raphanistrum* and sometimes difficult to distinguish from it. The combination of fruit width and number of ovules should be diagnostic and plants agreeing with the descriptions can be recorded with confidence. Some populations in the south west in fields or disturbed ground near the sea show overlap in characters between *R. maritimus* and *R. raphanistrum* and are probably of hybrid origin; these are 50-60% pollen- and seed-sterile (Harberd & Kay 1975).

Raphanus landra Moretti ex DC. has been reported as introduced with foreign grain but I have been unable to confirm its occurrence in Britain (most material is large *R. raphanistrum*). It will key out to *R. maritimus* but has petals 10-15 mm, white or yellow, and distant lateral lobes to the leaves.

Raphanus maritimus 2

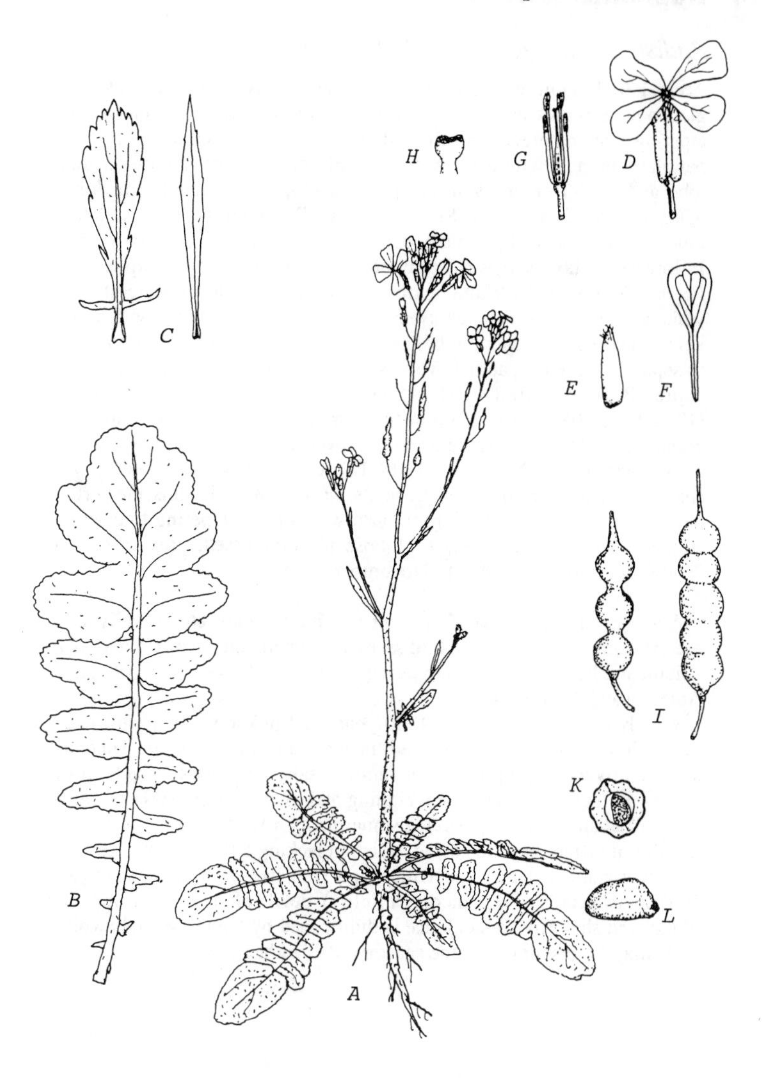

Ax0.2;B,Cx0.4;D-Gx1;Hx10;I,Kx1;Lx4.

3. **Raphanus sativus** L.

Radish, Garden Radish, Salad Radish

Annual 30-130 cm, with coarse, simple hairs below or glabrescent, often glaucous. Tap-root swollen, long, not or only moderately swollen and then tapering. Stems erect, branched above. Rosette leaves to 25(-35) cm, petiolate, pinnate with a large, broadly elliptic to ovate, obtuse terminal lobe and 0-3(-5 or rarely more) pairs of elliptic, smaller lateral lobes; margins coarsely toothed. Stem leaves similar, with 0-2 pairs of lateral lobes. Upper leaves lanceolate, acute; margins entire to acutely toothed. Inflorescence lax. Sepals 6-9 mm, oblong, saccate, green to purple, erect. Petals 12-21 x (4.5)5-9 mm; limb obovate, rounded to truncate at apex, white to pink or purple (exceptionally yellowish), sometimes with darker veins; claw ± longer than limb, linear, whitish. Petals about twice as long as sepals. Stamens 6; anthers yellow. Ovules (?1-)3-11. Stigma capitate, entire. Pedicels in fruit 4-21(-24) mm, stout, ascending to patent. Fruits (17-)20-54(-90) x 5-12(-15) mm (4-10 mm wide when dry), shape very variable, erect to patent, indehiscent. Lower segment 1-4 mm, with 0 or 1 seed. Terminal segment with 1-10(-12) seeds, ± terete, inflated or with shallow constrictions between the seeds, with *c.* 5-8 weak veins, the portion beyond the last seed *c.* 8-20 mm, narrowly conical, tapering to a sessile stigma. Seeds 2.7-3.7 mm, ± globose to compressed, grey to brown. 2n=18, 36. Flowering May to November.

A non-persistent casual of tips, waste ground, fields, etc., usually as an escape from cultivation or in bird seed. Infrequent, but recorded widely in Britain and Ireland close to habitation. A weed in Europe, Asia, N. and S. America and Australasia.

Widely cultivated for over 4000 years and unknown as a wild plant. Lewis-Jones *et al.* (1982) suggest it is derived from *R. landra.* There are many cultivars ranging from our familiar salad radishes to plants in the tropics with giant roots (some weighing 50 kg) or giant (to 100 cm) fruits. Fodder radish (?var. *oleiferus*) is sometimes grown for forage or a green manure; it differs from the salad radish in lacking a fleshy root.

Very variable in leaf shape, fruit shape, fruit size and seed set (Sampson 1957 gives details of self-incompatibility), petal colour and development of the "radish". Root succulence is diminished by long days and swollen roots may be absent late in the season (Al-Shehbaz 1985).

Raphanus sativus 3

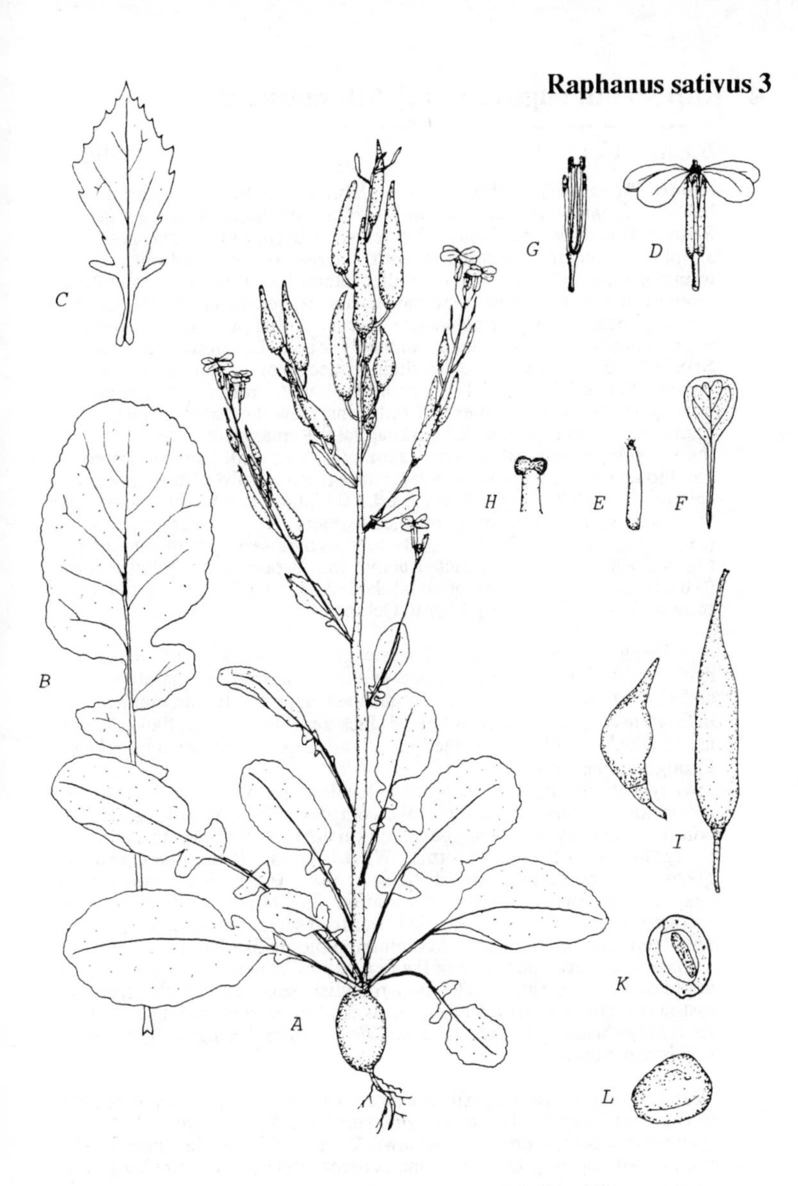

Ax0.2;B,Cx0.4;D-Gx1;Hx10;I,Kx1;Lx4.

4. Rapistrum rugosum (L.) All. *sensu lato*

Bastard Cabbage Map 4

Annual (? biennial) 15-100(-150) cm, glabrous or with coarse, simple hairs below. Stems erect, usually with widely spreading branches above. Rosette leaves not persisting. Lower stem leaves to 25 cm, petiolate, simple or pinnate to pinnatifid with a large, ovate to elliptic, obtuse terminal lobe and 0-5 pairs of smaller lateral lobes; margins irregularly toothed or lobed. Upper stem leaves smaller, sessile or shortly stalked, simple, linear, oblong or oblanceolate, cuneate below, acute to obtuse at apex; margins entire, toothed or sinuately lobed. Inflorescence crowded. Sepals 2.6-5 mm, ± linear, often slightly saccate or awned, green, erect to patent. Petals 5.7-11 x 2.4-4 mm; limb obovate, rounded to truncate at apex, pale yellow, sometimes with red veins; claw short or indistinct, paler. Stamens 6; anthers yellow. Stigma capitate, ± emarginate. Pedicels 1.5-5 mm, slender, appressed to erect. Fruits (3-)5-12 mm, hairy or glabrous, erect to appressed with two indehiscent segments (valves absent). Lower segment 1-3 x 0.8-1.5 mm (0.7-1.2 x 0.3-0.9 mm if sterile), ellipsoid, with 1(-3) seed(s), often remaining attached to plant. Terminal segment 1.3-3.5 (excluding style) x 1-2.8 mm, globose to ovoid, weakly ribbed to deeply rugose, with 1 seed, constricted above into a persistent, ± linear style (0.8-)1-3.5(-5) mm; stigma often bilobed. Seeds 1.1-2.4 mm, ovoid, light brown. 2n=16. Flowering May to October.

A locally abundant casual of waste ground, arable fields, roadsides, docks, rubbish tips, breweries, bird seed, etc. Introduced mainly with S. and C. European seed and well established and persistent in many places in England and Wales, rare in Ireland (Jackson 1981) and Scotland. Native around the Mediterranean to the Black Sea and a common weed introduced throughout the world.

Very variable in fruit shape and development which has led to the recognition of many taxa within the complex. There is little agreement whether distinct taxa can be distinguished or if there is complete intergradation between the forms. What is certain is that the characters given in many accounts are difficult to use. Provisional investigation suggests the two extremes of variation illustrated (*a, b*) may merit recognition, but plants with sterile lower fruit segments (*c*) do not. Plants with fruit type (*a*) are the most common in the British Isles.

Rapistrum rugosum is one of the few yellow crucifers with small fruits appressed to the stem. The two roundish segments of the fruit are distinctive (though the lower segments is often sterile) and distinguish it from *Sisymbrium officinale*, *Hirschfeldia* and *Brassica nigra*. The plant is self-incompatible.

5. **Rapistrum perenne** (L.) All. is a rare casual of waste ground, docks, etc. in England (Map 5). It is a perennial similar to *R. rugosum* but the two fruit segments are more alike, the lower 2-5 mm, elliptic, the upper 2.5-4.5 mm, ovoid, tapering to an indistinct, conical style 0.5-1.2 mm long. It is native C. and E. Europe.

Rapistrum rugosum 4

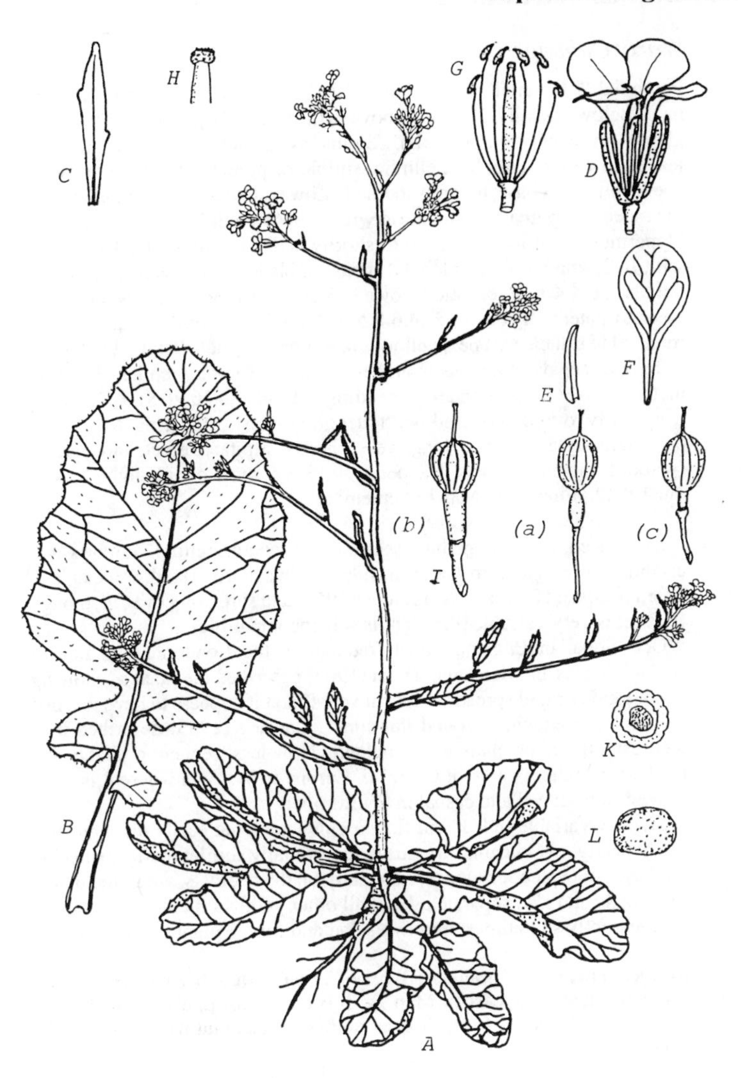

Ax0.2;B,Cx0.5;D-Gx3;Hx10;I,Kx3;Lx10.

6. **Bunias orientalis** L.

Warty Cabbage Map 6

Perennial (? biennial) 40-120 cm, softly hairy with simple and branched hairs below, usually glabrous above with many, large, yellow to black glands (especially on pedicels). Stems erect, much branched. Rosette leaves to 45 cm, petiolate, elliptic, simple or pinnatifid at base, acute at apex, cuneate at base; margins toothed. Lower stem leaves similar, elliptic to lanceolate, pinnate to pinnatifid with 1-3(-4) small lateral lobes at base. Uppermost stem leaves sessile to shortly stalked, lanceolate, ± entire or with 1(-2) small triangular lobe(s) at base of blade. Inflorescence crowded. Sepals 2.6-4.4 mm, elliptic to ovate, scarcely saccate, yellowish-green, erect to patent. Petals (4.2-)4.6-8.5 x 2.4-4.2 mm; limb broadly ovate, rounded to truncate at apex, yellow; claw short, distinct. Petals about twice as long as sepals. Stamens 6; anthers yellow. Stigma ± entire. Pedicels in fruit 10-18 mm, slender, ascending. Fruits (5-)6-7(-8) x 3-5 mm, irregularly ovoid, covered with irregular warts, tapering to a stout persistent style *c.* 1 mm long, very shortly stipitate, erect, indehiscent (without valves). Seeds 1-2 per fruit, 3-3.5 mm, ovoid, light brown. 2n=14, 42. Flowering May to September.

Introduced with grain, bird seed and other agricultural imports and established on waste ground, roadsides, railways, riverbanks, docks, etc. Often persistent for many years (Jones 1959) and spreading by seed, though fruit set is very variable (the plant is self-incompatible).

Occasional in England (most frequent in the London area), rare in Ireland, Wales and Scotland. Dunn (1905) described the plant as showing a remarkably rapid spread "in recent years" and it appears to have reached its greatest frequency around this time. Lousley (1953) described it as well-established on chalk in S. England. Now less frequent probably due to cleaner grain. Native in C. and E. Europe to W. Asia where it is often a weed, introduced and casual in W. Europe.

Slightly variable in leaf and flower characters, but easily distinguished by the warty, ovoid fruits. *Bunias* is superficially similar to *Rapistrum* but the large, conspicuous glands, especially on the pedicels, are immediately diagnostic and also separate it from all other yellow crucifers.

Apparently sometimes used as a salad and forage plant.

7. **Bunias erucago** L., *Southern Warty Cabbage*, has been recorded as a non-persistent, rare, casual (Map 7). It is easily distinguished by the four crested wings on the fruit. It is an annual or biennial native around the Mediterranean.

Bunias orientalis 6
*B. erucago 7

Ax0.2;B,Cx0.3;D-Gx4;Hx10;I,Kx2;Lx4;A*x0.2;*I,*Kx2.

8. **Neslia paniculata** (L.) Desv. *sensu stricto*

Ball Mustard Map 8

Annual (? biennial) 20-80 cm, with simple and branched hairs below, green. Stems erect, branched above. Rosette leaves to 10 cm, not persisting, oblanceolate with a broad petiole, obtuse; margins entire or sparsely toothed. Lower stem leaves to 8(-9) cm, oblong to lanceolate, with acute auricles clasping stem, obtuse to acute at apex; margins entire to sparsely toothed. Upper stem leaves similar but smaller, lanceolate, acute. Inflorescence crowded. Rarely the lowest 1-3 flowers bracteolate. Sepals 1.5-2 mm, oblong, green to yellow, erect. Petals 2.2-3.2 x 0.8-1.7 mm, narrowly obovate-oblanceolate, obtuse, yellow, indistinctly clawed. Petals *c.* 1.5-2 times as long as sepals. Stamens 6; anthers yellow. Stigma entire to ± emarginate. Pedicels in fruit 5-13 mm, slender, ascending to patent. Fruits 2.3-3.5 (including style) x 2.2-3.5 mm, ± globose, very shortly stipitate, erect to patent, indehiscent (valves absent). Lower segment 1.5-2.5 mm, rugose when dry with 2 inconspicuous, longitudinal ribs, with 1 seed. Persistent style 0.7-1.1 mm, linear, often deciduous. Seeds 1.5-2 mm, ovoid, brown. 2n=14. Flowering June to September.

A non-persistent casual of arable fields, waste ground, docks, chicken runs (rarely in bird seed, Hanson & Mason 1985), breweries, etc., characteristically introduced with grain and other agricultural imports. Formerly not infrequent, but now rarely reported. Occasionally recorded throughout England but only rarely so in Ireland, Scotland and Wales. A frequent weed in Europe, Asia and the Far East, widely introduced and often a common weed elsewhere in the world.

Little variable. Closely related to *N. apiculata* (see below; Ball 1961).

Easily distinguished by the little, round, wrinkled fruits, small yellow flowers and clasping leaves. Sometimes confused with immature *Lepidium draba* (which has white flowers), *Camelina* (which has many ovules or seeds in each fruit) or *Rorippa austriaca* (which is more robust and has dehiscent valves). Only separable in fruit from the following:

9. **Neslia apiculata** Fischer, C.A. Meyer & Avé-Lall. also occurs in similar places to *N. paniculata* but is much rarer (Sandwith & Sandwith 1935; Map 9). It has slightly larger fruits which are apiculate above and below and have 4 longitudinal ribs (examine 5-6 mature fruit from different parts of the inflorescence). It has a more southerly distribution in Europe. Intermediate plants occur and the taxa are sometimes treated as subspecies (e.g. Ball 1961).

Neslia paniculata 8

Ax0.3;B,Cx0.5;D-Lx12.

10. **Crambe maritima** L.

Sea Kale

Perennial 30-50 cm, forming low bushes, glabrous, green to purple, glaucous. Rootstock branched with many aerial stems. Leaves very variable, fleshy, undulate. Lower leaves to 50 cm, petiolate, ovate to elliptic in outline, cuneate to cordate at base, obtuse at apex, sinuately lobed to pinnatisect with 0-2 small lateral lobes at base of the terminal lobe; margins irregularly toothed to entire. Upper leaves smaller, ovate to lanceolate, acute, shortly stalked. Inflorescence a dense corymb. Sepals 4-7.5 mm, oblong, green often with purple tips, ascending to inclined. Petals (6-)6.5-11.1 x 5.5-10 mm; limb ± circular, white with purple veins; claw distinct, short, greenish. Petals *c.* 1.5 times as long as sepals. Stamens 6; inner 4 filaments with a lateral tooth; anthers yellow to purple. Stigma entire, sessile. Inflorescences remaining contracted in fruit. Pedicels in fruit (11-)14-32 mm, thick, ascending (to inclined). Fruits 6-15 mm, with 2 segments, erect. Lower segment 1-3 x *c.* 1.5 mm, ± oblong, sterile. Terminal segment (6-)7-14 x 6-11 mm, globose to ellipsoid, corky, indehiscent, with 0-6 indistinct veins; stigma sessile. Seeds 1-2 per fruit, 4-7 mm, ellipsoid, pale brown to black. 2n=60*(?30). Flowering May to August.

A native maritime plant found above the strand line on sandy or shingle beaches, rarely on sea cliffs. Inland, a rare escape from cultivation. Locally very abundant in suitable places (e.g. Dungeness) but more often in small populations. Said by some authors (e.g. Scott & Randall 1976) to be decreasing everywhere whilst others (e.g. Philp 1982, Mennema 1973) describe it as increasing. Scott & Randall (1976) describe the ecology in detail in the Biological Flora.

Most frequent on the south and west coasts of Britain north to the Clyde, scarcer and infrequent on the east coast north to Orkney. Formerly scattered around the Irish coast but now rare (map in Perring & Walters 1962). Coasts of Europe from N. Spain to the Baltic and C. Norway. A disjunct, distinct race occurs in the Black Sea, the plant surprisingly being absent from the Mediterranean.

A distinct plant variable in leaf shape and dissection. The large, glaucous leaves (similar to cabbage leaves) with dense corymbs of white flowers and round fruits distinguish it.

Rarely cultivated as a vegetable, the stalks and young shoots are prepared and eaten like asparagus (see Grigson 1955).

11. Crambe cordifolia Steven, a garden plant, native in the Caucasus, is rarely naturalised (Mullin 1988, 1989); Map 11. It is a tall plant 100-200 cm, hairy at least on pedicels, with thin leaves, lax racemes, petals to 7 mm and fruits 4-8 x 2.6-6.8 mm.

Crambe maritima 10

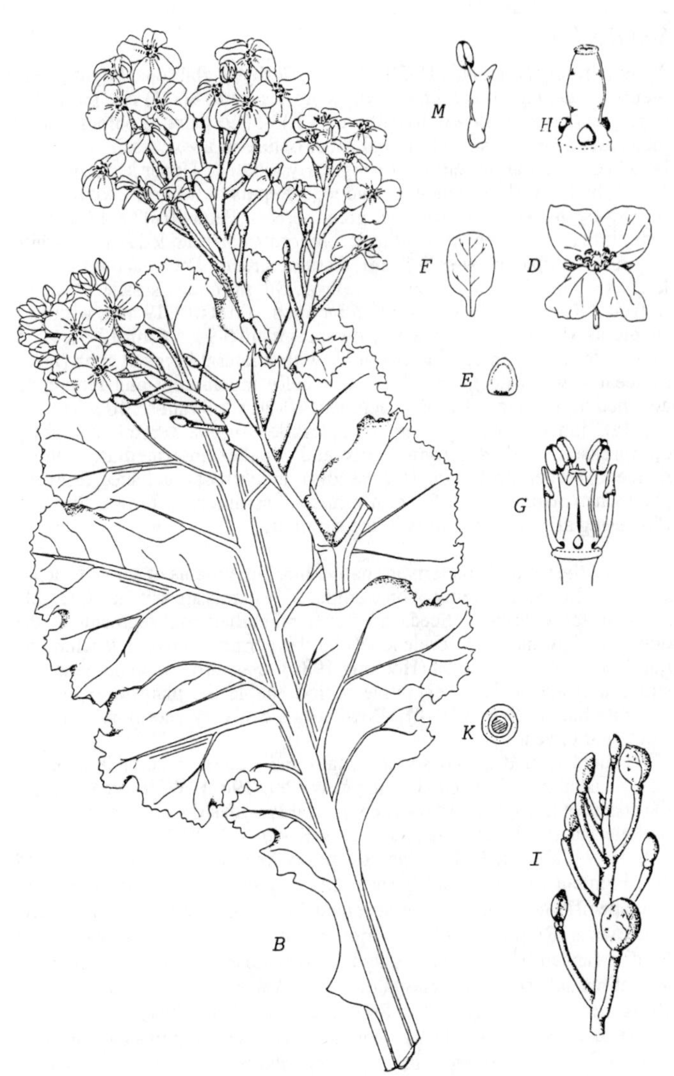

Ax0.3;D-Fx1;Gx2;Hx5;I,Kx0.7.

12. **Cakile maritima** Scop.

Sea Rocket

Annual (rarely perennial) 15-30(-45) cm, fleshy, ± glabrous, bright green. Stems decumbent, much branched. Stem leaves to 8(-11) cm, pinnatifid (rarely entire or shallowly lobed) with (2-)3-5(-9) pairs of oblong lateral lobes and a similar-sized terminal lobe, obtuse at apex, sessile or with a broad petiole; margins entire or irregularly toothed. Upper leaves usually less lobed. Inflorescence crowded. Sepals 3.9-6 mm, oblong to oblong-ovate, ± awned, saccate, green, erect. Petals (6-)7.5-13 x 3-7.5 mm; limb broadly obovate to elliptic, emarginate to rounded at apex, white to pink or purple; claw *c.* 3-5 mm, distinct, green. Petals about twice as long as sepals. Stamens 6; anthers yellow. Stigma entire, sessile. Pedicels in fruit 3-7 mm, thick, ascending to inclined. Fruits 12-19 mm, with two segments, fleshy, erect to patent. Lower segment 6-9 x 4-6 mm (2-5 x 1.5-5 mm if sterile), cylindrical to obovoid, with 2 lateral horns at apex (more noticeable when dry), with 1 seed, usually indehiscent and remaining attached to the plant. Terminal segment 10-20 x 5-9 mm (*c.* 10 x 3 mm if sterile), lanceoloid, tapering to a sessile stigma, terete to slightly compressed, with 4 strong veins and weaker intermediates (more noticeable when dry), with 1(-2) seed(s), indehiscent and shed from the plant when mature. Seeds 5-7 mm, ellipsoid, compressed, brown. 2n=18*. Flowering all year but mainly in the summer.

One of the most characteristics native maritime plants of sandy beaches and open dunes, less frequent on shingle. The plants are self-incompatible and fruit set is variable. Seeds are dispersed in their fruit segments by the tide, and high sodium chloride levels inhibit germination until leaching by rain lowers the salt content (Hocking 1982). Fast growth on nutrient-rich strandlines and a short vegetative period enable the plant to exploit an unstable habitat (Novo 1976). Populations can vary enormously in size from year to year.

Frequent around the coasts of Britain and Ireland in suitable habitats, rarely casual inland (map in Perring & Walters 1962). Widespread around the coasts of Europe and introduced, probably with ship's ballast, to N. and S. America and Australasia (Barbour & Rodman 1970, Rodman 1986).

Very variable in leaf lobing, flower colour, and in fruit shape and development of the lateral horns. Divided into a disputed number of infraspecific taxa but only represented in our area by subsp. *maritima* (following Rodman 1974), which occurs from C. Norway to the Mediterranean (Ball 1964 separated N. Atlantic and Mediterranean plants as distinct subspecies). Reports of the N. American *C. edentula* (Bigel.) Hooker from Britain are errors for *C. maritima* (Ball 1964).

The distinct fruit and fleshy, pinnatifid leaves distinguish *Cakile* and it is unlikely to be confused with any other species.

Cakile maritima 12

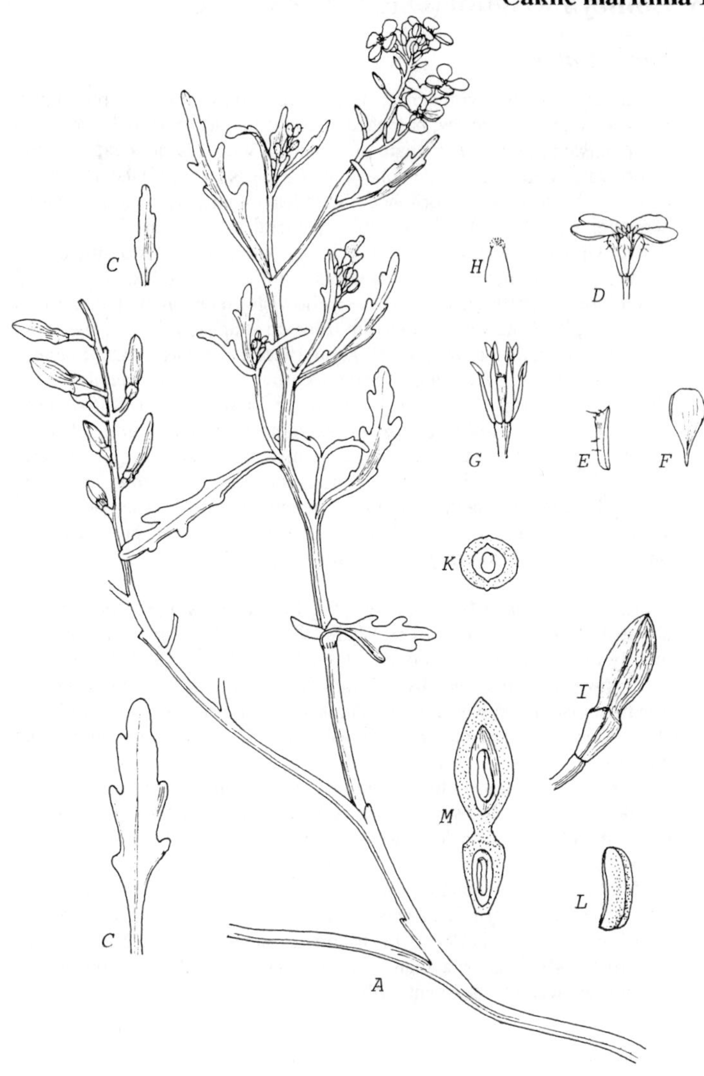

A,Cx0.5;D-Fx1;Gx2;Hx5;I,Kx1.5;Lx2;M=fruit L.S.x2.

13. Coincya wrightii (O.E. Schulz) Stace

Lundy Cabbage

Perennial 30-90 cm, with dense, simple hairs, somewhat glaucous above. Stems erect, woody below, branched. Rosette and lower stem leaves to 20 cm, persistent, petiolate, pinnate to pinnatifid with a large, ovate, elliptic or obovate, obtuse terminal lobe and 0-2 pairs of small lateral lobes; margins obtusely toothed or lobed. Stem leaves similar, the terminal lobe narrower with 1-4(-5) pairs of oblong lateral lobes, acutely toothed or lobed. Upper stem leaves to 8 cm, with an oblong to linear terminal lobe and 0-3 pairs of linear, acute, lateral lobes. Inflorescence lax. Sepals 7-13 mm, oblong, slightly saccate at base, yellowish-green, erect. Petals 15-20 x 5-8 mm; limb obovate, rounded to truncate at apex, yellow; claw about as long as limb, linear, pale. Petals about twice as long as sepals. Stamens 6; anthers yellow. Ovary hairy. Stigma entire. Pedicels in fruit 8-12 mm, slender, ascending to inclined. Fruits 30-65(-80) x 1.7-3(-4) mm, cylindrical, with a lower, dehiscent, valvular portion and an indehiscent terminal segment, terete or slightly angustiseptate, ascending to reflexed. Valves 21-55 mm, with 3 veins, glabrous, covering numerous, uniseriate seeds. Terminal segment 8-15 mm, ensiform, tapering to a sessile stigma, with 3 veins and 1-3 seeds. Seeds 1.2-1.6 mm, globose, grey-brown to black. 2n=24*. Flowering May to August.

Endemic to Lundy Island in the Bristol Channel, where a total of only about 300 plants are known (Cassidi 1981) on cliffs and open ground, mainly in the south-eastern part of the island (Marren 1972, 1973). The species was first described by O.E.Schulz in Wright (1936) and is known nowhere else in the world, though closely related species occur in Iberia (J.R. Akeroyd, pers. comm.). Pugsley (1936) gives some interesting discussion.

On Lundy only likely to be confused with *Sinapis arvensis* which has smaller flowers with ascending to reflexed sepals, and ovate, ± simple upper stem leaves. The dense hairs and perennial habit distinguish it from *C. monensis*.

Members of the genus *Coincya* will be better known to British botanists under the name *Rhynchosinapis*. *Coincya* has recently been revised by Leadlay & Heywood (1990). In Britain, *Coincya* can be distinguished by the fruits, which have a many-seeded, lower valvular portion and an indehiscent terminal segment with 1-5 seeds.

Coincya wrightii 13

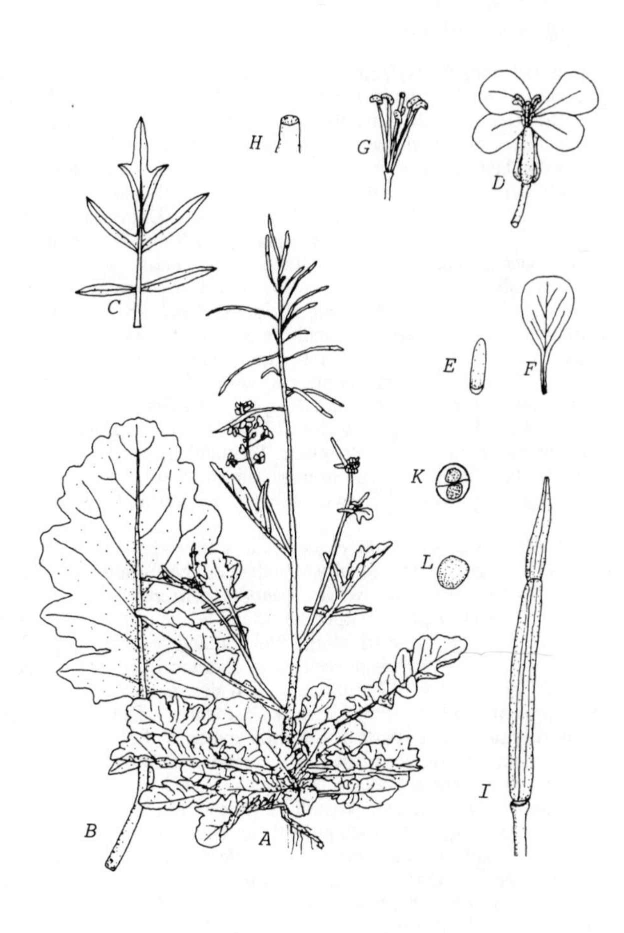

Ax0.2;B,Cx0.3;D-Gx1;Hx3;I,Kx1.5;Lx3.

14a. **Coincya monensis** (L.) W. Greuter & Burdet subsp. **monensis**

Isle of Man Cabbage Map 14a

Biennial (? annual) 15-50 cm, glabrous or with sparse, simple hairs below, glaucous above. Stems prostrate to ascending, branched mainly below. Rosette leaves to 20 cm, long-petiolate, pinnate to pinnatisect or bipinnatisect with an oblong, obtuse to acute, terminal lobe and 4-8 pairs of similar lateral lobes; margins entire or acutely or obtusely toothed, sparsely hairy (surfaces glabrous). Stem leaves 0-4(-6), similar to basal leaves but smaller with 3-6 pairs of lateral lobes. Upper stem leaves with linear, entire lobes. Inflorescence lax. Sepals (6-)7-11 mm, linear to oblong, saccate, hairy, yellowish-green, erect. Petals (11-)14-21 x (4-)5-8 mm; limb obovate, rounded, yellow to pale yellow, sometimes with darker veins; claw linear, greenish to reddish, longer than limb. Petals about twice as long as sepals. Stamens 6; anthers yellow. Ovary glabrous. Stigma entire. Pedicels in fruit 6-12 mm, stout, ascending to inclined. Fruits 35-70(-80) x 1.5-2.5 mm, cylindrical, with a lower, dehiscent, valvular portion and an indehiscent terminal segment, ± terete, ascending to patent, rarely with a 1 mm stipe. Valves (7-)20-45(-55) mm, with 3-5 veins, covering numerous, uniseriate seeds. Terminal segment 6.5-23(-26) mm, ensiform, tapering to a sessile stigma, with (0-)1-4(-5) seeds. Seeds 1.3-2 mm, subglobose to ovoid, brown. 2n=24*. Flowering May to September.

First discovered over 300 years ago at Ramsey, Isle of Man, and still present there today. Very locally abundant on the coast in open dunes and sandy shores, typically with *Ammophila arenaria* and *Raphanus maritimus*. Less frequent in open fields and hedgebanks near the sea, and rarely inland in Scotland (Stirling 1984); inland records for England are errors for *C. monensis* subsp. *recurvata* (Kiernan 1971). From the Clyde south to Cheshire, Isle of Man, rare in S. Wales. Surprisingly absent from Ireland. Apparently declining locally (lost from N. Devon and Mull) but still locally plentiful in parts of Cumbria and S. W. Scotland.

An endemic taxon confirmed to the west coast of Britain. Formerly regarded as a distinct species but now included with 4 others into an aggregate of closely related western European taxa (Leadlay & Heywood 1990). Very difficult to distinguish from one of these introduced to the British Isles, subsp. *recurvata*; the characters in the key must be used in combination. Inland records of subsp. *monensis* require voucher specimens if conservation considerations allow. For characters distinguishing the genus *Coincya*, see under *C. wrightii*.

C. wrightii is a densely hairy perennial. *C. monensis* subsp. *recurvata* is usually distinctly hairy but may be glabrescent; it also has more erect stems and smaller seeds. It can also occur on sand dunes but mixed populations have not been reported.

Coincya monensis subsp. **monensis 14a**

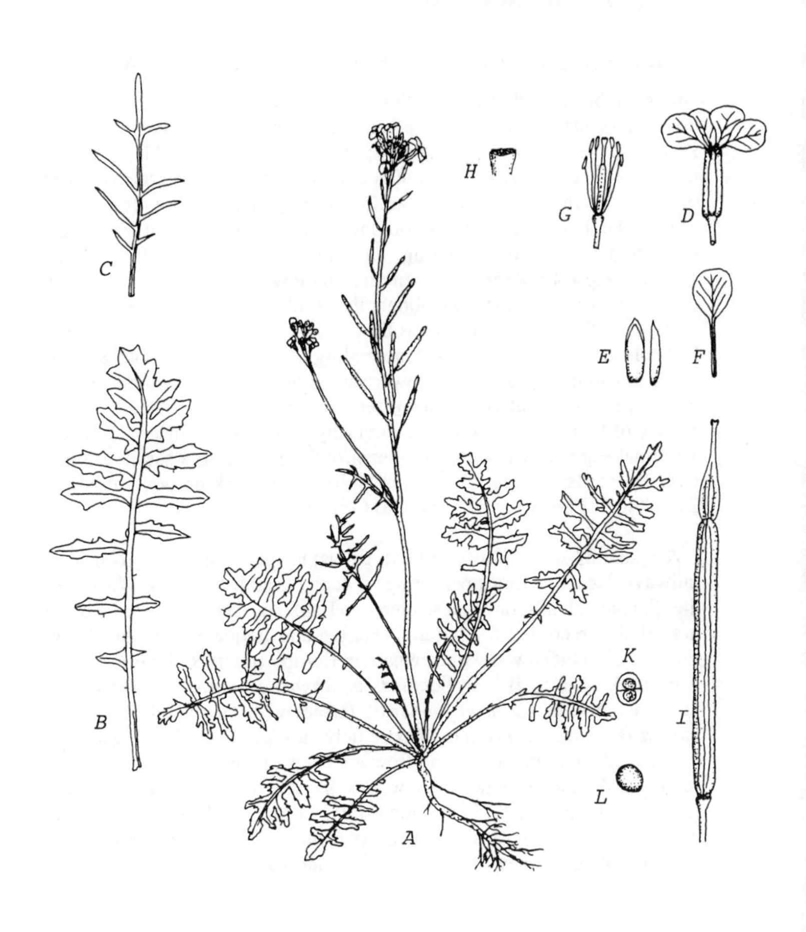

Ax0.2;B,Cx0.3;D-Gx0.7;Hx3;I,Kx1.5;Lx3.

14b. **Coincya monensis** (L.) W. Greuter & Burdet subsp. **recurvata** (All.) Leadlay

Wallflower Cabbage, Tall Wallflower Cabbage Map 14b

Annual or biennial 20-50 cm, with simple hairs below (rarely glabrescent), often glaucous above. Stems erect (rarely prostrate), branched below. Rosette leaves to 25 cm, pinnate to pinnatisect with a variable, ± elliptic terminal lobe and 4-6 pairs of smaller oblong lateral lobes; margins toothed or lobed. Stem leaves (2-)3-6, variable, deeply pinnatisect with 3-5 pairs of irregular, linear, lanceolate or oblong, acute, lateral lobes. Inflorescence lax. Sepals 7-10 mm, oblong, saccate, hairy, green, erect. Petals 12.5-22(-26) x 4.5-8 mm; limb circular to obovate, rounded, yellow with darker veins; claw linear, pale, longer than limb. Petals about twice as long as sepals. Stamens 6; anthers yellow. Ovary glabrous. Stigma entire. Pedicels in fruit 5-11 mm, stout, ascending to inclined. Fruits 45-80(-85) x 1.2-2.5 mm, linear with a lower dehiscent, valvular portion and an indehiscent terminal segment, ± terete, ascending to patent. Valves 30-55(-65) mm, with 3 veins, covering numerous, uniseriate seeds. Terminal segment 7-23(-34) mm, ensiform, tapering to a sessile stigma, with 1-4 seeds. Seeds 1.1-1.6 mm, subglobose, dark brown, uniseriate. 2n=24,48*. Flowering from May to October.

A casual of docks, waste ground, ballast heaps, sand dunes, roadsides, railways, etc, sometimes persisting. Locally established in S. Wales and the Channel Islands, rare and scattered in England. Extinct in Scotland and not reliably recorded for Ireland. Native and frequent in S. and S. W. Europe, introduced widely elsewhere in Europe and in N. America.

Better known to British botanists as *Rhynchosinapis cheiranthos* but now included in *Coincya monensis* (Leadlay & Heywood 1990). *C. monensis* subsp. *recurvata* is an extremely variable taxon showing a range of hairiness, leaf dissection and flower size. Leadlay & Heywood (1990) recognise 4 varieties which are not considered here. Glabrous or prostrate (beware of damaged plants) material is often confused with subsp. *monensis*. Many old records for subsp. *monensis* away from the Irish Sea refer to subsp. *recurvata* which was not widely known at the time.

Coincya monensis subsp. **recurvata 14b**

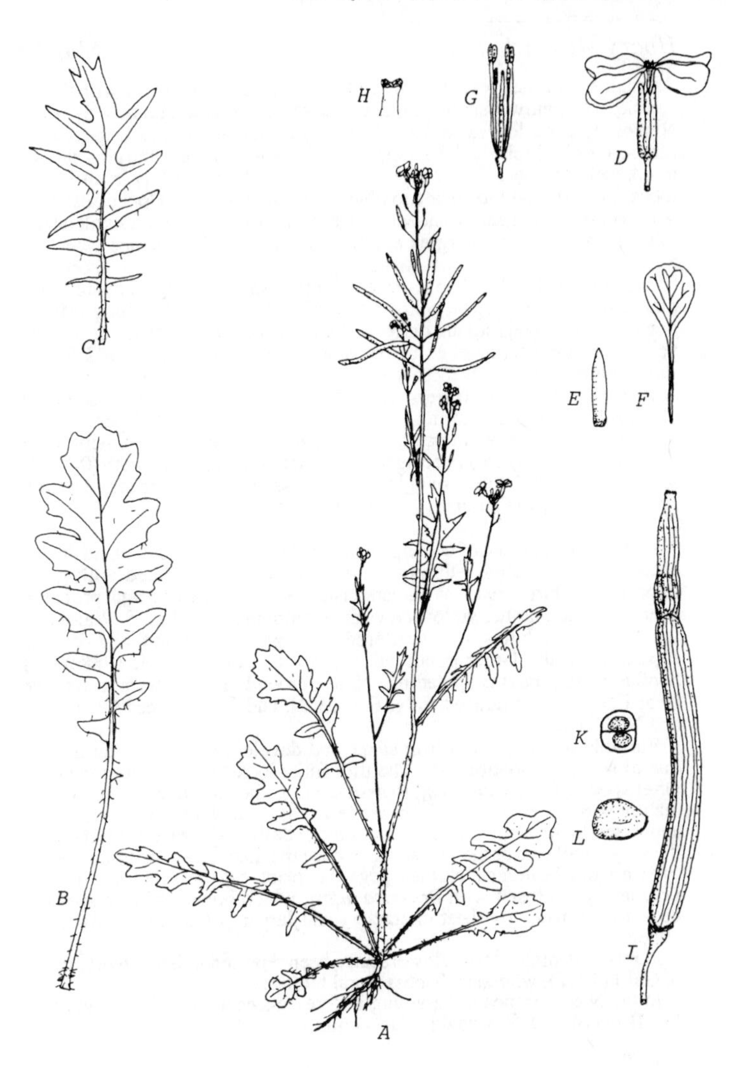

Ax0.2;B,Cx0.3;D-Gx1;Hx3;I,Kx1.5;Lx5.

15. **Hirschfeldia incana** (L.) Lagrèze-Fossat

Hoary Mustard Map 15

Annual, sometimes perennial 50-130(-200) cm, glabrescent or with dense, simple hairs below, usually glaucous above. Stems erect, branched mainly above. Rosette leaves to 35 cm, petiolate, pinnate with a large, ovate, obtuse terminal lobe and 2-9 pairs of small, lateral lobes; margins sinuate to coarsely toothed. Lower stem leaves similar, with 0-5 pairs of lateral lobes, the terminal lobe ovate to oblong, acute to obtuse. Uppermost stem leaves petiolate, linear to lanceolate, simple or sometimes with small lateral lobes at base, acute; margins entire to toothed. Inflorescence crowded. Sepals 2.6-5.2 mm, oblong, awned, green to reddish, erect to ascending (rarely inclined). Petals 5.3-10 x 2-5 mm; limb broadly obovate, apex obtuse to emarginate, pale yellow; claw distinct, about half as long as limb, linear, pale. Petals about twice as long as sepals. Stamens 6, anthers yellow. Stigma capitate, entire to emarginate. Pedicels in fruit 2-5 mm, stout, appressed. Fruits 7-16 x *c.* 1.3-1.7 mm, ± linear, terete, with two segments, appressed (rarely spreading obliquely). Lower segment (4-)4.5-9(-11) x *c.* 1(-1.5) mm, dehiscent, seeds (0-)2-6(-?9) per loculus. Valves with 0-3 veins (most obvious when dry). Terminal segment 3-6.5 x 1-1.8 mm, lanceolate, tapering to a persistent style *c.* 1 mm, with (0-)1(-2) seeds. Seeds 0.9-1.4 mm, ovoid to subglobose, brown, uniseriate. 2n=14*. Flowering May to October (-December).

An introduced weed of waste ground, docks, railways, fields, sand dunes, power stations, rubbish dumps, cities, etc., often associated with grain imports and bird seed. Well-established in many places and probably overlooked as another yellow crucifer. Common in London, S. Wales and the Channel Islands and scattered throughout England and Wales, especially in the larger conurbations, rare in Ireland (Rich 1988c) and Scotland and probably spreading. Native around the Mediterranean to the Near East, and introduced to N. Europe, N. and S. America, Australasia and N. Africa.

Variable in size, and in fruit shape and development (the latter mainly due to self-incompatibility). The thin fruits look like immature fruits of other species (e.g. *Rapistrum*, *Brassica*, etc.), and careful examination for mature seeds is essential. There are few other yellow crucifers with fruits appressed to the stem, and *Hirschfeldia* can be distinguished from these by the very small size of the fruits. Also, once the distinctive fruits are known (like an old-fashioned clothes peg) the plant can be identified easily. Frequently confused with *Brassica nigra*, which however has a seedless beak to the fruit (the best character) and larger petals and fruits (Rich 1988c).

Plants with male-sterile flowers have been reported in Israel by Horovitz & Galil (1972), who also discuss floral biology.

In the Mediterranean, the young inflorescences are collected and eaten like Broccoli (J.J. Zawadzki, pers. comm.).

Hirschfeldia incana 15

Ax0.1;B,Cx0.3;D-Gx3;Hx10;I,Kx5;Lx10;M=hairs on stem.

16. Brassica nigra (L.) Koch

Black Mustard Map 16

Annual 30-200(-310) cm, glabrous or sparsely hairy with coarse simple hairs below, often glaucous. Stems erect, branched mainly above, the branches widely spreading. Leaves very variable in dissection, all petiolate. Lower stem leaves to 20 cm, coarsely and irregularly pinnatifid to pinnate, the terminal lobe ovate, obtuse, truncate to cordate below, coarsely sinuate, with 1-3 pairs of smaller lateral lobes; margins acutely toothed. Upper stem leaves smaller and less divided, lanceolate, acute to obtuse at apex, with 0-1 pair of lateral lobes; margins entire to serrate. Uppermost leaves to 5 cm, linear to oblong, entire. Inflorescence crowded, rarely lowest 1-6 flowers bracteolate. Sepals 4-7 mm, narrowly oblong, green, ascending to patent. Petals (7-)9.5-13 x (3-)4-5.5 mm; limb obovate, obtuse, yellow to pale yellow; claw *c.* 5 mm, linear. Petals about twice as long as sepals. Stamens 6; anthers yellow. Stigma capitate, entire to emarginate. Pedicels in fruit 3-8 mm, usually appressed. Fruits 8-25(-33) x (1.5-)2-4.5 mm, linear to elliptic, compressed (angustiseptate, the septum (1-)1.5-2 mm wide), dehiscent. Valves (6-)7-24(-28) mm, strongly keeled with a prominent mid vein and finer laterals. Beak 2-5(-6) mm, linear (rarely narrowly conical), sterile. Seeds 2-5(-8) per loculus, 1.3-2 mm, globose, grey, brown or black, uniseriate. 2n=16. Flowering May to September.

A native species of riverbanks, sea cliffs and shingle but also widely introduced and casual on roadsides, waste ground, towns and cities, fields, etc. Locally very abundant in England and Wales, rare in S. Scotland and scattered mainly around the coast in Ireland. The precise native range is unclear in Britain, Ireland and elsewhere. Probably native through most of W. Europe to Turkey and C. Europe, southern Scandinavia. Introduced in Africa, N. and S. America and Australasia.

Somewhat variable in leaf and fruit shape.

Few other crucifers have appressed fruits and yellow flowers. *Sisymbrium officinale, Rapistrum* and *Barbarea* species are all quite different plants, but *Hirschfeldia* and *Sinapis arvensis* in particular are often confused with *B. nigra. Hirschfeldia* has smaller, thinner fruits which usually have a seed in the beak, and is often densely hairy below and has smaller petals and seeds (Rich & Rich 1988). *Sinapis arvensis* may also have a seed in the beak, or longer fruits and ovate upper stem leaves.

Formerly cultivated for its seeds which were used in mustard and for oil used in soap and medicine.

For characters identifying the genus *Brassica*, see p.36.

Brassica nigra 16

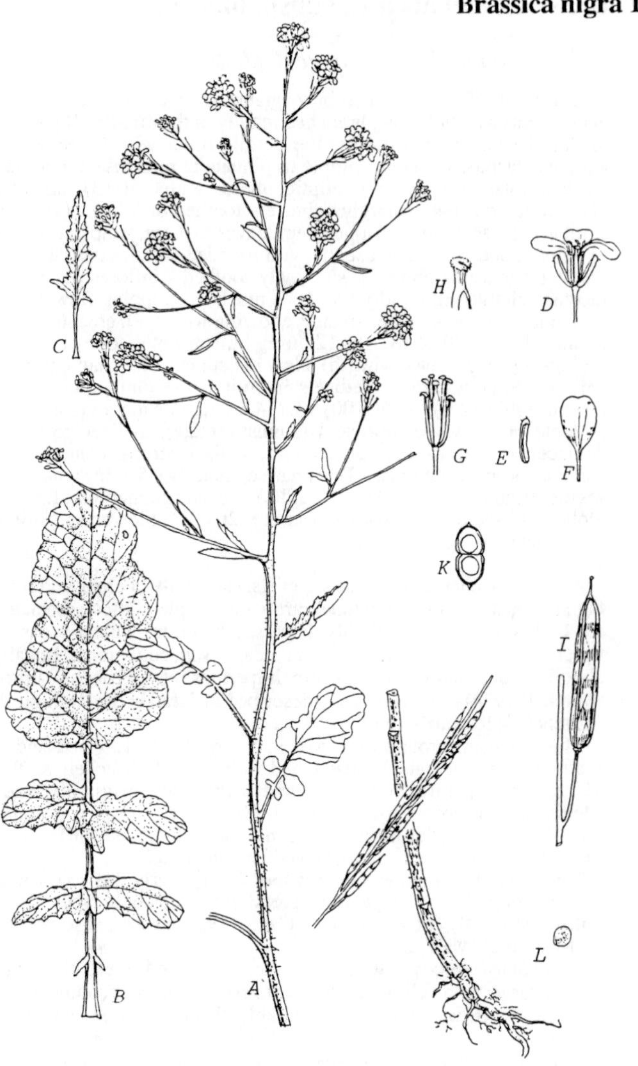

Ax0.2;B,Cx0.4;D-Gx1;Hx5;Ix1.5;Kx2;Lx2.

17. **Brassica oleracea** L. subsp. **oleracea**

Wild Cabbage, Sea Cabbage, Cabbage

Perennial 30-60(-150) cm, robust, glabrous, glaucous. Stems erect or decumbent, woody below, branched mainly in the middle. Rosette leaves to 20(-30) cm, very variable, the petiole often with broad wings and auricles clasping the stem, simple or pinnate to pinnatisect, undulate, the terminal lobe large, ovate to elliptic, obtuse, with 0-5(-13) pairs of small, obovate lateral lobes; margins obtusely toothed or lobed. Lower leaves similar but less divided. Upper stem leaves to 7 cm, simple, oblanceolate to oblong, obtuse, sessile usually with rounded auricles partially clasping the stem; margins entire or shallowly toothed. Inflorescence very lax, racemes elongating rapidly and buds much overtopping flowers. Sepals 8-16 mm, oblong, saccate, green (yellowing with age), erect to ascending. Petals (16-)18-30 x (8.5-)9-12 mm; limb elliptic to ovate, rounded, (yellow-) pale yellow (-whitish); claw indistinct, pale. Stamens 6; anthers yellow. Stigma entire. Pedicels in fruit 10-33 mm, stout, inclined to patent. Fruits (25-)35-85(-100) x 2.5-4.5 mm, linear, variable in seed set, ± terete or weakly compressed (angustiseptate), inclined to ascending, dehiscent. Valves (22-)30-76 mm, with 1 strong central vein and sometimes weak laterals. Terminal segment 3-9.5 mm, conical with a sessile stigma, with (0-)1(-2) seed(s). Seeds numerous, 1.5-2.5 mm, globose, black to dark brown, uniseriate. 2n=18*. Flowering April to June (-October).

A calcicole characteristic of sea cliffs, especially of chalk and limestone but also on shale and sandstone, rarely on shingle. Often associated with sea bird colonies and probably dispersed by birds. Inland occasional on tips, roadsides, etc. as a casual or an escape from cultivation. Locally very abundant but many colonies are impermanent. Probably introduced (Mitchell 1976). The ecology is described in detail in the Biological Flora (Mitchell & Richards 1979).

Most frequent around the coast in S. and S. W. England and Wales, scattered elsewhere in England and S. Scotland (Mitchell & Richards 1979). Not naturalized in Ireland. Probably originally native on Mediterranean coasts from Spain to Greece (as subsp. *robertiana*) but widely introduced elsewhere, and naturalized on sea cliffs in France, Germany, N. America and introduced in Australasia.

Very variable in many characters (particularly leaves) within and between populations suggesting reversion towards wild types from plants cultivated locally for vegetables. Gates (1953) describes some effects of cultivation of "wild" material.

The robust woody habit, thick, fleshy lower leaves, clasping upper leaves, large, pale yellow flowers and very lax racemes should distinguish it from *Brassica napus*. It is unlikely to be confused with any other crucifer.

Grown for centuries as a vegetable with numerous cultivated races such as the many forms of Cabbage, Brussels Sprouts, Broccoli, etc.

Brassica oleracea 17

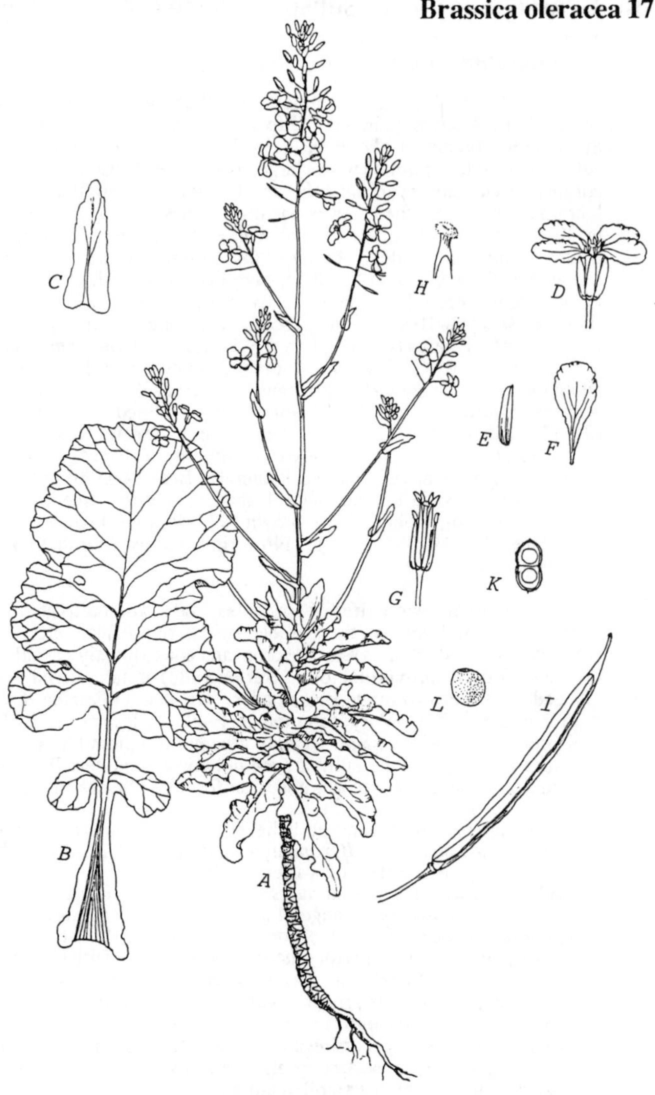

Ax0.2;B,Cx0.4;D-Gx0.8;Hx5;I,Kx0.7;Lx3.

18a. **Brassica napus** L. subsp. **oleifera** (DC.) Metzger

Oil-seed Rape, Rape, Cole, Cole-seed

Annual or biennial (rarely perennial) 30-130 cm, glabrous or with coarse simple hairs below, glaucous. Tap root cylindrical to Carrot-shaped. Stems erect, branched above. Rosette leaves to 40 cm, glaucous, long petiolate, simple to pinnate with a large, ovate (often lobed at base), obtuse, cordate to cuneate, terminal lobe and 0-6 pairs of small lateral lobes; margins obtusely toothed; not persisting. Lower stem leaves similar but to 30 cm, the petiole often winged with clasping auricles. Upper stem leaves simple, lanceolate to oblong, obtuse to apiculate, sessile with broad, rounded auricles clasping the stem; margins entire to shallowly toothed. Inflorescence crowded, the buds overtopping (rarely equalling) the flowers. Sepals 6-10 mm, oblong, scarcely saccate, green to yellow, erect to ascending. Petals (10.5-)11-18 x 6.3-11 mm, limb obovate, rounded at apex, pale yellow to yellow; claw short, ± distinct, triangular, paler. Petals about twice as long as sepals. Stamens 6; anthers yellow. Stigma capitate, ± entire. Pedicels in fruit 15-30 mm, slender, inclined to reflexed. Fruits (20-)35-95(-110) x (2.5-)3-5 mm, linear, terete to compressed (angustiseptate), inclined to reflexed, dehiscent. Valves (14-)18-85 mm, with 1 strong central vein and weak laterals. Beak (4-)5-16 mm, conical, tapering to a sessile stigma, with 1 seed or sterile. Seeds numerous 1.3-2.7(-3.1) mm, globose, dark brown to black, light brown or reddish, uniseriate. 2n=4x=38*(57,76). Flowering mainly March to June (to October).

A common yellow crucifer of roadsides, waste and cultivated ground, docks, cities and towns, tips, arable fields, riverbanks, etc. Widely cultivated for seed oil or as a forage crop, and consequently casual ("fallen off the back of the proverbial lorry" Burton 1983) or naturalized wherever oil-seed rape is grown in the British Isles. The map in Perring & Walters (1962) is unreliable. Very common in England, less so in Wales and Scotland, scattered in Ireland. Widely grown throughout the world.

Somewhat variable, the cultivars are not considered here because their recognition depends on growth under uniform conditions in agricultural trials.

Unknown as a wild plant, and believed to be an allotetraploid derived from a *Brassica oleracea* x *B. rapa* hybrid, probably in Europe in the 16th century. A great number of varieties are now known and cultivated throughout the world (see Bunting 1988 for a recent review). Oil extracted from the seeds is used for a range of domestic, chemical, engineering and commercial purposes. Seedlings are used as "cress" (Rich 1988b).

Unlikely to be confused with other yellow crucifers with clasping stem leaves apart from *Brassica oleracea* and *B. rapa* (for details, see under these species). In the past, much confused with *B. rapa*, not least due to confusion between the English and Latin names.

18b. Brassica napus L. subsp. **rapifera** Metzger, the *Swede* or *Swedish Turnip*, may rarely occur as an escape or relict of cultivation. It is distinguished from subsp. *oleifera* by the swollen tap-root.

Brassica napus 18

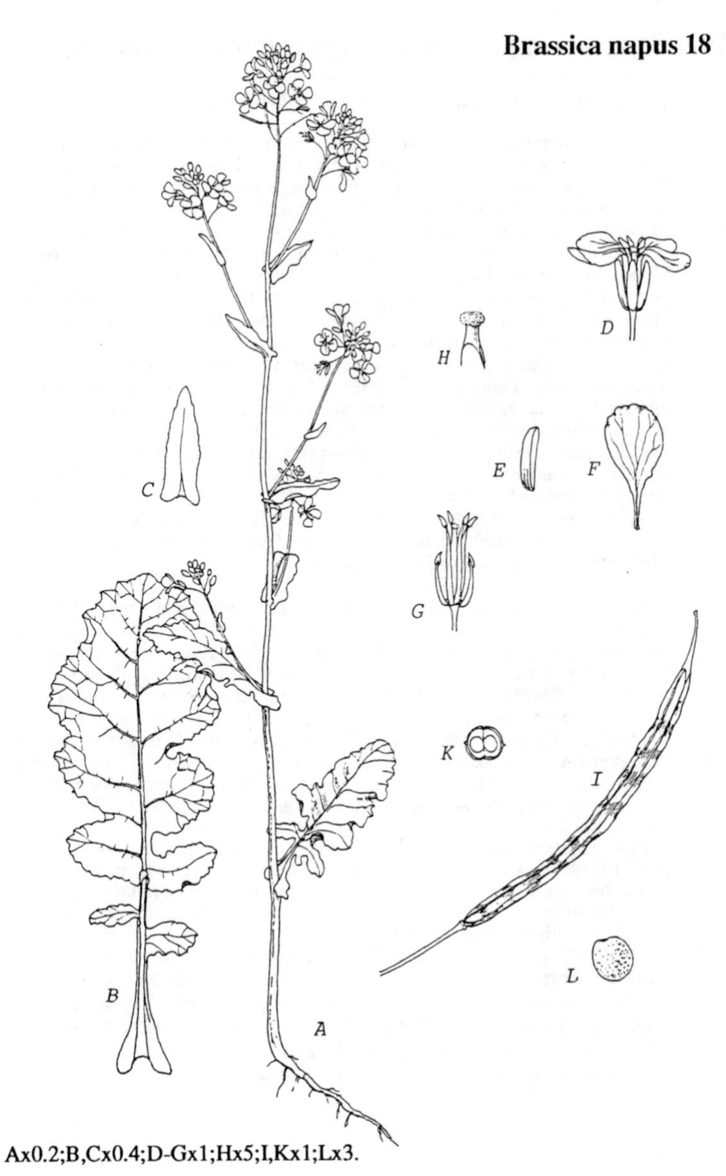

Ax0.2;B,Cx0.4;D-Gx1;Hx5;I,Kx1;Lx3.

19a. **Brassica rapa** L. subsp. **sylvestris** (L.) Janchen

Wild Turnip, Navew, Bargeman's Cabbage

Annual or biennial 30-100 cm, usually with coarse, simple hairs below (rarely glabrous), green to slightly glaucous. Rosette leaves to 40 cm, petiolate, pinnate to pinnatisect with a large ovate, obtuse terminal lobe and 2-6 pairs of small, ovate, lateral lobes; margins sinuate or obtusely toothed; leaves not persisting. Lower stem leaves similar but to 20 cm, petiole winged, with auricles clasping stem and 1-4 pairs of lateral lobes. Upper stem leaves lanceolate to ovate, acute to obtuse at apex, sessile, perfoliate or with rounded, clasping auricles. Inflorescence crowded, the open flowers equalling or overtopping buds. Sepals 4.5-7(-8) mm, oblong to ovate, green to yellow, erect to inclined. Petals 6-12(-14) x (3.2-)4-6(-8) mm; blade elliptic, rounded at apex, deep yellow to yellow; claw short, triangular, pale. Petals *c.* 1.5-2 times as long as sepals. Stamens 6; anthers yellow. Stigma capitate, entire to emarginate. Pedicels in fruit 5-30 mm, slender, ascending to patent (rarely reflexed). Fruits (20-)30-68 x 2-4 mm, linear, terete to compressed (angustiseptate, septum 2-3 mm wide), rarely shortly stipitate, erect to patent, dehiscent. Valves (13-)18-51 mm, with 1 strong central vein and weak laterals. Beak 7-22 mm, conical, tapering to a capitate, sessile stigma, sterile (rarely 1-seeded). Seeds numerous, 1.1-2 mm, globose to ovoid, black, brown or reddish, uniseriate. 2n=20*(?40). Flowering mainly May to July (to September).

Probably native but widely introduced (e.g. with bird seed). Locally abundant on roadsides, railways, gardens, tips, arable fields, waste ground, etc. and particularly characteristic in England of river and canal banks. Reasonably common throughout Britain and Ireland, though rare in N. Scotland. Much less frequent in England than *Brassica napus*, but probably commoner elsewhere. The map in Perring & Walters (1962) is unreliable. The native range is unclear but it is probably native in most of Europe to Asia, introduced in N. and S. America, Australasia and Africa.

Variable in size and leaf shape. Varieties based on annual or biennial forms do not merit recognition as this character is inconsistent. Two other subspecies are given below.

Few crucifers have yellow flowers and clasping stem leaves like *B. rapa*, and of these the only likely confusion is with *B. napus*. The most reliable characters are given in the key; *B. rapa* also has darker petals and green first year rosettes (note these are absent at flowering; Wigginton & Graham 1981). For discussion see Rich (1987a).

19b. Brassica rapa L. subsp. **rapa**, the cultivated *Turnip*, may occur rarely as an escape or relict of cultivation. It is distinguished by the swollen tap-root.

19c. Brassica rapa L. subsp. **oleifera** (DC.) Metzger, *Turnip-rape*, is apparently grown as a fodder or oil-seed crop and is said to have larger (?1.5-2 mm) reddish-brown seeds, but I have been unable to ascertain its exact identity or how widely it occurs in Britain.

Brassica rapa 19

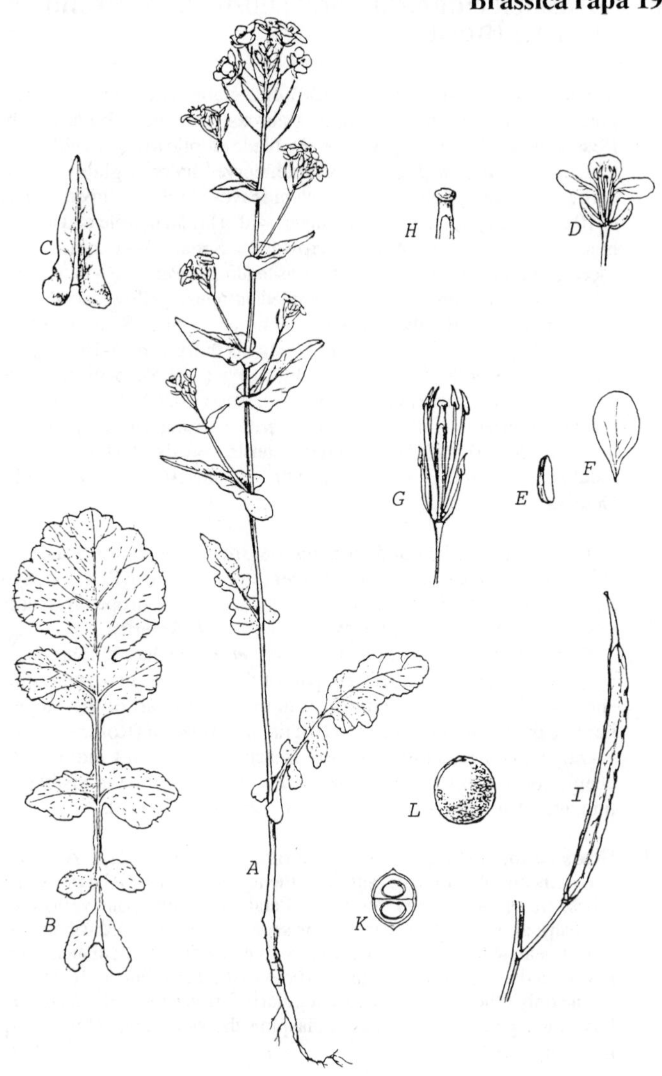

Ax0.2;B,Cx0.4;D-Gx1;Hx5;I,Kx1;Lx10.

20. **Brassica elongata** Ehrh. subsp . **integrifolia** (Boiss.) Breistr.

Map 20

Biennial or perennial 50-100(-130) cm, glabrous or with simple hairs, green. Stem erect to ascending, glabrous, branched above and below. Rosette leaves to 30 cm, petiolate, the blade elliptic to oblanceolate, obtuse to acute, cuneate, with coarse, simple, curved hairs or glabrous; margins entire to sinuately lobed. Lower stem leaves similar. Upper stem leaves to 10 cm, linear-oblanceolate to linear, with a broad petiole and acute apex; margins entire. Inflorescence crowded. Sepals 3-4.6 mm, oblong, ± saccate, green to yellow, erect to ascending. Petals (6-)6.5-8.5(-10) x 2.5-3.7(-4) mm; limb elliptic, rounded at apex, yellow; claw *c.* 3 mm, linear, pale. Petals about twice as long as sepals. Stamens 6; anthers yellow. Stigma capitate, ± entire. Pedicels in fruit (6-)8-18 mm, slender, ascending to inclined. Fruits 15-40(-50) x (1-)1.5-2.5 mm, ± terete to compressed (angustiseptate), torulose, stipe (0.8-)1.5-5 mm, erect to patent, dehiscent. Valves 12-38 mm, with 1 strong central vein and weak laterals. Beak 0.5-2(-3) mm, linear, sterile. Seeds (2-)4-10(-12) in each loculus, 1-1.6 mm, globose, grey to brown, uniseriate. Flowering May to October.

An introduced plant of docks, quarries, roadsides, fields, waste ground, etc., sometimes persistent for a number of years. Rare in England, Scotland and Wales, not recorded in Ireland. Native in S. E. Europe, W. Russia and the Near East. Introduced to W. Europe and N. America.

Two subspecies are known: subsp. *elongata* has sinuate-pinnatifid leaves and occurs in the western part of the range. Subsp. *integrifolia* has entire to sinuate leaves and occurs in the eastern part of the range. This latter is the taxon introduced to Britain and America (Rollins 1980).

Only *Diplotaxis tenuifolia* also has a stipe more than 1 mm long but has deeply lobed stem leaves, larger petals, etc. The stipe is visible even in immature fruit which helps to identify non-fruiting plants.

21. Brassica tournefortii Gouan is a rare casual (Map 21). Annual 10-50 cm, hairs simple. Leaves petiolate, the lower with (0-)4-10 pairs of lateral lobes; margins strongly toothed. Petals 4-7 mm, pale yellow (fading whitish), about 1-2 times as long as sepals. Fruits (30-)35-65 mm x 1.5-5 mm, linear with a narrower beak. Valves 22-50 cm, dehiscent, torulose, many seeded. Terminal segment (beak) 10-20 cm, linear, 1(-3) seeded.

The only other *Brassica* with a regularly fertile beak is *B. oleracea* which has larger petals, fleshy leaves clasping the stem, etc. *Coincya* species have larger petals.

Brassica elongata 20

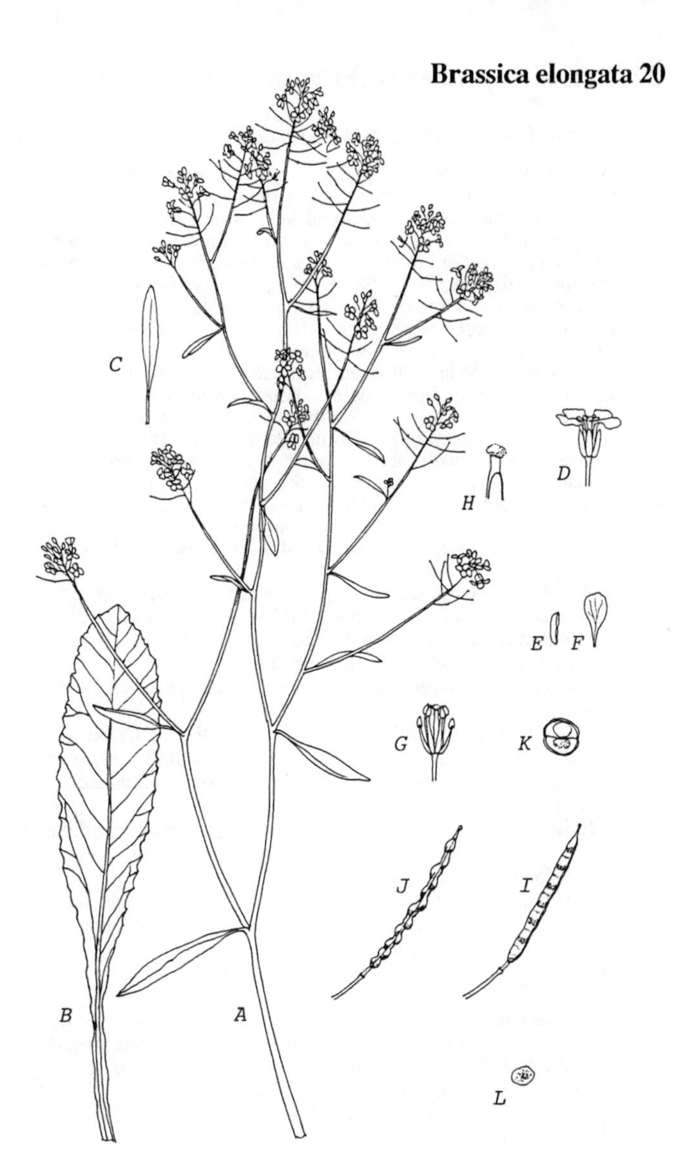

Ax0.2;B,Cx0.3;D-Gx1;Hx5;I,Kx1;Lx3.

22. **Brassica juncea** (L.) Czern.

Brown Mustard, Indian Mustard, Chinese Mustard Map 22

Annual 20-100(-150) cm, glabrous, green, sometimes slightly glaucous. Stems erect, branched above. Lower stem leaves to 20 cm, long petiolate, pinnate to pinnatifid, the terminal lobe ovate, obtuse at apex, cordate to truncate at base, with 1-3 pairs of much smaller lateral lobes; margins irregularly lobed or crenate to dentate. Upper stem leaves shortly stalked (rarely sessile), narrowly oblong, oblanceolate or lanceolate, acute at apex, cuneate at base, sometimes with 1-2 pairs of lateral lobes; margins entire or toothed. Inflorescence crowded or lax. Sepals 4.6-7 mm, linear (inrolled) to oblong, green to yellow, ascending to patent. Petals 9-13 x 4.8-7.5 mm; limb broadly ovate, entire to emarginate at apex, yellow; claw distinct, *c.* 5 mm, linear, paler. Petals about twice as long as sepals. Stamens 6; anthers yellow. Stigma capitate, entire. Pedicels in fruit 5-14 mm, slender, ascending. Fruits (20-)28-50(-?75) x 3-5 mm, linear, compressed (angustiseptate, septum 2-3 mm wide), erect to ascending, dehiscent. Valves 21-40 mm, weakly torulose with 1 strong central vein (often ± keeled) and weak laterals. Beak (4-)5-9(-12) mm, conical tapering to a sessile stigma, sterile. Seeds numerous, 1.2-1.7 mm, globose, light to dark brown or grey, uniseriate. 2n=4x=36. Flowering May to October.

A casual of fields, roadsides, paths, waste ground, docks, tips and cities, introduced with foreign grain and bird seed. Of unpredictable occurrence in England, rare in Ireland, Scotland and Wales. Probably much overlooked as yet another yellow crucifer. Possibly native in Asia and E. Africa but widely cultivated. Introduced to Europe, N. and S. America and Australasia.

British material is little variable, but in India and China where it is an important vegetable and oil-seed crop, it is particularly polymorphic in leaf shape (see also Vaughan *et al.* 1963). Commercial mustard is now made by mixing seed flour with that of *Sinapis alba*.

Believed to be an allotetraploid derived from a *B. rapa* x *B. nigra* hybrid (Harberd 1975) and may have originated independently in both E. Africa and Asia (Vaughan & Gordon 1973).

Difficult to distinguish from *Sinapis arvensis* which has inclined to reflexed sepals, and valves with 3-7, ± equally strong veins and a beak 7-16 mm, 0- or 1-seeded. Also confused with the following species. For details see Rich (1987c).

23. Brassica carinata A. Braun, *Abyssinian Mustard* or *Ethiopian Rape*, is similar to *B. juncea* but is annual and glaucous; leaves with 0-1 pairs of lateral lobes; petals 13-17 mm, pale yellow; fruits 29-60 x 5-9 mm, beak 2.5-6(-7) mm. It is thought to be an allotetraploid (2n=4x=34) derived from a *B. oleracea* x *B. nigra* hybrid. A rare casual (Map 23). For more details, see Rich (1987c).

Brassica juncea 22
*B.carinata 23

Ax0.2;B,Cx0.4;D-Gx1;Hx10;I,Kx1;Lx3;*Cx0.4;*Ix1.

24. Sinapis arvensis L.

Charlock, Kilk, Wild Mustard, Brassocks

Annual (5-)20-100(-220) cm, dark green to purplish, densely to sparsely hairy below with simple, coarse, deflexed hairs. Stems erect, much branched. Lower stem leaves to 20 cm, petiolate, pinnate to pinnatifid or simple, the terminal lobe ovate, elliptic or obovate, obtuse at apex, with 1-4 pairs of small lateral lobes; margins coarsely and irregularly toothed to lobed. Upper stem leaves smaller, sessile or with petiole to 5(-10) mm, ovate to lanceolate, acute at apex, sometimes with a pair of shallow, acute lobes at base; margin entire to acutely toothed. Inflorescence crowded. Sepals (4.5-)5-7.5 mm, linear (inrolled), usually awned, hairy or glabrous, green to yellow, inclined to reflexed. Petals (7.5-)8-17 x 4-7.5 mm; limb obovate, rounded at apex, yellow to pale yellow; claw about as long as limb, linear, pale. Petals about twice as long as sepals. Stamens 6; anthers yellow. Stigma capitate, entire to emarginate. Pedicels in fruit (2-)3-24 mm, stout, erect to patent. Fruits 22-57 x 2-4 mm, linear, with a lower valvular portion and a terminal segment (beak), ± terete to compressed (angustiseptate, septum 2-3.5 mm wide), appressed to patent, dehiscent. Valves (6-)10-43 mm, often torulose, with 3-7 strong veins (the central slightly stronger), hairy or glabrous. Terminal segment 7-16 mm, conical, with 3-5 veins, a sessile stigma and 0 or 1 seed. Seeds 2-12 per loculus, 1.6-2.1 mm, globose, black to dark brown, uniseriate. 2n=18*. Flowering most of the year but mainly spring and early summer.

A very common weed of fields, roadsides, riverbanks, railways, gardens, tips and waste ground, virtually ubiquitous in Britain and Ireland (map in Perring & Walters 1962). The ecology is described in the Biological Flora (Fogg 1950) and by Mulligan & Bailey (1975). Probably native in Europe to Asia and N. Africa, introduced to N. and S. America, Australasia and S. Africa.

One of the most variable crucifers with a large range in size and in leaf and fruit characters. Plants with hairy valves (var. *orientalis* (L.) Koch & Ziz) are not considered to merit recognition as the plants otherwise resemble typical *S. arvensis*. Variable in fruit set, probably due to self-incompatibility (Ford & Kay 1985).

It is difficult to describe in a few words how *S. arvensis* can be distinguished. Few other yellow-flowered crucifers have a beak 7 mm or more long; *Coincya* species have erect sepals and 1-5 seeds in the terminal segment; *Brassica* species (excluding those with clasping leaves) are less easy to separate but have valves with 1 strong central vein and weaker laterals, and usually have a sterile beak. The linear, reflexed sepals (which should be examined in fresh mature flowers) are distinctive of *Sinapis* but can usually only be seen clearly on *c.* 50% of the plants (other sepals are inclined or patent).

Once one of the most serious weeds of crops, the seeds being persistent in the soil for over 50 years. Now easily controlled by herbicides and much reduced in overall frequency.

Sinapis arvensis 24

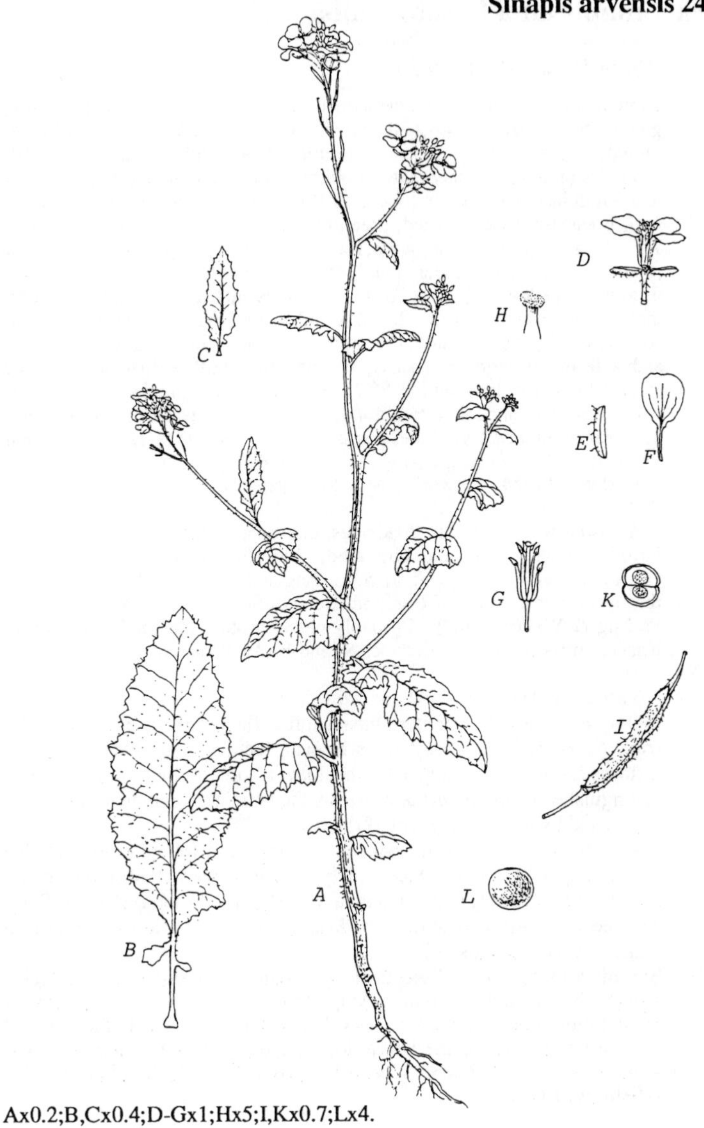

Ax0.2;B,Cx0.4;D-Gx1;Hx5;I,Kx0.7;Lx4.

25a. **Sinapis alba** L. subsp. **alba**

White Mustard, Mustard

Annual 20-100(-220) cm, glabrescent or with simple, deflexed hairs below, green. Stems erect, branched above. Stem leaves to 15 cm, petiolate, pinnate to irregularly pinnatisect, terminal lobe 3-fid, ± lanceolate, with 1-3 pairs of smaller, lanceolate to oblong, lateral lobes; margins coarsely and irregularly toothed or lobed. Inflorescence crowded. Sepals 4.2-8 mm, linear (inrolled), awned, green to yellow, patent to reflexed. Petals 7.5-14 x 4.2-7 mm; limb obovate, rounded at apex, pale yellow; claw about as long as limb, linear, paler. Petals about twice as long as sepals. Stamens 6; anthers yellow. Stigma entire. Pedicels in fruit 5-15 mm, slender, inclined to patent. Fruits 20-42 x 2-5.5 mm, with a lower, terete to compressed (angustiseptate, the septum 3-4 mm wide), valvular segment and a terminal segment (beak), inclined to patent, dehiscent. Valves (5-)7-17 mm, torulose, with 3-5(-7) veins, hairy. Terminal segment 10-24 x 4-6 mm, flat, lanceolate to oblong, with many veins, straight or curving upwards, with 0 or 1 seed and a sessile stigma. Seeds (1-)2-3(-4) in each loculus, 1.7-2.9 mm, globose, pale brown or grey to reddish brown, uniseriate. 2n=24. Flowering May to September.

A casual weed of fields, roadsides, quarries, waste ground, cities, etc., introduced with grain and bird seed. Most frequent and persistent on calcareous soils. Frequent in S. E. England, occasional and scattered through the remainder of England, Ireland, Scotland and Wales (map in Perring & Walters 1962). Probably native around the Mediterranean to Russia, introduced in N. Europe, N. and S. America, Australasia and the East.

Variable in leaf dissection and fruit shape.

Only *Eruca* and *Carrichtera* have similar, flat terminal segments to the fruit; *Eruca* has seeds in 2 rows in each loculus, and *Carrichtera* has inflorescences borne opposite the leaves (not in their axils). Best distinguished from *Sinapis arvensis* by the petiolate, pinnate to pinnatifid leaves and by the flat segment of the fruit.

Widely cultivated and introduced as a seed and forage crop, and also used as a green manure. Once widely used as a green salad (mustard and cress) but now largely discarded in favour of *Brassica napus* (Rich 1988b). The seeds are mixed with those of *Brassica juncea* (or rarely *B. nigra*) to make commercial mustard.

25b. Sinapis alba L. subsp. **dissecta** (Lag.) Bonnier is a rare casual of docks, tips etc, in England and Wales (Map 25b). It differs from subsp. *alba* in having more finely divided leaves (often ± bipinnatifid with the terminal lobe of similar size to the lateral lobes), fruits 4-7 mm wide and valves glabrous or sparsely hairy. It probably evolved by selection in flax fields (Hjelmqvist 1950).

Sinapis alba 25

Ax0.2;Bx0.5;D-Gx0.1;Hx5;I,Kx0.7;Lx2.

26. **Eruca vesicaria** (L.) Cav. subsp. **sativa** (Miller) Thell.

Garden Rocket, Salad Rocket Map 26

Annual 30-80 cm, sparsely hairy below with simple hairs or glabrous, green, foetid. Stem erect, simple or branched above. Leaves very variable. Lower leaves to 20 cm, petiolate, oblanceolate in outline, simple to pinnatisect, terminal lobe ovate to elliptic, obtuse at apex, and (2-)3-9 pairs of smaller, triangular to oblong lateral lobes; margins entire or irregularly toothed or lobed. Upper leaves less divided, sessile or shortly stalked, the terminal segment oblong to oblanceolate, with 0-4 lateral lobes, the basal lobes rarely semi-auriculate. Sepals 6-11.5 mm, oblong, ± saccate and awned, green to purple, erect, deciduous. Petals 13-26 x 5-9 mm; limb obovate, rounded, truncate or emarginate at apex, cream or pale yellow with dark veins; claw about as long as the limb, linear, pale. Petals about twice as long as sepals. Stamens 6; anthers yellow. Stigma 2-lobed. Pedicels of fruit 2-8(-10) mm, stout, erect to ascending. Fruit 15-36(-40) x 2.5-5 mm, with a lower valvular portion and a terminal segment (beak), ± terete or slightly compressed (angustiseptate, the septum 2.5-4 mm wide), erect, dehiscent. Valves (7-)10-28(-38) mm, with 1 strong central vein (± keeled), glabrous or hairy. Beak (4-)5-11 mm, triangular, flat, with 3 veins, sterile. Seeds numerous, 1.6-2.5 mm, subglobose to ovoid, pale or greyish-brown, biseriate. 2n=22. Flowering May to November.

A plant of flax and corn fields, docks, waste places, chicken runs, etc., introduced with foreign grain and bird seed and also escaping from cultivation. Once frequent in England, occasional in Wales and Scotland and rare in Ireland, but now rarely recorded. It occurs in W., C. and E. Europe, N. Africa and S. W. Asia but the native range is unclear due to widespread cultivation. Introduced to N. America (a noxious weed in Mexico), Asia and Australasia.

Very variable in size, leaf shape and flower colour. The flat, long terminal segment of the fruit and seeds in 2 rows in each loculus should distinguish it from other crucifers. *Sinapis alba* can look similar but has spreading sepals and uniseriate seeds. The plant is self-sterile (Verma *et al.* 1977) and isolated plants do not set fruit.

Cultivated for many centuries as a salad, pot herb, animal feed, and seed oil crop.

Eruca vesicaria subsp. *vesicaria* has been reported but no voucher material has been seen. It is best distinguished by the sepals which are persistent at least until the fruits ripen. It is native in Spain and N. Africa.

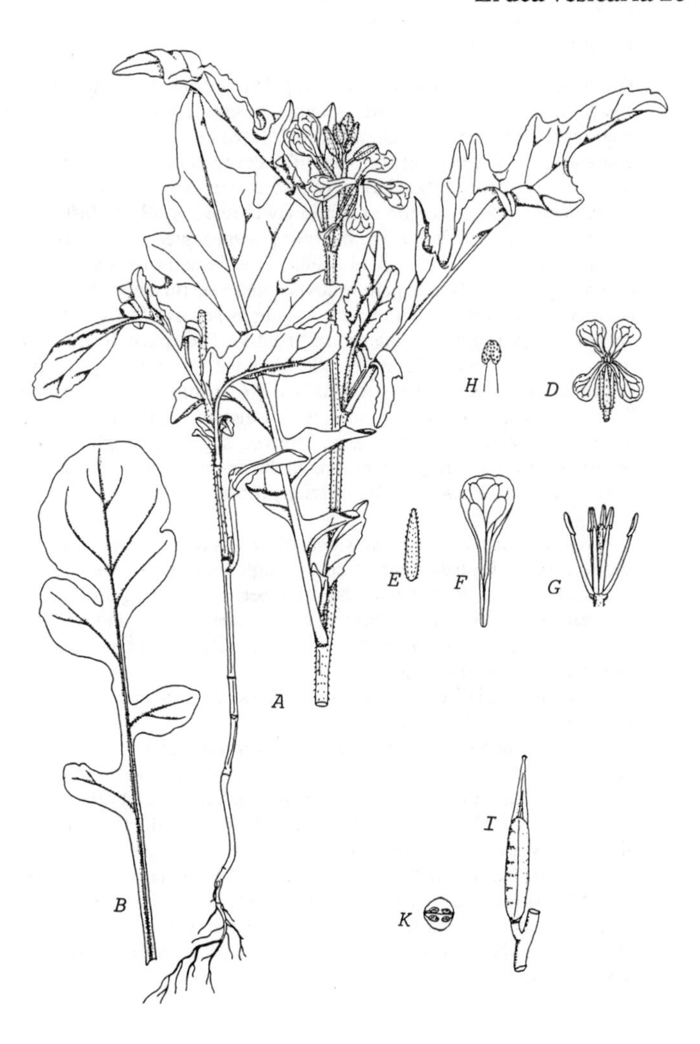

Ax0.2;Bx0.4;Dx0.7;E-Gx1;Hx5;I,Kx0.8.

27. Diplotaxis tenuifolia (L.) DC.

Perennial Wallrocket

Perennial 30-80(-?130) cm, glabrous, pale glaucous-green, foetid. Stems erect, branched mainly above, tough. Rosette leaves absent except in seedlings. Stem leaves numerous, to 15(-18) cm, petiolate or ± sessile, entire and linear-lanceolate or pinnatifid and lanceolate to oblanceolate in outline, the terminal lobe oblong, obtuse at apex, with 1-4 pairs of similar lateral lobes; margins entire or shallowly toothed or lobed. Inflorescence lax or crowded. Sepals 5-7.5 mm, oblong, awned, green to yellowish, erect to inclined. Petals 8-15 x 5-11.5 mm; limb broadly obovate to elliptic, rounded at apex, yellow to pale yellow (sometimes flushed red); claw short, distinct, paler. Petals about twice as long as sepals. Stamens 6; anthers yellow. Stigma capitate, emarginate to entire. Pedicels in fruit 11-40(-50) mm, slender, inclined. Fruits 14-50(-60) x 1-2.5 mm, linear, compressed (latiseptate), stipitate (the stipe (0.3-)0.5-6.5 mm), erect to ascending, dehiscent. Valves 10-48(-58) mm, with 1 distinct central vein. Persistent style 1.2-3(-3.5) mm, cylindrical, sterile; stigma emarginate. Seeds numerous 0.9-1.2 mm, ± ovoid, brown, biseriate (sometimes uniseriate at ends). 2n=22*. Flowering May to September.

In docks, waste ground, roadsides, old walls (especially castles), cities and roadsides. Probably introduced, though thought by some to be native in S. E. England. Widespread and often persistent in England and Wales, frequent in the London area but less common elsewhere, and less frequent overall than *Diplotaxis muralis*. Very rare in Scotland and Ireland (map in Perring & Walters 1962).

Native in S. and C. Europe to Syria, introduced to N. and S. America and Australasia.

The large, leafy stems and large, pale yellow flowers distinguish it immediately from *Diplotaxis muralis*. The combination of seeds in 2 rows in each loculus with a stipe should distinguish it from all other crucifers, but as the plant is largely self-incompatible and sets little fruit this may not always be apparent. Somewhat variable in leaf shape.

Diplotaxis is easily distinguished from most other yellow crucifer genera by the 2 rows of seeds in each loculus of the fruit. *Rorippa, Eruca, Arabis, Erysimum* and *Draba* also may have this character but are quite different plants. A number of other species of *Diplotaxis* have been recorded as very rare casuals.

Diplotaxis tenuifolia 27

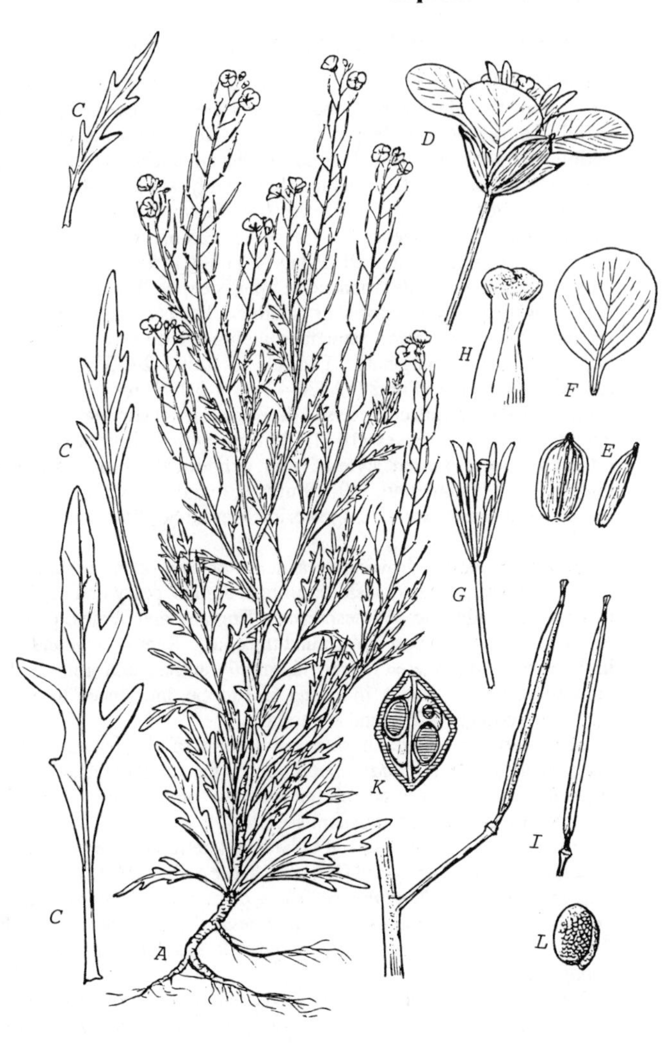

Ax0.2;Cx0.5;D-Gx2;Hx10;I,Kx1;Lx10.

28. **Diplotaxis muralis** (L.) DC.

Annual Wallrocket, Wall Mustard, Stinkweed

Annual (rarely perennial) (5-)10-60 cm, sparsely hairy below with simple hairs, dark to light green, foetid. Stems erect to ascending, branched mainly below. Rosette leaves to 13(-19) cm, with a narrowly winged petiole, narrowly oblanceolate in outline but sinuately lobed to pinnatisect with 3-6 broad, rounded lateral lobes, rarely simple, obtuse at apex; margins entire to toothed. Stem leaves 0-4(-6), smaller, less lobed and with sharper teeth. Inflorescence lax. Sepals (3-)3.5-5.1 mm, oblong to ovate, awned, green, erect to inclined. Petals (4-)5.8-8(-8.5) x (2.5-)2.8-5.8(-6.2) mm; limb broadly ovate, rounded at apex, deep yellow to yellow (often reddish when pressed); claw short, distinct. Petals *c.* 1.5 times as long as sepals. Stamens 6 (lateral stamens sometimes reduced); anthers yellow. Stigma capitate, emarginate. Pedicels in fruit 5-24(-35) mm, slender, ascending to patent. Fruits (12-)20-42 x 1.5-2.7(-3) mm, linear, tapering somewhat at each end, compressed (latiseptate), not stipitate, ascending to patent, dehiscent. Valves (10-)19-40 mm, with 1 distinct central vein, often torulose. Persistent style 1-3 mm, cylindrical, sterile, emarginate to bilobed. Seeds numerous, 0.9-1.3 mm, ± ovoid, brown, biseriate (often uniseriate at ends). 2n=42*. Flowering May to September (-November).

Doubtfully native, growing in a range of dry, open habitats such as railways, docks, walls, roadsides, rubbish dumps, peaty and sandy fields, flowerbeds, paths, etc. Widespread in England and Wales, particularly frequent in the south east and around the coast. Rare in Scotland (mainly in the east), and scattered in Ireland, mainly in the south (map in Perring & Walters 1962). Native in S. and C. Europe, introduced to N. and S. America, southern Africa and Australasia.

Little variable except in size and lobing of the leaves and easily distinguished from *Diplotaxis tenuifolia* by the small petals, absence of a stipe and leaves in a basal rosette with few on the stem. The plant is self-compatible and sets abundant seed.

Diplotaxis muralis is believed to be an allotetraploid derived from a *D. tenuifolia* x ? *viminea* hybrid (Harberd & McArthur 1972).

D. muralis var. *babingtonii* Syme is ± perennial with many stem leaves but has flowers and fruit like typical *D. muralis.* It is not known whether the differences are genetically based. It has been mistaken for *D. tenuifolia.*

The combination of biseriate seeds, small flowers and small fruit distinguishes it from most other yellow crucifers except *Rorippa* species. These usually have petals less than 6 mm and ± terete fruits, or are perennial.

Diplotaxis muralis 28

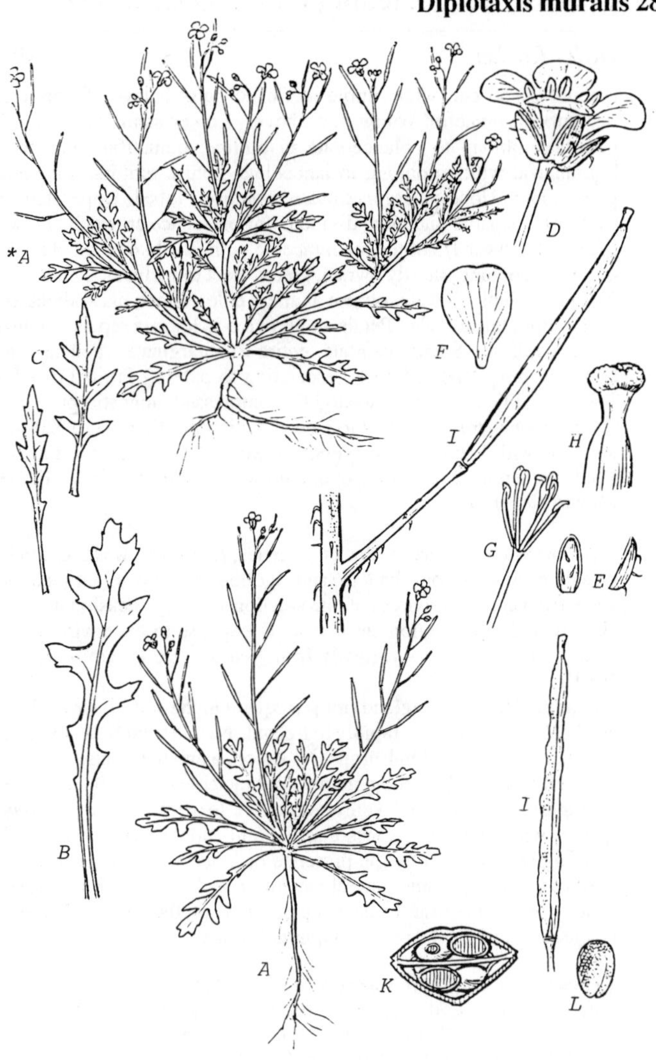

Ax0.2;B,Cx0.5;D-Gx2;Hx10;I,Kx1:Lx10.

29. **Erucastrum gallicum** (Willd.) O.E. Schulz

Hairy Rocket Map 29

Annual 20-60 cm, with simple, deflexed hairs below. Stems erect to decumbent, branched below and above. Lower stem leaves to 25 cm, petiolate, oblong to oblanceolate in outline, pinnatifid, pinnatisect or bipinnatifid with an oblong to lanceolate, obtuse terminal lobe and 3-8 pairs of lateral lobes; margins sinuate to obtusely lobed. Upper stem leaves similar but smaller and more divided. Inflorescence crowded. Flowers in at least the lower 1/3 of the main raceme with leaf-like bracteoles. Sepals 4-6.5 mm, oblong, shortly awned, green, erect to inclined. Petals 5.8-10 x 2-3.5 mm; limb obovate, obtuse at apex, pale yellow; claw about as long as the limb, linear, pale. Petals about twice as long as sepals. Stamens 6; anthers yellow. Stigma capitate, entire to emarginate. Pedicels in fruit (4-)5-19 mm, slender, (erect-) ascending to patent. Fruits (16-)20-45 x 1-2.5 mm, linear, subterete to slightly compressed (angustiseptate, septum 0.7-1.5 mm wide), ascending to patent, dehiscent. Valves (15-)20-45 mm, torulose, with 1 central vein. Persistent style (1.5-)2-3.5(-3.7) mm, linear. Seeds numerous, 0.9-1.5 mm, ovoid, brown, uniseriate. 2n=30. Flowering May to November.

An occasional casual of docks, quarries, roadsides, waste ground, etc., on a range of soils from chalk to sand. It often occurs as solitary individuals away from apparent sources of introduction and only occasionally persists. Growth and reproduction are plastic in response to the environment and seed production can vary greatly from year to year (Klemow & Raynall 1983).

Scattered through England (most frequent in the south), rare in Scotland and Wales but spreading rapidly in Ireland. Native from S. W. to C. Europe but widely introduced in Europe, N. America and the Urals.

Variable in leaf dissection, number of bracteoles and fruit size, but easily distinguished from other yellow crucifers (except two rare *Sisymbrium* species which have fruits less than 20 mm and *Arabis turrita* which has clasping stem leaves) by the presence of bracteoles in at least the lower third of the main raceme (lateral racemes may have few or no bracteoles). Other species may rarely have up to 4(-6) of the lowest flowers with bracteoles but then the usually bipinnatifid leaves should separate it.

30. **Erucastrum nasturtiifolium** (Poiret) O.E. Schulz is a rare casual with 2 post-1950 records. It is distinguished from *E. gallicum* by the absence of bracteoles, fruit with a short (0.3-0.7 mm) stipe and a beak 3-6.5 mm with (0-)1(-2) seeds. It may key out unsatisfactorily as a *Barbarea* or *Sinapis*.

Erucastrum gallicum 29

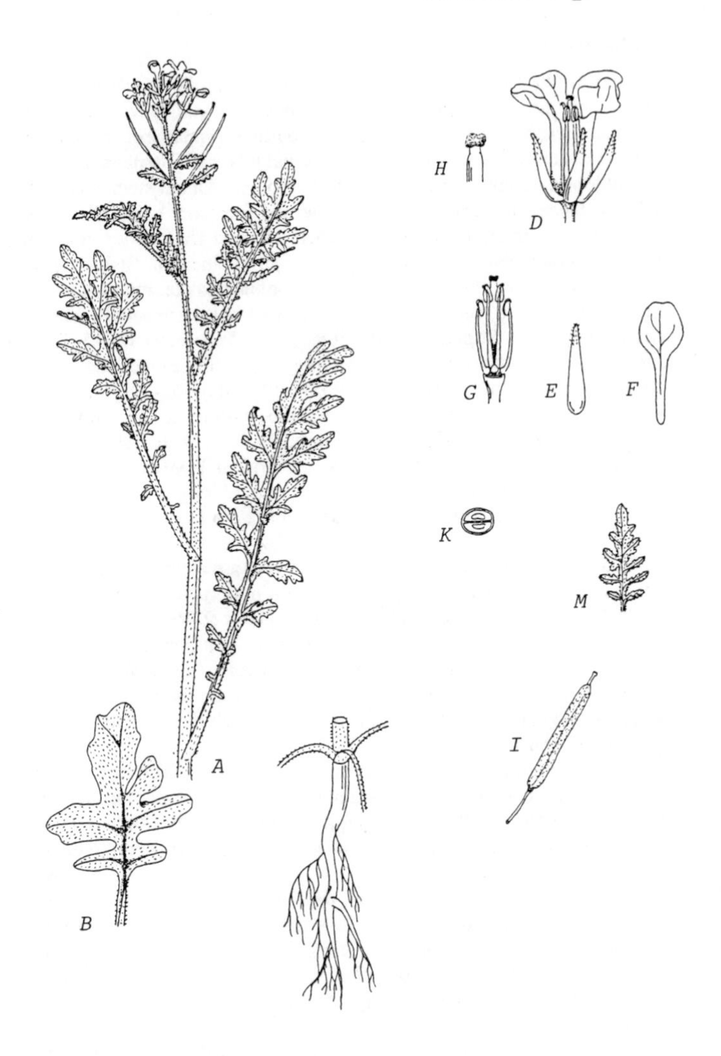

A,Bx0.5;D-Gx2;Hx5;Ix0.6;Kx2.5;M=bracteole,x1.

31. Sisymbrium irio L.

London Rocket

Annual 10-60(-130) cm, glabrous or with sparse, simple hairs below, bright green. Stems erect, much branched above. Basal leaves to 16 cm, petiolate, spathulate to oblanceolate in outline, pinnatisect to pinnatifid with an ovate to lanceolate, obtuse, terminal lobe and 2-6 pairs of smaller lateral lobes; margins usually irregularly toothed. Upper stem leaves with a linear-lanceolate, acute terminal lobe and 0-3 pairs of smaller lateral lobes at base; margins entire or toothed. Young fruits overtopping the flowers, ebracteolate. Sepals 1.6-2.5(-?3.5) mm, linear, not awned, green, erect. Petals 2.5-4(-?6) x 0.5-1 mm, linear-oblanceolate, rounded at apex, indistinctly clawed, yellow. Petals *c.* 1-2 times as long as sepals. Stamens 6; anthers yellow. Stigma capitate, emarginate. Pedicels in fruit 5-12(-20) mm, slender (0.2-0.4 mm wide in middle), ascending to inclined. Young fruits glabrous. Fruits (20-)25-53(-?65) x (0.7-)0.8-1.2(-1.3) mm, linear, terete to compressed (angustiseptate, the septum *c.* 0.5 mm wide), dehiscent. Valves (20-)25-52(-?65) mm, torulose, with 1 strong central vein and 2 weaker lateral veins. Persistent style 0.3-0.9 mm, stout, entire to emarginate. Seeds numerous, 0.7-1.1 mm, oblong, yellowish-brown, uniseriate. 2n=14,21,28,42,56. Flowering May to December.

An introduction found on waste ground, roads, pavements, docks, walls and banks, etc., often associated with grain imports and often as a wool alien. Persistent in London and Dublin (Bangerter & Welch 1952; Brunker 1952) but usually casual elsewhere. Rare in England and Ireland, not recorded recently in Wales or Scotland (map in Perring & Walters 1962). Probably native from S. Europe and N. Africa to India but introduced widely elsewhere in Europe, N. and S. America (abundant in the Colorado Desert) and Australasia.

Variable in leaf shape and hairiness in Britain. The species is an autopolyploid complex and is variable in many characters throughout its range (Khoshoo 1958 and subsequent papers). The diploid form is widespread and is the only taxon confirmed for Europe; polyploids occur in Asia (Khoshoo 1966a).

The very slender (*c.* 1 mm wide), torulose fruits are distinctive but are also found in *Descurainia* (which has finely divided leaves), *Erysimum repandum* (which has appressed forked hairs) and other *Sisymbrium* species. It is distinguished from the latter taxa which have fruits over 20 mm long by the small petals (less than 4.5 mm) combined with the long, slender pedicels.

Called the London Rocket after its abundance following the Great Fire of London in 1666.

Sisymbrium irio 31

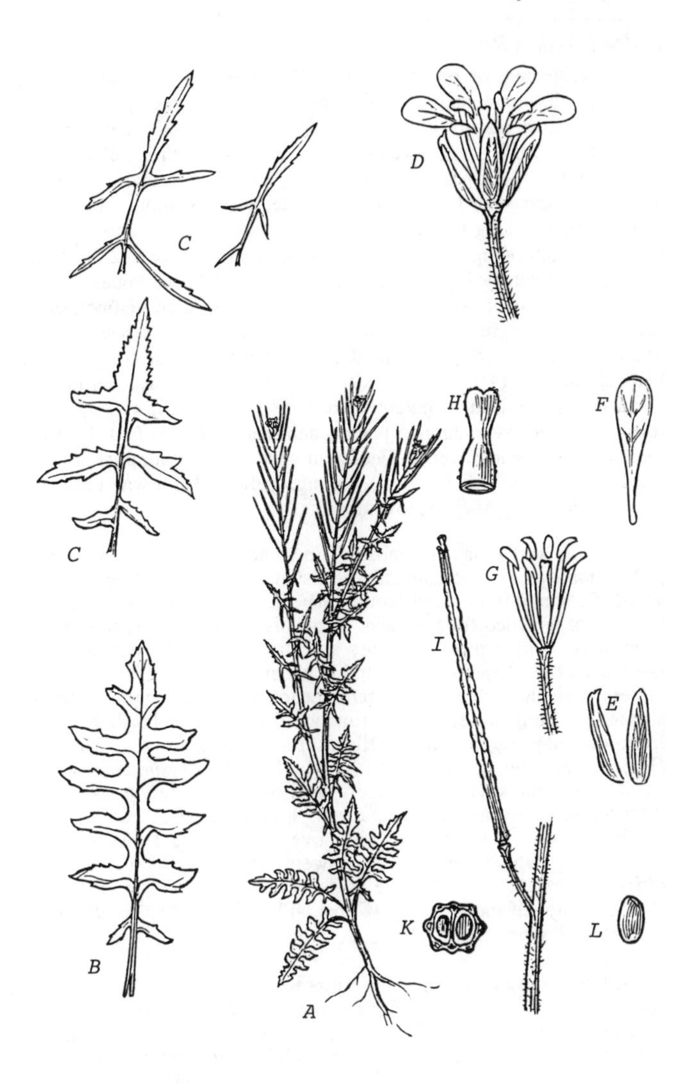

Ax0.2;B,Cx0.5;D-Gx5;Hx10;Ix1.2;K,Lx10.

32. Sisymbrium loeselii L.

False London Rocket Map 32

Annual (?biennial or perennial) 30-150(-?180) cm, with simple, long, deflexed hairs below, sparsely hairy above (rarely glabrous), green. Stems erect, branched mainly above. Basal leaves to 8(-10) cm, petiolate, pinnatisect with a large ovate-hastate or triangular, acute to obtuse, terminal lobe and 1-6 pairs of small, oblong lateral lobes; margins coarsely toothed. Upper stem leaves with a hastate, acute terminal lobe and 1-3 pairs of long, oblong to triangular lateral lobes. Inflorescence crowded. Flowers overtopping young fruits, ebracteolate. Sepals 2.5-4.5 mm, oblong, shortly awned, yellow to green, erect to inclined. Petals 4.5-7.5 x 1.9-3.7 mm; limb obovate, rounded at apex, yellow; claw distinct, shorter than limb, pale or greenish. Petals *c.* 1.5-2 times as long as sepals. Stamens 6; anthers yellow. Stigma capitate, emarginate. Young fruit glabrous. Pedicels in fruit (4-)5-14 mm, slender (0.15-0.5 mm wide in middle), (?ascending) inclined to patent. Fruits (7-)11-31(-?46) x 0.6-1.1 mm, linear, ± terete, ascending to patent, dehiscent. Valves (7-)11-30 mm, torulose or not, with 3 veins. Persistent style 0.3-1.5 mm, thick, bilobed. Seeds numerous, 0.6-1.1 mm, oblong, yellowish-brown, uniseriate. 2n=14. Flowering May to November.

An introduced casual of railway lines, docks, waste ground, walls, pavements, tips, etc., occasionally persistent (especially in London) in the south. Scattered in England (commoner than *S. irio*), rare in Ireland and Wales, not recorded for Scotland. Probably more frequent now than 50 years ago. Native from Germany and Italy east to C. Asia and India, introduced elsewhere in Europe and in N. and S. America.

Very variable in leaf shape, pubescence and fruit length. The very slender (*c.* 1 mm wide), patent fruits are distinctive and will separate it from the majority of other yellow crucifers (as for *S. irio*) except *Descurainia* (which has finely divided leaves), *Erysimum repandum* (which has appressed, stellate hairs) and other *Sisymbrium* species, but there are no easy diagnostic characters. It is similar to *S. irio* (which has smaller flowers and longer fruits overtopping the flowers) and *S. erysimoides* (which has smaller flowers and short, fat pedicels). *S. altissimum* and *S. orientale* have longer fruits.

Apparently self-incompatible (Khoshoo 1966b) but usually setting good fruit, though short, atypical fruits are sometimes seen.

33. **Sisymbrium erysimoides** Desf. is a rare wool alien (Map 33). It is annual, sparsely hairy; leaves pinnatifid; racemes ebracteolate; petals 1-3.5 mm, pale yellow; pedicels in fruit 1-3 mm, as wide as fruits; fruits 18-38(-?50) x 0.7-1.3 mm, subulate, patent; valves with 3 veins, not torulose.

Sisymbrium loeselii 32

Ax0.2;B,Cx0.5;D-Gx5;Hx10;Ix1.2;K,Lx10.

34. **Sisymbrium officinale** (L.) Scop.

Hedge Mustard

Annual or biennial 15-100(-140) cm, glabrous or with simple hairs, dark green, often purple. Stems erect, branched above. Rosette leaves to 15(-?20) cm, petiolate, pinnate to pinnatisect with a broadly ovate, obtuse to truncate terminal lobe and 2-6 pairs of smaller, ovate to oblong or triangular lateral lobes; margins sinuate, irregularly toothed or lobed. Lower stem leaves with an ovate to lanceolate terminal lobe and 2-4 pairs of lateral lobes. Upper stem leaves smaller, the terminal lobe lanceolate or hastate with 0-2 pairs of smaller, oblong lateral lobes; margins entire to irregularly toothed. Inflorescence crowded. Flowers overtopping fruits, ebracteolate. Sepals 1.7-2.5 mm, oblong, not awned, green, erect. Petals 3.1-4.2 x 0.9-1.4 mm, obovate, obtuse or rounded at apex, tapering to an indistinct claw, pale yellow. Petals *c.* 1.5 times as long as sepals. Stamens 6; anthers yellow. Stigma ± entire to emarginate. Young fruits glabrous or hairy. Pedicels in fruit 1-3 mm, ± as wide as fruits, appressed. Fruits (7-)10-16(-18)(-?20) x 0.9-1.6 mm, linear, ± terete, cylindrical-conical, tapering to an indistinct style 0.5-2 mm, dehiscent. Valves (6-)8-15(-16) mm, with 3 veins, not torulose. Seeds *c.* 10 per loculus, 1-1.7 mm, oblong, pale to dark brown, uniseriate. 2n=14. Flowering all year but mainly in the summer.

A weed of gardens, roadsides, arable land, waste ground, etc, usually associated with habitation or man. One of the commonest crucifers in Britain and Ireland, though rare in the upland regions (map in Perring & Walters 1962). Native in Europe, N. Africa and the Near East, but widely introduced and nearly cosmopolitan.

Not very variable except in size and branching (small, simple annuals contrast with tall, branched biennials) and hairiness. Hairiness varies within and between populations and it is doubtful if infraspecific taxa describing the variation (e.g. var. *leiocarpum* DC. with glabrous fruits) merit recognition.

The small, appressed fruits are very distinctive. *Hirschfeldia incana* is the only other yellow crucifer with appressed fruits of similar size but these typically have swollen, fertile tips; in any case it has larger petals and different leaves and the two are unlikely to be confused. *Sisymbrium polyceratium* and *S. runcinatum* also have small, often appressed fruits but have bracteolate inflorescences.

Sisymbrium officinale 34

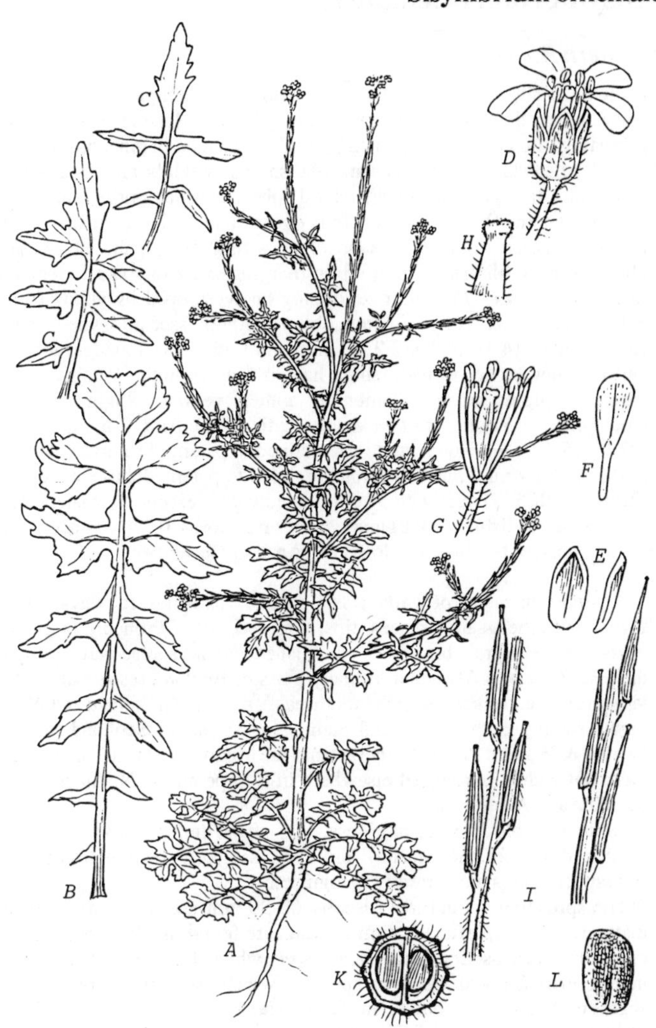

Ax0.2;B,Cx0.5;D-Gx5;Hx10;Ix1.2;K,Lx10.

35. Sisymbrium orientale L.

Eastern Rocket

Annual 10-80(-100) cm, softly hairy below with simple hairs, sparsely hairy above. Stems erect, branched mainly below. Rosette leaves lax, not persisting. Lower leaves to 15 cm, petiolate, pinnate to pinnatisect with a large, ovate to lanceolate or hastate, obtuse, terminal lobe and (0-)1-4 pairs of broadly triangular to linear, lateral lobes; margins entire, sinuate or shallowly toothed. Upper leaves smaller, the terminal lobe linear-lanceolate to linear-oblanceolate with 0-1(-2) pairs of smaller, oblong (rarely filiform) lateral lobes; margins entire or sparsely toothed. Inflorescence lax. Flowers overtopping or overtopped by young fruits, ebracteolate. Sepals (3-)4.5-6 mm, oblong, slightly saccate, awned, green, erect. Petals (4-)7-10.5 x 2.2-4.1 mm; limb obovate, rounded at apex, yellow; claw broad, whitish, about half as long as limb. Petals *c.* 1.5-2 times as long as sepals. Stamens 6; anthers yellow. Stigma entire to emarginate. Young fruits hairy. Pedicels in fruit 2-6 mm, thick ((0.3-)0.5-1.6 mm wide in middle), inclined to patent. Fruits (25-)50-120 x (0.5-)1.1-1.7 mm, linear, ± terete, inclined to patent, dehiscent. Valves (22-)47-100(-117) mm, with 3 veins, usually glabrescent. Persistent style 0.3-3.5 mm, indistinct, emarginate. Seeds numerous, 0.7-1.4 mm, oblong, brown, uniseriate. 2n=14. Flowering April to December.

A persistent casual of walls, waste ground, roadsides, railways, docks, towns and cities, sandy places, tips, bird seed, etc. Probably originally introduced with grain but widely established and has spread particularly in the last 50 years. Abundant in many parts of England (especially in the larger towns and cities and in the south-east), scattered in Ireland and Wales (again mainly in the cities) and mainly in the east of Scotland (map in Perring & Walters 1962). Native around the Mediterranean and in the Near East, but widely introduced elsewhere in Europe and in N. America, the Far East and Australasia.

Quite variable in size, hairiness, leaf shape (entire leaves are often produced late in the season) and fruit set (self-incompatible). No infraspecific taxa are currently recognised.

The spreading, cylindrical (usually 60-90 mm x *c.* 1.5 mm) fruits are distinctive but superficially similar fruits are found in a few other yellow crucifers, such as *Erysimum repandum* (which has appressed, stellate hairs), *Brassica* and *Coincya* species (which have beaks or terminal segments), and, of course, in *Sisymbrium.* Often confused with *S. altissimum* which has 2-5 filiform lobes on the upper leaves and young fruits usually (but not always) glabrous, and *S. irio* which has smaller flowers, and thinner fruits and pedicels.

Sisymbrium orientale 35

Ax0.2;B,Cx0.5;D-Gx5;Hx10;Ix1.2;K,Lx10.

36. Sisymbrium altissimum L.

Tall Rocket

Annual 30-100(-120) cm, hairy below with long, simple hairs, glabrous above. Stems erect, branched mainly above. Rosette leaves not persisting. Lower stem leaves to 30 cm, petiolate, the terminal lobe ovate to oblong, obtuse to acute with 5-10(-11) pairs of narrowly triangular to oblong lateral lobes; margins entire to coarsely lobed and toothed. Uppermost stem leaves smaller, very finely divided with a linear terminal lobe and 2-5(-6) pairs of linear lateral lobes, sessile or shortly stalked. Inflorescence lax, flowers overtopping or overtopped by young fruits, ebracteolate. Sepals 3.6-6.5 mm, oblong (often inrolled), awned, green, erect to patent. Petals 5.7-11 x 2-3.5 mm; limb obovate, rounded at apex, pale yellow (rarely white); claw broad, whitish, slightly shorter than limb. Petals about twice as long as sepals. Stamens 6; anthers yellow. Stigma capitate, emarginate. Young fruits usually glabrous, sometimes glabrescent. Pedicels in fruit 4.5-10(-12) mm, as wide as fruits, ascending to patent. Fruits (30-)40-90(-110) x 1-1.5 mm, linear, terete, ascending to patent, dehiscent. Valves (29-)38-89(-100) mm, with 1-3 obscure veins, glabrous. Persistent style 1-2.5 mm, indistinct, swollen. Seeds numerous, 0.7-1.2 mm, ovoid, brown, uniseriate. 2n=14. Flowering May to October.

A casual of walls, waste ground, roadsides, railways, docks, towns and cities, sandy places, tips, bird seed, etc., often abundant and well established, but less frequent and less persistent than *S. orientale*. Frequent in England, rare and scattered in Ireland, Wales and Scotland (map in Perring & Walters 1962). Probably native from Austria to India but widely introduced in Europe, Asia, the Far East, N. and S. America and Australasia.

Relatively constant except for minor differences in leaf dissection, and variable in overall size.

The very fine, linear lobes of the upper leaves are very distinctive; similar leaves may be found in *Descurainia sophia* and *Rorippa sylvestris* (which, however, have much smaller flowers and fruits), and *Coincya* (which has a fertile terminal segment to the fruits and larger petals). Often confused with *Sisymbrium orientale*, which sometimes has similar linear lobes on the upper leaves; the lower leaves must then be examined which are typically hastate with 1-4 pairs of lateral lobes in *S. orientale*, much less divided than in *S. altissimum*.

Sisymbrium altissimum 36

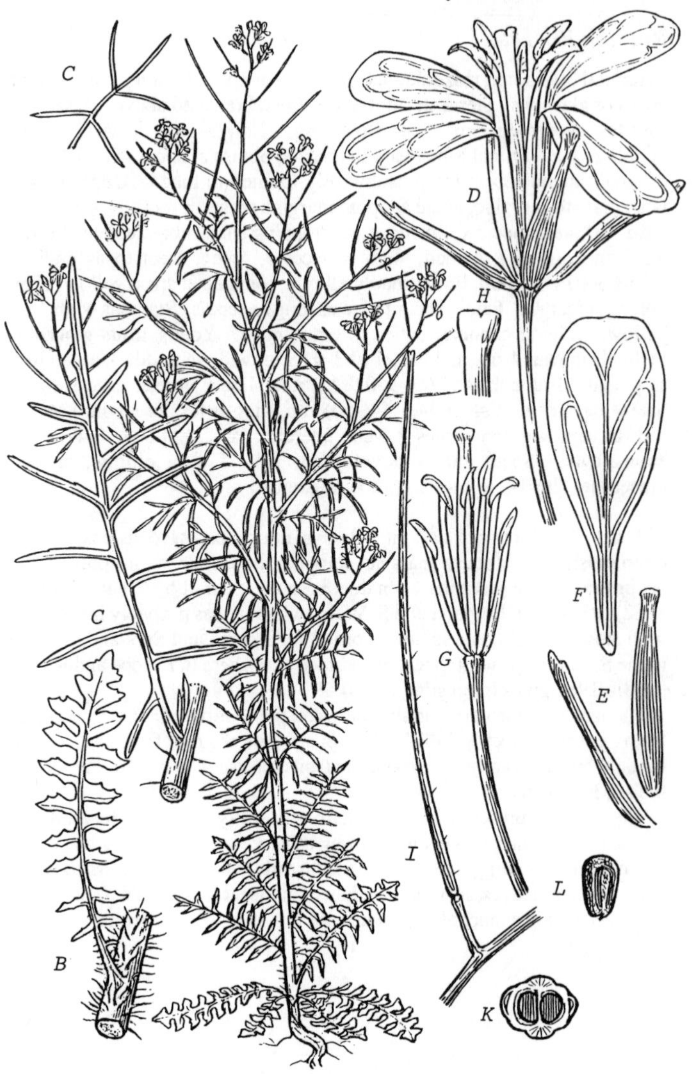

Ax0.2;B,Cx0.5;D-Gx5;Hx10;Ix1.2;K,Lx10.

37. **Sisymbrium volgense** Bieb. ex E. Fourn.

Russian Rocket Map 37

Perennial 30-75 cm, rhizomatous, glabrous or sparsely hairy below, dark green or glaucous. Stems decumbent to erect, branched above. Basal leaf rosettes evergreen, some not flowering. Lower stem leaves 15 cm, petiolate, the terminal lobe ovate to hastate, obtuse at apex, with 0-4 pairs of small lateral lobes; margins coarsely toothed to lobed. Upper leaves lanceolate to rhombic, acute at apex; margins entire or toothed or lobed at base. Inflorescence lax, flowers ± overtopping fruits, ebracteolate. Sepals 3-5.2 mm, oblong, awned, slightly saccate, green, erect to ascending. Petals 6-10 x 4-5 mm; limb broadly elliptic, rounded at apex, yellow; claw oblong, greenish. Petals about twice as long as sepals. Stamens 6; anthers yellow. Stigma capitate, entire to emarginate. Young fruits glabrous. Pedicels in fruit 4-6 mm, slender (0.3-0.5 mm wide in middle), ascending to patent. Fruits 15-45(-60) x 0.9-1.2 mm, linear, terete, ascending to patent, dehiscent. Valves 14.5-44(-60) mm, slightly torulose, with 1 strong central vein and sometimes 2 weak laterals. Persistent style 0.4-1 mm, emarginate. Seeds rarely developing, 1.3-1.5 mm, elliptic, brown, uniseriate. 2n=14. Flowering June to August.

A rare introduction found on railways, docks, paths, roadsides, canals, often persistent for many years but showing little tendency to spread. The occurrence of the plant has been documented in detail by Clement (1979, 1982); it is rare in England and S. Wales where it was probably introduced with Russian grain, and is not recorded for Ireland and Scotland. Native in the S. E. European Steppes, introduced elsewhere in Europe and Russia (Jehlik 1981 gives interesting notes).

The leaf shape is very variable between populations, the leaves ranging from entire to deeply lobed. The plant does set ripe fruit and seed in Britain, but only rarely; otherwise the fruits are usually only 10-25 mm and the seeds abortive.

The lack of fruit set makes it difficult to identify the genus, let alone the species. There are, however, very few other yellow crucifers which form patches with creeping rhizomes rather than stolons; *Rorippa austriaca* has clasping stem leaves, *R. sylvestris* has pinnatisect leaves, and *Sisymbrium strictissimum* has lanceolate, acuminate leaves and larger fruits.

Sisymbrium volgense 37

Ax0.2;B,Cx0.5;D-Gx5;Hx10;Ix1.2;K,Lx10.

38. Sisymbrium strictissimum L.

Perennial Rocket Map 38

Perennial 60-120(-?150) cm, softly hairy with long, simple hairs. Stems clumped, erect, branched above. Lower leaves to 17 cm, petiolate, lanceolate, acute or acuminate at apex, rounded to cuneate below (sometimes shallowly lobed); margins ± entire to shallowly toothed. Upper leaves similar but smaller, shortly petiolate or sessile. Inflorescence crowded. Young fruits overtopping flowers, ebracteolate. Sepals 4.5-6.5 mm, oblong, awned, green, inclined. Petals 8.4-10 x 2.4-3.1 mm; limb oblanceolate, rounded at apex, yellow; claw indistinct, whitish. Stamens 6; anthers yellow. Stigma ± entire. Young fruits glabrous. Pedicels in fruit (4-)8-15 mm, thinner than fruit (0.3-0.5 mm wide at middle), erect to recurved. Fruits 50-86 x 0.7-1.2 mm, linear, 4-angled, latiseptate, sometimes stipitate (the stipe 0-0.7 mm), erect to recurved, dehiscent. Valves 51-74 mm, torulose, with (1-)3(-5) veins. Persistent style 0.3-2 mm, capitate. Seeds numerous, 1.5-2.5(-3) mm, cylindrical, yellow to pale brown, uniseriate. 2n=28. Flowering June to August.

A rare casual or garden escape on waste ground, path sides, docks, etc. Scattered in England, not recorded for Ireland, Scotland or Wales. Native in S. Europe to Russia, occasionally introduced elsewhere in Europe.

European material is more variable than British material, and may have narrower, less toothed leaves and be glabrous.

The lanceolate, acuminate leaves with crowded flowers and fruits make it a distinct plant; no other yellow crucifer has all leaves ± simple and similar long, thin fruits.

39. Sisymbrium polyceratium L. is a rare casual (Map 39). It is a glabrous annual 10-30 cm; stems decumbent to ascending; stem leaves petiolate, ovate, lobed to pinnatifid, acute; racemes bracteolate, the flowers in clusters of (1-)2-5 in the bracteole axils; petals 1.5-2 mm, pale yellow, *c.* 1.5 times as long as sepals; fruits 9-20(-?25) x 0.7-1.2 mm, linear with a short persistent style *c.* 2 mm. The most distinctive feature is the bracteolate racemes (but see also **29.** *Erucastrum gallicum* which has longer petals and fruits).

40. S. runcinatum Lag. ex DC. is very similar to *S. polyceratium* but has flowers solitary in the bracteole axils and petals 2.5-3.5 mm. There are 2 post-1950 records. For details of both species and illustrations, see Clement (1982).

Sisymbrium strictissimum 38

Ax0.2;B,Cx0.5;D-Gx5;Hx10;Ix1.2;K,Lx10.

41. **Descurainia sophia** (L.) Webb ex Prantl

Flixweed

Annual (rarely biennial) 10-100 cm, densely to sparsely covered with stellate, forked and simple hairs at least below, grey-green to green. Stems erect, branched above. Lower leaves to 8(-10) cm, oblong to ovate, variably but finely divided, bipinnate to quadripinnate, the lobes ovate to linear-oblong, acute, the lowest pair often clasping the stem. Upper stem leaves pinnate to bipinnate. Inflorescence crowded. Sepals 2-3 mm, linear, green to yellow, erect to ascending. Petals 1.4-1.8 x *c.* 0.3 mm, spathulate, pale yellow. Petals ± as long as or shorter than sepals. Stamens 6; anthers yellow. Stigma ± capitate. Pedicels in fruit 5-14(-50) mm, slender, inclined to patent. Fruits 10-26(-?35) x 0.5-1.1 mm, linear, ± compressed (angustiseptate) to terete, erect to ascending, dehiscent. Valves 10-26(-35) mm, torulose, ± keeled with 1 strong central vein and weak lateral veins. Persistent style 0.2-0.3 mm, ± sessile, capitate. Seeds numerous, 0.8-1.1 mm, ellipsoid to oblong, pale brown, uniseriate. 2n=28(?56). Flowering May to October (November).

Probably a long-established introduction. Particularly characteristic and persistent as an arable weed on sandy or peaty soils in E. England, but elsewhere an uncommon casual of waste ground, docks, cities, etc., usually associated with grain imports. Susceptible to herbicides and an easily controlled weed (Best 1977). If Salisbury (1961) is to be believed, the plant is much less common now than in the 16th and 17th centuries.

Common in E. England, scattered elsewhere throughout England and Wales (map in Perring & Walters 1962). Mainly around the Firth of Forth and the Moray Firth in Scotland. Rare in Ireland, mostly in the larger cities. Native in Europe from Scandinavia to S. W. Asia and N. Africa. Introduced to N. and S. America, Africa and Australasia.

Although somewhat variable in leaf dissection, British material is relatively constant. Elsewhere variations in petal size, pubescence and fruit length are noted. No infraspecific taxa are accepted here. The plant is self-compatible and usually sets abundant seed.

Easily distinguished by the very finely divided leaves. Very few other yellow crucifers (*Sisymbrium irio* and *Rorippa* spp.) may have the petals more or less equalling the sepals. *D. sophia* was originally included in *Sisymbrium*, and the genus *Descurainia* is now separated from *Sisymbrium* by the finely divided leaves and stellate hairs.

Descurainia sophia 41

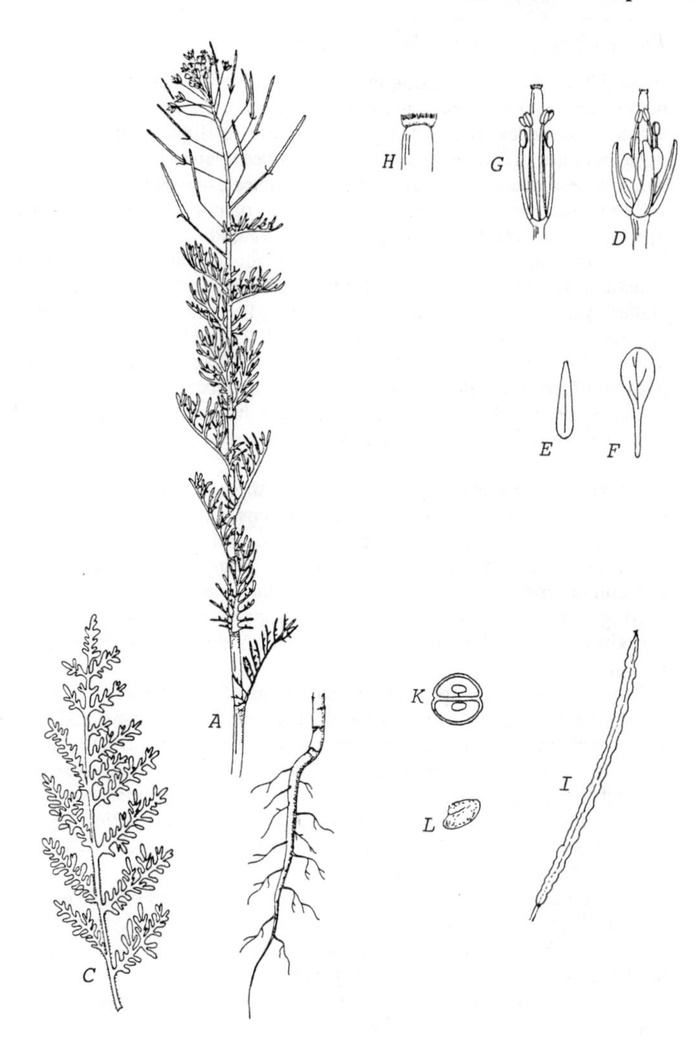

A,Cx0.5;D,Ex5;Fx10;G,Hx15;Ix2;Kx10;Lx8.

42. **Arabidopsis thaliana** (L.) Heynh.

Thale Cress, Common Wall Cress

Annual 2-30(-50) cm with simple, and sometimes forked and stellate hairs below, grey-green, often ± glaucous. Stems erect, branched above and below. Rosette leaves to 4 cm, petiolate, ovate to elliptic, obtuse at apex; margins entire to coarsely toothed (rarely lobed). Stem leaves few, to 1.5 cm, sessile, narrowly lanceolate, elliptic or linear; margins usually entire. Inflorescence crowded. Sepals 1.3-2.5 mm, oblong, green, with purple tips, yellowing rapidly with age, erect to ascending. Petals 2.5-4.5 x 0.6-1.5 mm; limb spathulate to oblanceolate, rounded at apex, white; claw indistinct, yellowish. Petals 1.5-2 times as long as sepals. Stamens 4-6; anthers yellow. Stigma capitate, ± entire. Pedicels in fruit 4-10 mm, slender, inclined to patent. Fruits 6-16(-18) x 0.3-0.7 mm, linear and slender, flattened (latiseptate), ascending to inclined, dehiscent. Valves 6-15.5(-17.5) mm, with 1 strong central vein. Persistent style 0.2-0.5 mm. Seeds numerous, 0.3-0.5 mm, ellipsoid, pale brown, uniseriate. 2n=10. Flowering nearly all year but mainly April to July.

Native on rocks and ledges, dunes, river shingle, open sandy ground, etc., and also a frequent weed of paths, gardens, waste ground, roads and characteristically railways. In open, dry places and most frequent on light, sandy soils in the spring but rarely abundant. Throughout Britain though uncommon in the hilly areas, surprisingly infrequent in Ireland (map in Perring & Walters 1962). Native from Europe to the Far East, E. Africa. Introduced to N. America, S. Africa and Australasia and becoming ubiquitous.

Variable in size in response to the environment but otherwise relatively constant at first glance. At a finer level, populations tend to differ slightly from each other and there are numerous, simple, genetically-based mutations which when combined with the small size and short life cycle, make it an excellent experimental species (e.g. Estelle & Somerville 1986). The genus even has its own journal, the *Arabidopsis Information Service*. Largely self-fertilized, though it will outbreed if cross-pollinated. No infraspecific taxa are recognised.

Distinguished by its small size, small, linear fruits and entire to toothed leaves. The contrast between the white petals and yellowing sepals is often distinctive. Somewhat like *Arabis hirsuta* but this has many stem leaves with ± clasping bases and erect, longer fruits.

Arabidopsis is a poorly defined genus and there is no currently accepted consensus as to whether to include it in *Arabis* or not.

Arabidopsis thaliana 42

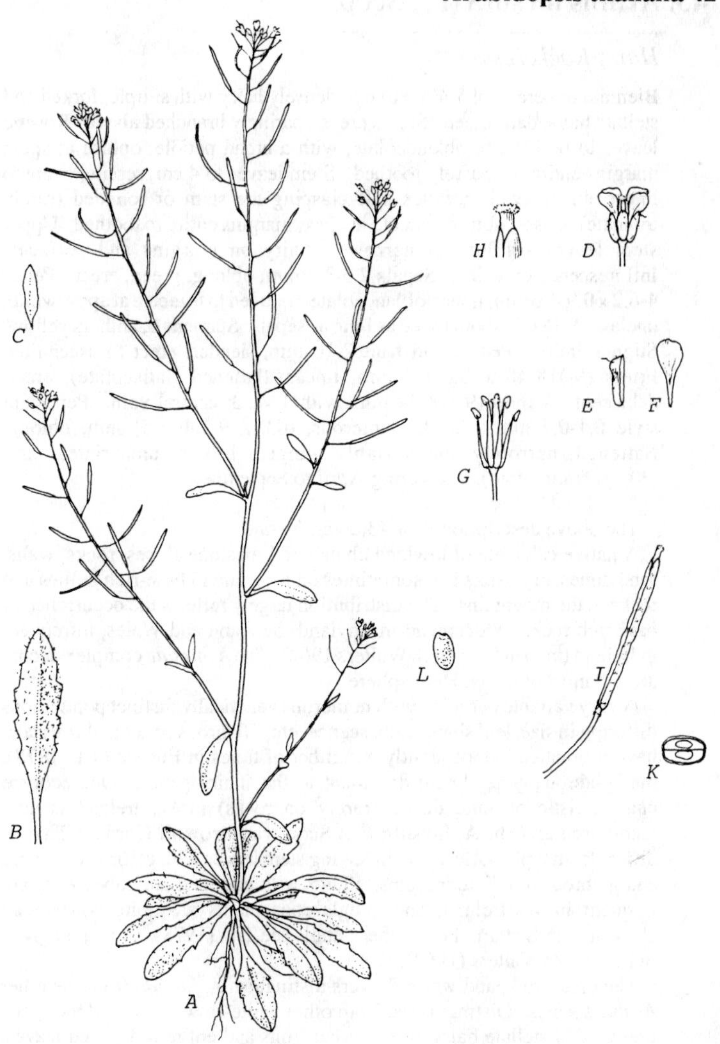

A-Cx1;D-Gx3;Hx10;Ix2;K,Lx10.

43. **Arabis hirsuta** (L.) Scop.

Hairy Rockcress

Biennial or perennial 5-40(-80) cm, densely hairy with simple, forked and stellate hairs, dark green. Stems erect, sparingly branched above. Rosette leaves to 6(-12) cm, oblanceolate, with a broad petiole, obtuse at apex; margins entire or sparsely toothed. Stem leaves to 4 cm, sessile, ovate to linear-oblong, with auricles half clasping the stem or rounded (rarely truncate) at base, obtuse to acute at apex; margins entire to toothed. Upper stem leaves similar but narrower, hairy on margins and surfaces. Inflorescence crowded. Sepals 2.5-3.7 mm, oblong, green, erect. Petals 4-6.2 x 0.9-1.6 mm, linear-oblanceolate, rounded to truncate at apex, white, unclawed. Petals about twice as long as sepals. Stamens 6; anthers yellow. Stigma entire. Pedicels in fruit 2-10 mm, slender, erect to ascending. Fruits (9-)18-48 x 0.6-1.5 mm, linear, flattened (latiseptate), erect, dehiscent. Valves (9-)18-48 mm, with 1 weak central vein. Persistent style 0.1-0.7 mm. Seeds numerous, (0.8-)0.9-1.4(-1.5) mm, oblong, flattened, narrowly and variably winged, brown, uniseriate. 2n= (30-)32*(tetraploid). Flowering April to September.

The above description is of **43a.** var. *hirsuta*.

A native calcicole of lowland chalk and limestone slopes, rocks, walls, sand dunes, dry banks and sometimes on peat, and in base-rich gullies and rocks in the mountains. The distribution largely reflects the occurrence of base-rich rock. Widespread in England, Scotland and Wales, infrequent in Ireland (map in Perring & Walters 1962). The *A. hirsuta* complex occurs around most of the N. Hemisphere.

A very variable complex with numerous genetically distinct populations differing in size, leaf shape, pubescence, etc. Glabrous or subglabrous taxa have originated independently a number of times in Europe (Titz 1978), the "glabrous gene" being dominant to the "hairy gene". One ecotype characteristic of sand dunes (rarely on rocks) in W. Ireland is now recognised as **43b. A. hirsuta** (L.) Scop. var. **brownii** (Jordan) Titz. It differs from typical *A. hirsuta* in having stem leaves with glabrous surfaces and glabrous or ciliate margins. Plants with intermediate pubescence are frequent in W. Ireland, and subglabrous plants are quite widespread elsewhere in Britain. For further details see Titz (1978). A map is given in Perring & Walters (1962).

The erect fruits and white flowers distinguish *A. hirsuta* from the other *Arabis* species. Distinguished from other white-flowered crucifers by the presence of stellate hairs, linear, erect fruits and entire to toothed leaves. *Arabidopsis* is similar but altogether more delicate and has spreading fruits.

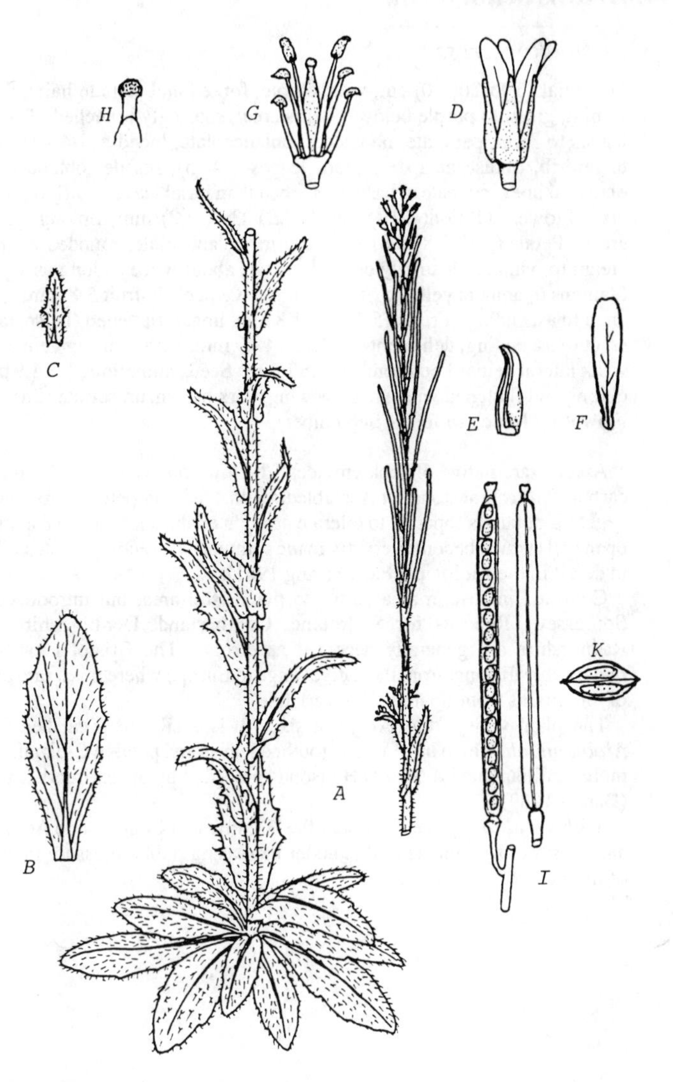

Ax0.5;B,Cx0.7;D-Gx5;Hx10;Ix2;Kx10.

44. Arabis scabra All.

Bristol Rockcress

Perennial (3-)5-20(-30) cm, with simple, forked and stellate hairs, deep, shining green to purple below. Stems erect, sparingly branched. Rosette leaves to 3 cm, petiolate, obovate to oblanceolate, lobed *c.* 1/4 - 1/2 way to midrib, obtuse at apex. Stem leaves 1-4(-5), sessile, oblanceolate, obtuse at apex, cuneate at base, less lobed than basal leaves. Inflorescence lax. Flowers (1-)3-8(-12). Sepals (2.7-)3-4(-4.2) mm, oblong, green, erect. Petals (5-)5.5-8(-8.8) x 1.8-3 mm, oblanceolate, rounded at apex, cream to white, indistinctly clawed. Petals about twice as long as sepals. Stamens 6; anthers yellow. Stigma entire. Pedicels in fruit 5-8 mm, stout, erect to ascending. Fruits 15-50 x 1-1.8 mm, linear, flattened (latiseptate), erect to ascending, dehiscent. Valves 14-49 mm, with 1 strong vein and 2 weak lateral veins. Persistent style *c.* 1 mm. Seeds numerous, 1.2-1.9 mm, oblong and flattened with a narrow wing, dark brown, uniseriate. 2n=16*. Flowering March to June (-September).

A very rare native of rock crevices, shallow soil pockets and scree on carboniferous limestone. It is unable to withstand competition from other vegetation, but its capacity to tolerate drought enables it to survive in more open soils which become very dry in the summer. The ecology is described in detail in the Biological Flora (Pring 1961).

Confined in Britain as a native to the Bristol area, but introduced in Somerset. Records for S. Ireland, Cumberland, Denbighshire and Radnorshire are generally accepted as errors. The Bristol locality is remarkably disjunct from the native range in Europe where it is a rare plant of mountains from Spain to Switzerland.

The plant varies little except in size. It is easily distinguished from *Arabis hirsuta* which has entire to toothed leaves and petals 4.2-6 mm. The more familiar name *A. stricta* Hudson is preceded by the earlier *A. scabra* (Dandy 1969).

Arabis scabra is protected under the Wildlife and Countryside Act 1981 and must not be collected. It is under increasing public pressure in many of its sites.

Arabis scabra 44

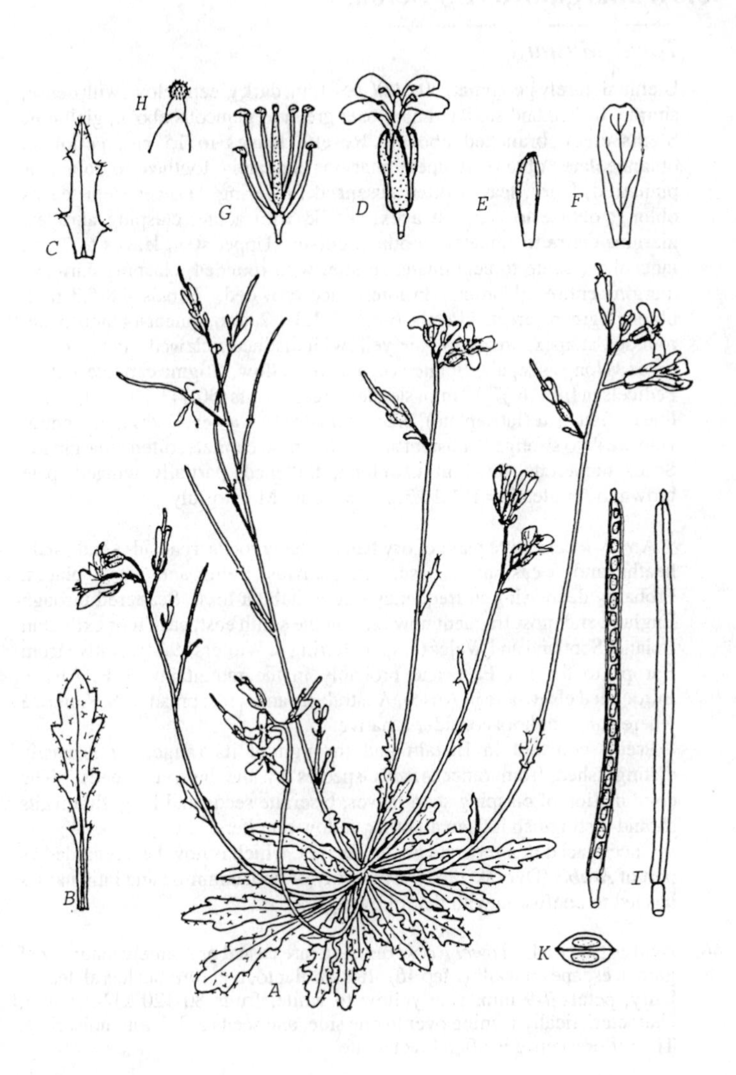

Ax0.5;B,Cx1;D-Gx4;Hx10;Ix2;Kx6.

45. **Arabis glabra** (L.) Bernh.

Tower Mustard

Biennial (rarely perennial) 30-100(-150) cm, dark green below, with dense, simple, forked and stellate hairs, pale green to glaucous above, glabrous. Stems erect, branched above. Rosette leaves to 15 cm, petiolate, oblanceolate, obtuse at apex; margins sinuately toothed to lobed or pinnatifid, hairy; leaves often absent at flowering. Lower stem leaves oblong, obtuse to acute at apex, sessile with acute, clasping auricles; margins entire to sinuately toothed, ciliate. Upper stem leaves to 7 cm, lanceolate, acute to acuminate, sessile, with rounded, clasping auricles; margins entire, glabrous. Inflorescence crowded. Sepals 4.8-5.3 mm, oblong, green, erect. Petals 6-8.4 x 1.3-1.7 mm, linear-oblanceolate, rounded at apex, cream to pale yellow, indistinctly clawed. Petals *c.* 1.5 times as long as sepals. Stamens 6; anthers yellow. Stigma capitate, entire. Pedicels in fruit (6-)7-19 mm, slender, erect. Fruits (30-)43-70 x 1-1.7 mm, linear, flattened (latiseptate), erect, dehiscent. Valves 42-66 mm, central vein weak to strong. Persistent style 0.6-1 mm, capitate, often emarginate. Seeds numerous, 0.6-1 mm, oblong, flattened, partially winged, pale brown, biseriate. 2n=12*,16,32. Flowering May to July.

A very local native plant of dry banks, shady rocks, roadsides and sandy heaths and occasionally casual in quarries, fields and waste places. Probably decreasing in frequency due to habitat loss. Scattered through England and most frequent now only in the south east, casual or extinct in Ireland, Scotland and Wales (map in Perring & Walters 1962). Native from Europe to the Far East, and probably in the mountains of E. Africa. Introduced elsewhere in Africa, Australasia and widespread in N. America where some authors consider it native.

Pretty constant in Britain and throughout its range. It is easily distinguished from other *Arabis* species by the biseriate seeds. The combination of clasping stem leaves, biseriate seeds and long, thin fruits should distinguish it from all other yellow crucifers.

Once placed in a separate genus *Turritis*, which is now best regarded as part of *Arabis* (Dvořák 1967). Similarity of both common and latin names has led to confusion with the following species.

46. Arabis turrita L., *Tower Rockcress*, is a rare casual or scarcely naturalized garden escape on walls (Map 46). It is similar to *A. glabra* but has all leaves hairy; petals 7-9 mm, pale yellow to white; fruits 80-120 x 2-2.5 mm, characteristically leaning over to one side, and seeds 2-2.7 mm, uniseriate. The inflorescence is often bracteolate.

Arabis glabra 45

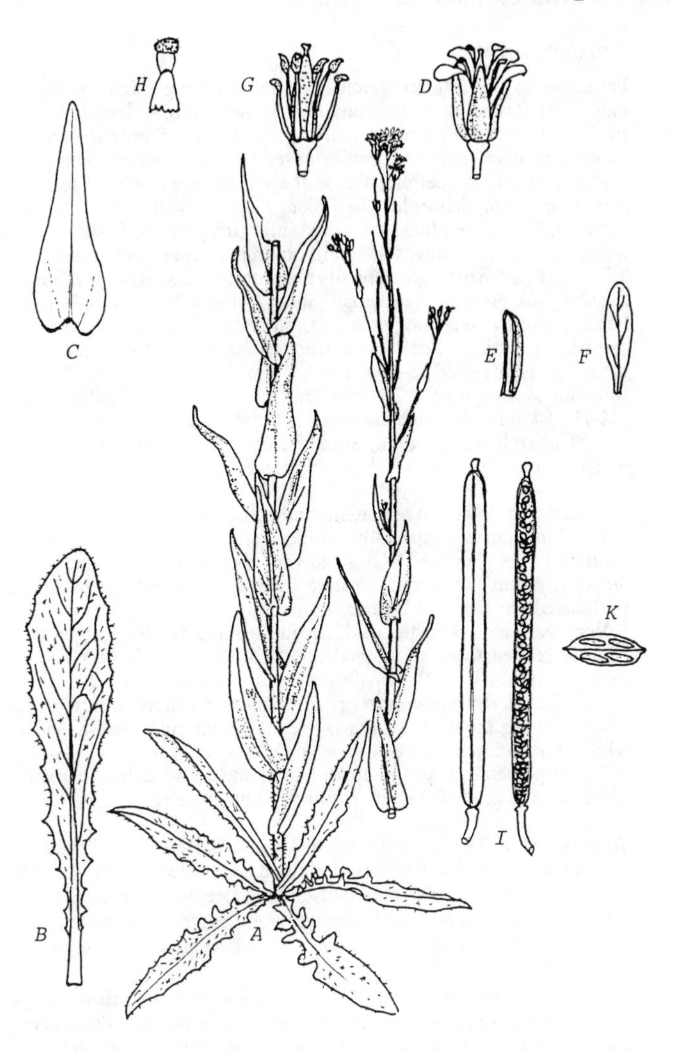

Ax0.5;B,Cx0.7;D-Gx3;Hx10;Ix1.5;Kx10.

47. **Arabis caucasica** Willd.

Garden Arabis

Perennial 15-25(-50) cm, much branched, forming spreading mats with many non-flowering rosettes and thin, brown stolons, densely hairy with stellate, forked and simple hairs, grey-green. Stems decumbent to ascending, unbranched. Rosette leaves to 6 cm, oblanceolate, cuneate below, with a broad petiole, obtuse at apex; margins with 2-3 teeth. Stem leaves to 4 cm, lanceolate to oblong, sessile with rounded, clasping auricles, acute to obtuse at apex; margins with 2-6 shallow teeth. Inflorescence lax. Sepals 4-7 mm, oblong, saccate, green, erect. Petals 9.5-16(-18) x 4-8 mm; limb broadly obovate, rounded at apex, white (rarely flushed pink-purple); claw broad, about twice as long as limb, greenish. Petals about twice as long as sepals. Stamens 6; anthers yellow. Stigma capitate, ± entire. Pedicels in fruit 10-20 mm, slender, ascending to inclined. Fruits (22-)32-55 x 1.5-2.5 mm, linear, flattened (latiseptate), sometimes with stipe 0.2-1 mm, erect to inclined, dehiscent. Valves (22-)31-54 mm, somewhat torulose, the central vein weak or absent. Persistent style 0.5-0.9 mm. Seeds numerous, 1.4-1.6 mm, flat, elliptic, winged, uniseriate. 2n=16. Flowering mainly March to June.

A very common plant of gardens and consequently occasionally thrown out and naturalised, especially on old walls, limestone rocks and cliffs. Scattered in England and Wales and very abundant in some localities (e.g. the Derbyshire limestone). Native in S. Europe and the Middle East, introduced elsewhere in Europe.

Very variable and often divided into a number of infraspecific taxa. Closely related to *A. alpina* and on a worldwide scale very difficult to distinguish from it. In Britain the plants cultivated in the garden have often been selected for horticultural performance and there is little difficulty distinguishing them: *A. alpina* is greener, with small petals and fruits, whilst *A. caucasica* is more robust and distinctly grey.

The large petals, clasping upper leaves and dense stellate hairs make it unlikely to be confused with any other white crucifer.

48. Arabis collina Ten. agg. (including *A. muralis* Bertol non Salisb.)

Perennial 10-30 cm, with dense, stellate hairs. Stems erect. Basal leaves to 10 cm, petiolate, obovate, obtusely and deeply toothed. Stem leaves oblong to ovate, sessile with clasping auricles, toothed or entire. Petals 6-10 x 2-4 mm, pink to purple or white, about twice as long as sepals. Fruits 20-70(-90) x 1-2.2 mm, erect.

There are 5 or 6 post-1950 records for pink- or purple-flowered garden plants on walls etc., which are referred here to the *A. collina* aggregate (including *A. muralis*). *A muralis* and *A. collina* are closely related and I have been unable to resolve satisfactorily the identities of naturalised plants, which have possibly been modified by horticultural selection.

Arabis caucasica 47

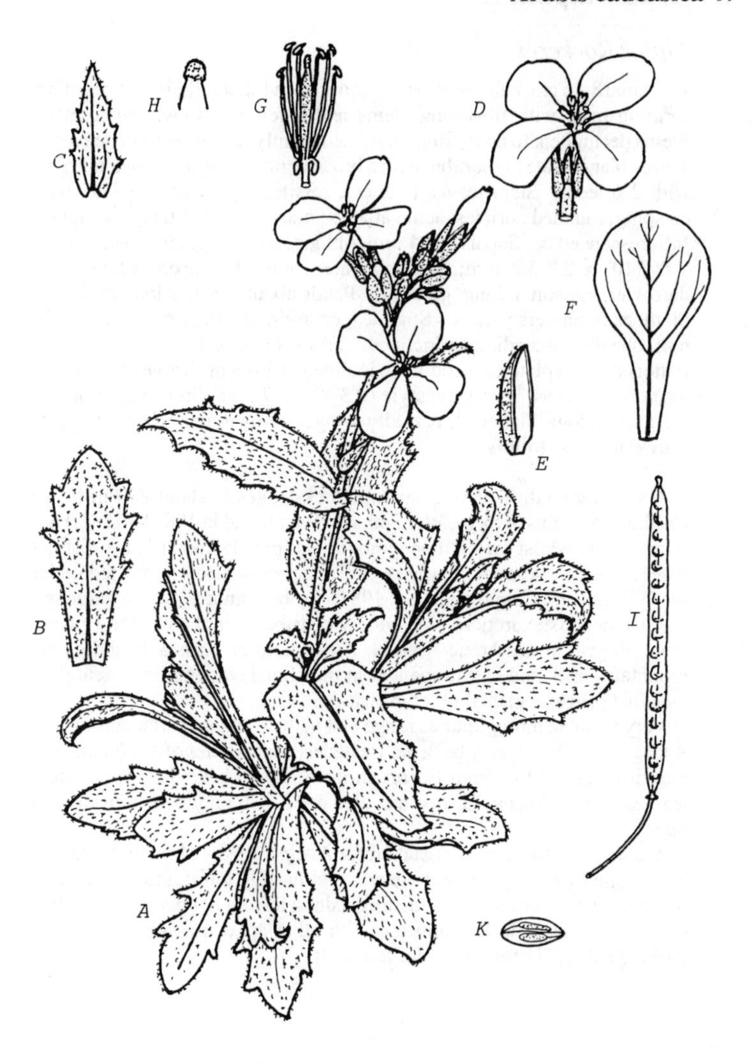

A-Cx0.5;Dx1;E-Gx3;Hx10;Ix1;Kx5.

49. Arabis alpina L.

Alpine Rockcress

Perennial 8-15 cm, hairy with simple, forked and stellate hairs but greenish, forming mats with flowering stems and a few non-flowering rosettes. Stems decumbent to ascending, branched mainly below. Rosette leaves to 4 cm, oblanceolate to rhombic with a broad petiole, obtuse at apex; margins with 2-6 teeth. Stem leaves to 3 cm, sessile, ovate to lanceolate, with clasping, rounded auricles, acute at apex; margins with 3-6 teeth or lobes. Inflorescence lax. Sepals 2.5-4 mm, oblong, saccate, green, erect. Petals 5-8(-?10) x 2.8-3.2 mm; limb obovate, rounded at apex, white; claw narrower, indistinct, long, greenish. Petals about twice as long as sepals. Stamens 6; anthers yellow. Stigma ± emarginate. Pedicels in fruit 5-10 mm, slender, ascending to inclined. Fruits 17-30 x 1.3-1.7 mm, linear, flattened (latiseptate), ascending to inclined, dehiscent. Valves 17-30 mm, torulose, veinless. Persistent style (0.3-)0.5-0.7 mm. Seeds numerous, *c.* 1.5 mm, oblong, flattened, partially winged, brown, uniseriate. 2n=16*. Flowering May to July.

A very rare native of wet, igneous rock ledges at about 800 m in the Cuillin Mountains of Skye, W. Scotland. First found in 1887 by H. C. Hart (Hart 1887) whilst on his honeymoon. Originally found "in 3 distinct places" but now known in only two. The species is now protected under the Wildlife and Countryside Act 1981. Probably not grown in gardens.

A characteristic arctic-alpine widely distributed in the N. Hemisphere, circumboreal in the Arctic from N. America to N. Russia, Scandinavia, mountains of S. and C. Europe and in the Himalayas and E. Africa. It is usually taken as a glacial relict in Britain.

Very variable throughout its range and in the southern part blending into *A. caucasica*, which may be better treated as a subspecies of *A. alpina*. The Scottish plants differ from European material and have quite lobed stem leaves. Populations in Europe etc. also often differ noticeably from each other.

A. petraea is frequently mistaken for *A. alpina* by wishful thinking, but has sessile (not clasping) stem leaves and usually lobed, much less hairy rosette leaves. For characters distinguishing it from *A. caucasica*, see that species. Unlikely to be confused with other species as you need to be looking especially for it in order to find it.

Arabis alpina 49

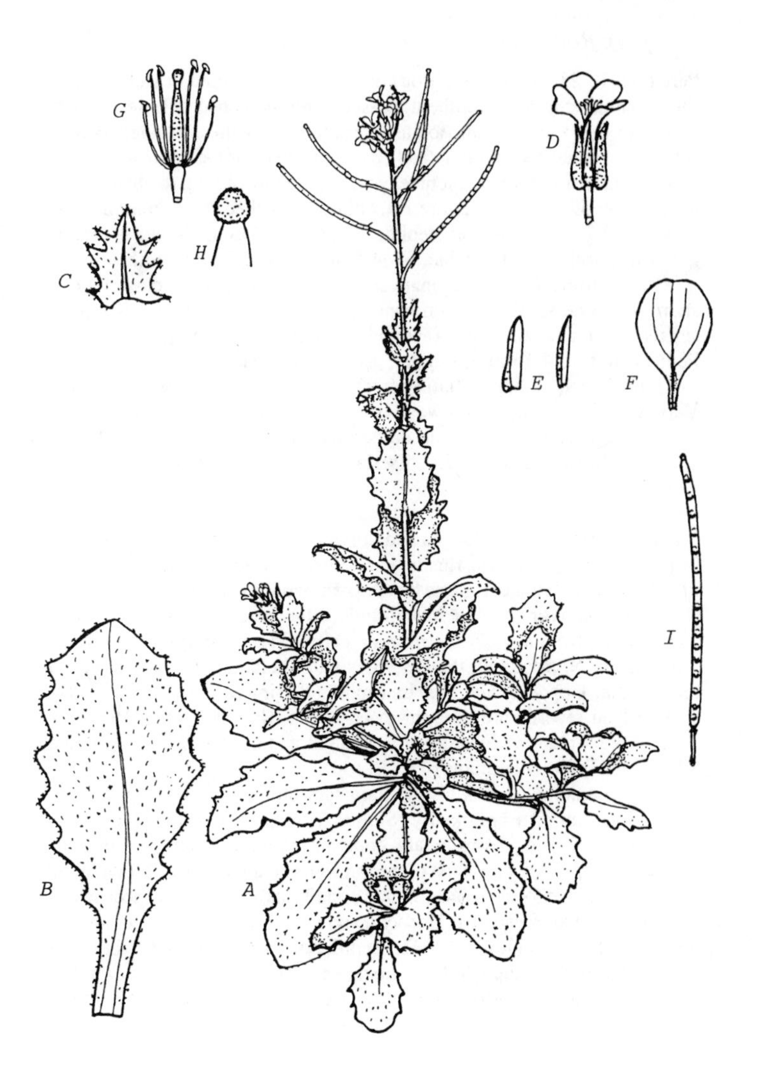

Ax0.5;Bx2;Cx1;Dx2;E-Gx3;Hx10;Ix1.5.

50. **Arabis petraea** (L.) Lam.

Northern Rockcress

Perennial 10-25(-45) cm, glabrous or with simple, forked or stellate hairs, rhizomatous. Stems decumbent to ascending, branched at base. Rosette leaves to 5(-9) cm, oblanceolate to spathulate, with a winged petiole, simple or pinnatisect with 1-6(-8) pairs of lateral lobes, obtuse at apex; margins entire or toothed. Stem leaves 2-9, to 2 cm, oblong to oblanceolate at apex, sessile; margins entire or shallowly toothed. Inflorescence lax. Lowest 2-3 flowers often bracteolate. Sepals 2-3.5 mm, oblong, ± saccate, green to purple, erect. Petals (4-)4.5-9(-9.1) x 2-5 mm; limb obovate, rounded to truncate or emarginate at apex, white to purple; claw distinct, short, with one small tooth on each side. Petals about twice as long as sepals. Stamens 6; anthers yellow. Stigma capitate, ± emarginate. Pedicels in fruit 4-15 mm, slender, ascending to patent. Fruits (5-)8-35 x 1-2 mm, linear, flattened (latiseptate), ascending to inclined, dehiscent. Valves (5-)7-34 mm, with a weak or strong central vein. Persistent style 0.3-1 mm, emarginate. Seeds numerous but variably set, 1.2-1.6 mm, ellipsoid, brown, unwinged, uniseriate. 2n=16*. Flowering June to August.

A local alpine of rock ledges and crevices, cliffs, screes, mountain tops (especially snowbeds and flushes) and ultrabasic serpentine soil; it has a broad ecological range, though it is intolerant of grazing and nutrient-poor soils. Also on river gravel at low altitude, either washed down from higher up or a relict from a wider distribution earlier in this interglacial period.

Mainly in N. Scotland, the Hebrides and Shetland, but also, rarely, in Snowdonia, Tipperary and Leitrim (map in Perring & Walters 1962). The Cumberland records are thought to be errors (Ratcliffe, D.A. 1959). A circumboreal plant of Scandinavia, Iceland and the Faroes, N. America, Siberia and in the mountains of C. Europe.

A very variable species in all characters throughout its range. Local populations can differ markedly (Stelfox 1970). Overall the plant shows continuous variation from small, compact plants in open, summer-dry habitats to large, lax, branched plants on wet, sheltered cliffs. Much of this variation is due to local environmental conditions and the infraspecific taxa describing it (var. *faeroensis* Hornem. and var. *grandiflora* Druce) are best treated as ecotypes (Jones 1963). Populations are partially out-breeding under favourable circumstances. Local vegetative reproduction may give morphologically uniform plants in otherwise heterogenous populations.

Arabis petraea 50

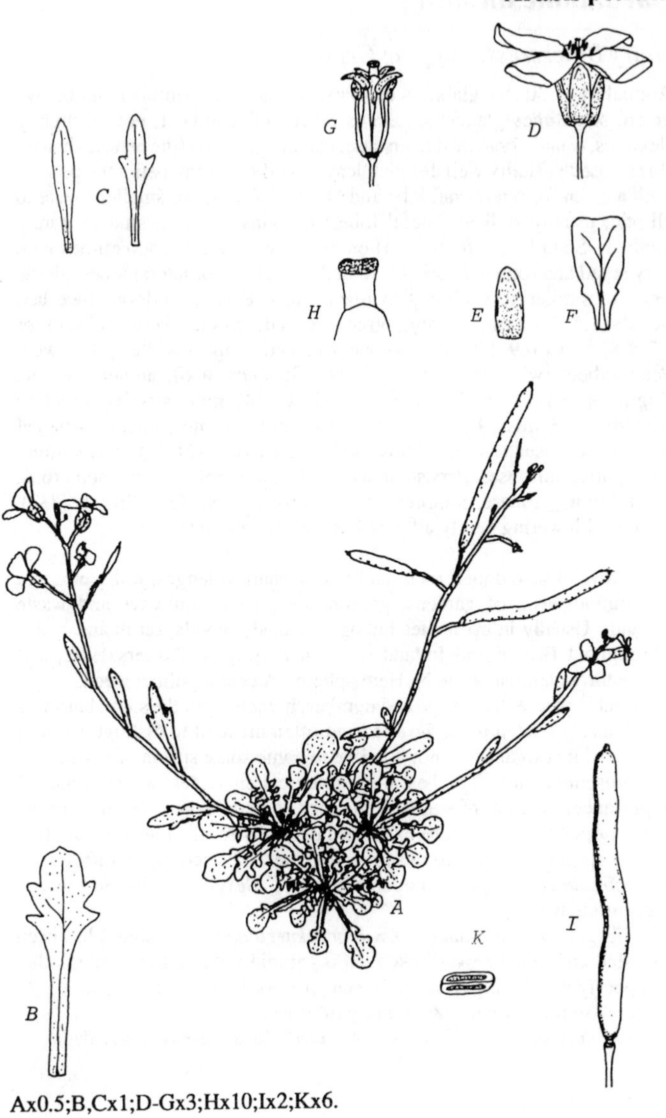

Ax0.5;B,Cx1;D-Gx3;Hx10;Ix2;Kx6.

51. Cardamine hirsuta L.

Hairy Bittercress, Popping Cress

Annual (3-)5-30 cm, glabrous or sparsely hairy with simple hairs below, green, sometimes glaucous. Stems erect to decumbent, not or slightly flexuous, usually branched from base, usually glabrous (the petioles hairy). Basal rosette usually well defined; leaves to 10(-13) cm, petiolate, pinnate with a ± reniform terminal lobe and (1-)2-6(-7) pairs of smaller, ovate to elliptic, usually stalked, lateral lobes; margins entire to sinuate, rarely toothed. Stem leaves (0-)1-4(-5) on main stem, smaller, sometimes with very small and rounded auricles, with 2-4(-5) pairs of lateral lobes, all the lobes of similar size, oblong to linear-oblanceolate. Inflorescence lax. Sepals 1.5-2.5 mm, oblong, green to red, erect. Petals absent or 2.7-4.8(-5.2) x 0.9-2.2 mm, obovate, rounded at apex, white, not clawed. Petals about twice as long as sepals. Stamens 4(-6); anthers yellow. Stigma ± entire. Pedicels in fruit 2-10(-14) mm, slender, erect to ascending. Fruits (9-)10-21(-25) x (0.6-)0.8-1.4 mm, linear, flattened (latiseptate), usually erect, dehiscent. Valves 10-20.5(-25) mm, without veins, often torulose. Persistent style 0.2-0.6(-1) mm. Seeds numerous, 0.7-1.2 mm, ± square, compressed, light brown, uniseriate. 2n=16*(?32), diploid. Flowering nearly all year but mainly March to July.

A native of sand dunes, rocks and screes, shallow ledges, walls, etc., and a common weed of gardens, greenhouses, paths, railways and waste ground. Usually in open sites but also in shady woods, scrub and rocks. Throughout Britain and Ireland (map in Perring & Walters 1962) and native through most of the N. Hemisphere. A cosmopolitan weed.

Variable in size, leaf shape and number, branching of the stem, hairiness etc. Closely related to *C. flexuosa* and often difficult to distinguish from it - indeed it is sometimes not possible to name some specimens without a chromosome count. The best morphological characters are (in order of importance); number of stamens, number of stem leaves, hairiness of the stem. Populations should be examined carefully for the range in variation and the species may sometimes grow together. The ecological differences of *C. hirsuta* being a plant of open sites and *C. flexuosa* of shaded sites are very unreliable.

C. flexuosa x *C. hirsuta* (*C.* x *zahlbruckneriana* O. E. Schulz) has been recorded and may be overlooked. It is triploid with highly sterile pollen and poorly developing fruits (1-2 seeds per pod). It looks more similar to *C. flexuosa* rather than being strictly intermediate.

The young stems and flowers make a delicious ingredient in salads.

Cardamine hirsuta 51

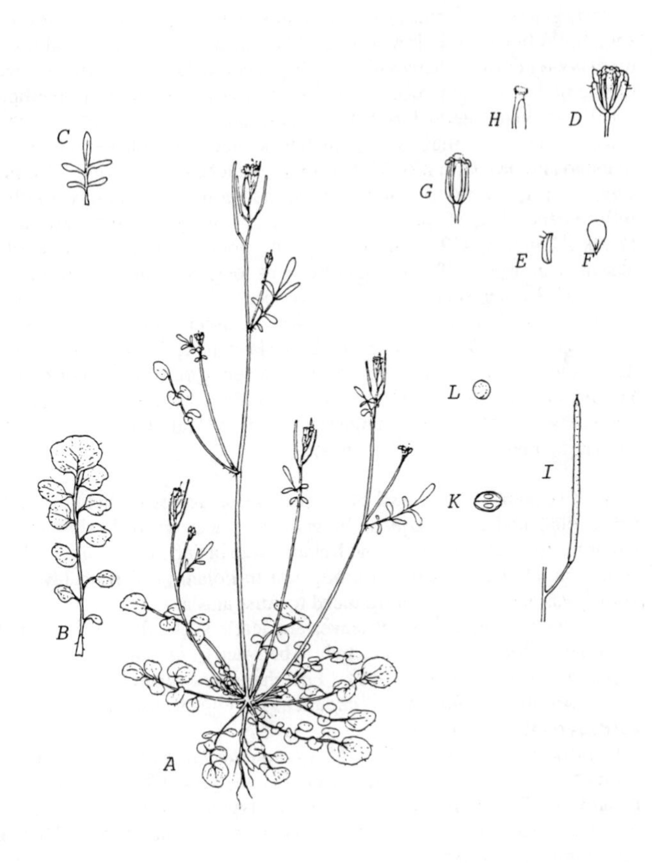

Ax0.5;B,Cx0.3;D-Gx3;Hx10;Ix1.5;Kx3.

52. **Cardamine flexuosa** With.

Wavy Bittercress, Wood Bittercress

Annual, biennial or perennial 7-50 cm, usually with simple hairs below (?rarely glabrous). Stems erect to decumbent, weakly to strongly flexuous, sometimes branched below, usually (90% of plants) hairy. Basal rosette often poorly defined; leaves to 15(-20) cm, petiolate, pinnate with an ovate to ± reniform terminal lobe and 2-7(-8) pairs of smaller, ovate to elliptic, usually stalked, lateral lobes; margins usually coarsely toothed. Stem leaves (3-)4-10 on main stem, smaller, sometimes with very small and rounded auricles, with 2-6(-7) pairs of lateral lobes, all the lobes of similar size, elliptic-ovate to oblong; margins entire or coarsely toothed. Inflorescence lax. Sepals 1.5-2.5 mm, oblong, green to purple, erect. Petals 2.2-4.3(-5) x 0.8-1.8 mm, obovate, rounded at apex, white, claw absent or indistinct. Petals about twice as long as sepals. Stamens (4-)6 (note the 2 outer stamens may be reduced); anthers yellow. Stigma ± entire. Pedicels in fruit 5-15 mm, slender, ascending to patent. Fruits (8-)12-26 x 0.9-1.5 mm, linear, flattened (latiseptate), ascending to patent, dehiscent. Valves (8-)11.5-25 mm, often torulose, without veins. Persistent style 0.2-1 mm; seeds numerous 0.9-1.4 mm, ± square, compressed, light brown, uniseriate. 2n=32*(allotetraploid), *c.* 50. Flowering mainly May to September.

A native plant of shady flushes, rocks, woods, ravines, mountains, stream sides, etc., and also frequent in gardens, waste ground, shingle, etc. Common throughout Britain and Ireland (map in Perring & Walters 1962). Native in W., C., N. and S. Europe east to Poland, and probably in the Himalayas and Far East. Introduced to Australasia.

Variable in size, number of leaves and leaflets, etc. It is believed to be an allotetraploid derived from a hybrid between *C. hirsuta* and possibly *C. impatiens* (Ellis & Jones 1969). For characters separating it from the former see under *C. hirsuta*. *C. impatiens* has conspicuous, acute, clasping auricles on the stem leaves.

C. flexuosa x *C. pratensis* (= *C. x haussknechtiana* O. E. Schulz) has been recorded in v.c. 5, 9, 14, 17, 21, 22, 35, 38 and 99, usually in open meadows. The hybrids are intermediate, typically with stems branched below and narrow (*c.* 10 x 4 mm), lilac (rarely white) petals. They are sterile but reproduce vegetatively.

Cardamine flexuosa 52

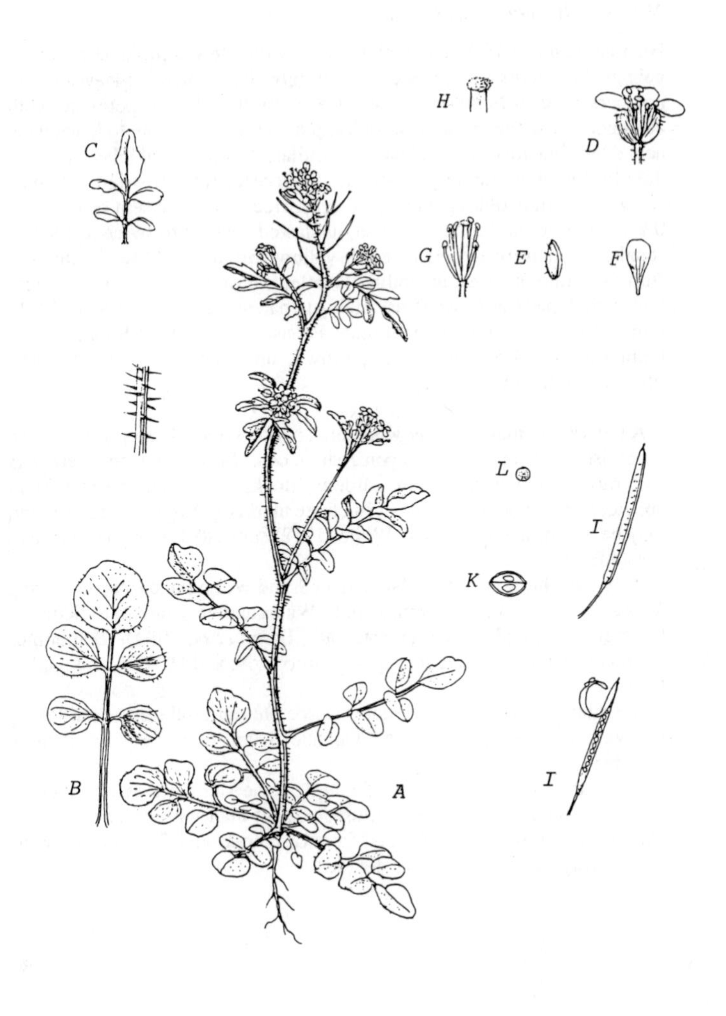

Ax0.5;B,Cx0.3;D-Gx3;Hx10;Ix1.5;Kx3.

53. Cardamine impatiens L.

Narrow-leaved Bittercress

Biennial (?annual) 20-80 cm, glabrous or with a few, simple hairs below, pale green. Stems erect, branched sparingly above. Rosette leaves small, not persisting to flowering. Stem leaves to 20(-23) cm, petiolate, with conspicuous, acute, clasping auricles, pinnate with an ovate to lanceolate, acute, terminal lobe and 6-11 pairs of similar, stalked lateral lobes; margins deeply lobed to acutely toothed. Inflorescence crowded. Sepals (?0.9-)1.2-2 mm, oblong, green to purple, erect. Petals absent or 1.5-3.6 x 0.5-1.3 mm, oblanceolate, whitish, unclawed. Petals to twice as long as sepals. Stamens 6; anthers greenish-yellow. Stigma ± entire. Pedicels in fruit 4-11 mm, slender, ascending to patent. Fruits 12-30 x 0.7-1.5 mm, linear, flattened (latiseptate), ascending to patent, dehiscent. Valves 11-29 mm, ± torulose, central vein absent. Persistent style 0.4-1.8 mm. Seeds numerous, 0.9-1.5 mm, oblong, brown, uniseriate. 2n=16*(diploid). Flowering May to August.

A native biennial of shady woodland, rocks, screes, banks, paths, often on moist, limestone soils in open Ash woods. Sometimes on riverbanks and rarely also a casual in gardens, docks, etc. Unpredictable in appearance, the populations can fluctuate markedly from year to year and may be absent in some years. White (1912) suggested it may be sensitive to late frosts.

Very local but sometimes abundant in S. and W. England (map in Perring & Walters 1962), not uncommon in E. Wales and very rare in S. Scotland (Corner 1988) and Ireland (Westmeath; Breen *et al.* 1984). Sometimes casual elsewhere. Native throughout Europe, N. and C. Asia, and the Far East.

In Britain, little variable. Petals may be present or absent. Leaf lobing can vary between populations. Outside Britain it is very much more variable.

The numerous pairs of lateral lobes on the stem leaves make this a distinct plant unlikely to be confused except than with other species of *Cardamine*. Easily distinguished from *C. hirsuta* and *C. flexuosa* by the acute, clasping auricles on the stem leaves.

Cardamine impatiens 53

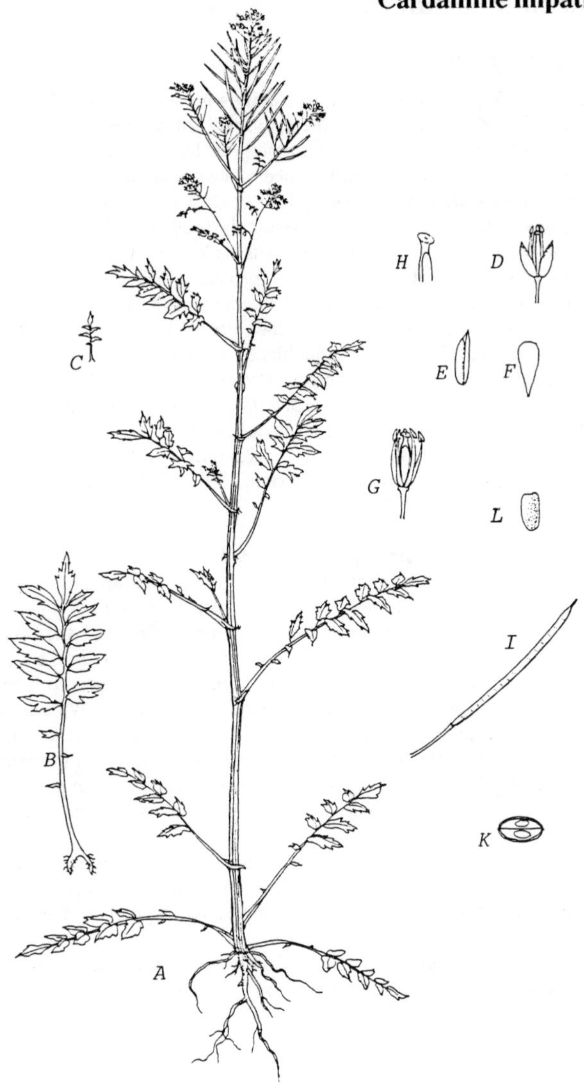

Ax0.5;B,Cx0.3;D-Gx3;Hx10;Ix1.5;Kx3.

54. Cardamine amara L.

Large Bittercress

Perennial 15-50(-60) cm, glabrous (rarely with sparse, simple hairs), light green. Stems creeping, decumbent to erect, often rooting at the nodes, sparsely branched above. No rosette present. Lower stem leaves to 15 cm, petiolate, pinnate with a terminal lobe ovate to orbicular, obtuse at apex, cordate to cuneate at base, and with 2-5 pairs of slightly smaller, stalked or sessile, lateral lobes; margins sinuately toothed. Upper stem leaves similar, with lobes elliptic to ovate (rarely oblong). Inflorescence lax. Sepals 2.7-4.3 mm, oblong to ovate, ± saccate, green, erect. Petals 5.4-12 x 3-6.5 mm; limb obovate, rounded to emarginate at apex, white (rarely pink or purplish); claw short, indistinct, greenish. Petals about twice as long as sepals. Stamens 6; undehisced anthers red to purple (?rarely yellowish-black). Stigma ± entire. Pedicels in fruit 5-20 mm, slender, ascending to patent. Fruits 14-35(-40) x 0.7-1.5 mm, linear, flattened (latiseptate), erect to ascending (-?patent), dehiscent. Valves 12-34 mm, central vein absent. Persistent style 0.5-2.5 mm, slender. Seeds numerous, 0.7-1.2 mm, ± oblong, brown, uniseriate. 2n=16*(?32). Flowering April to July.

Represented in Britain and Ireland by subsp. *amara*.

A locally abundant plant of swamps, fens, Alder carr, canal and riverbanks, springs and wet woodlands, often in slow moving or nearly stagnant water. Common and widespread in S. E. England, the Midlands, N. England and S. and E. Scotland but surprisingly rare S. E. of a line from Portsmouth to Liverpool and N. W. of a line from Glasgow to Inverness. Rare in Wales and in N. Ireland (map in Perring & Walters 1962). Generally lowland but not exclusively so. Throughout most of Europe to Asia Minor. Possibly declining in some areas due to river "improvements".

Variable in leaf shape and size, and in petal size. Plants in grassland are often more slender than those in swamps. Self-incompatible and fruit set is often poor.

Similar to *C. pratensis* but this has erect stems not rooting at the nodes, an obvious basal rosette and yellow anthers. Somewhat similar to Watercress species (*Rorippa*) but they have smaller petals and nearly terete fruits.

C. amara x *C. pratensis* (= *C.* x *ambigua* O. E. Schulz) has been recorded but is unlikely and requires verification.

Cardamine amara 54

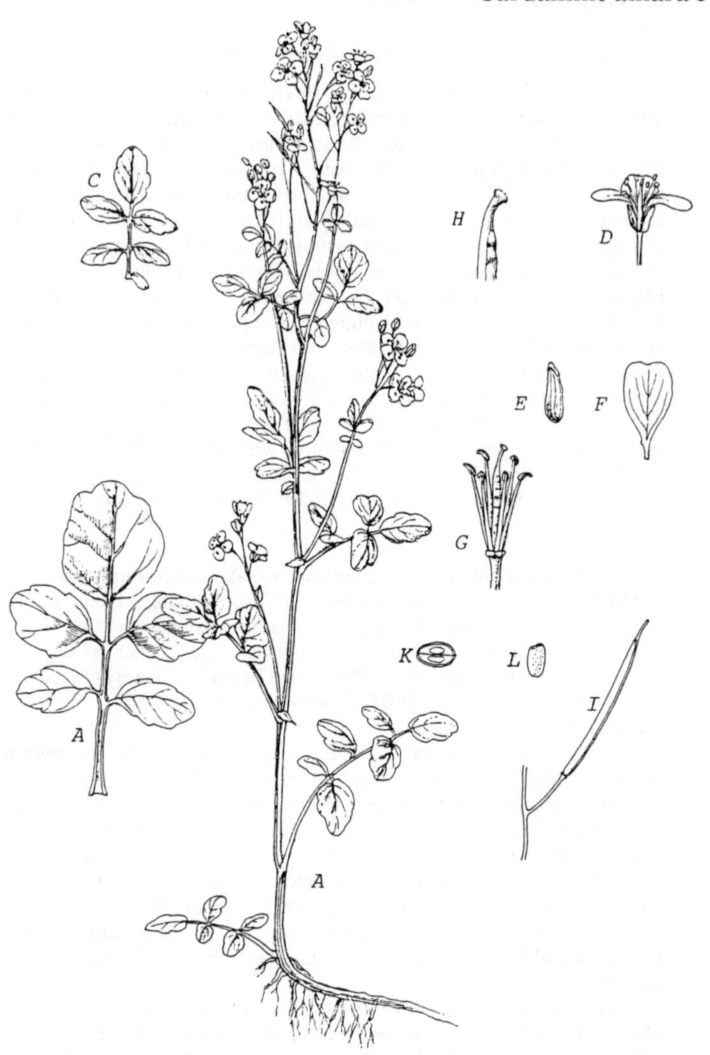

A-Cx0.5;Dx1;E-Gx1.5;Hx10;Ix1;Kx6;Lx5.

55. Cardamine pratensis L.

Cuckoo Flower, May Flower, Lady's Smock

Perennial 15-80 cm, glabrous (rarely with simple hairs below), dark green. Stems erect, not rooting at the nodes, usually unbranched. Rosette leaves to 16(-20) cm, petiolate, pinnate with an ovate to reniform, obtuse terminal lobe and (0-)2-8 pairs of stalked, ovate to obovate, lateral lobes; margins sinuate or coarsely toothed. Lower stem leaves similar, with ovate segments. Upper stem leaves with 2-7(-10) pairs of sessile or stalked, oblong to linear, lateral lobes and a similar terminal lobe; margins entire. Inflorescence crowded. Flower rarely "double". Sepals 2.7-5.3 mm, oblong to ovate, saccate, green (fading yellow), erect to patent. Petals 6.1-15.5(-18) x 2.8-10 mm; limb obovate to elliptic, rounded to emarginate at apex, white to purple or pink often with darker veins; claw short, green. Petals 2-3 times as long as sepals. Stamens 6; anthers yellow. Stigma capitate, ± entire. Pedicels in fruit 5-30 mm, slender, ascending. Fruits 23-50(-55) x (1.1-)1.4-2.3 mm, linear, flattened (latiseptate), erect to ascending, dehiscent. Valves 22.5-49 mm, without veins. Persistent style 0.3-1.8 mm, stout or slender. Seeds numerous, 1.3-2 mm, oblong, compressed, brown, uniseriate. 2n=16-30*-56*-96. Flowering March to July.

A locally abundant native of grassland, woods, fens, swamps, ditches, streamsides, road verges, tracks, mountain flushes and dune slacks, rarely casual in cities. Particularly frequent in damper habitats and tolerant of waterlogging though avoiding stagnant water. Somewhat resistant to herbicides. Common throughout Britain and Ireland and nearly ubiquitous (map in Perring & Walters 1962). Native in most of the N. Hemisphere. Introduced in Australasia.

Very variable between, and sometimes within, populations. The species aggregate consists of a partly aneuploid chromosome series in Europe, the ploidy levels correlating to a degree with morphological characters. It has not been possible to apply Lövkvist's (1956) treatment to the British Isles due to phenotypic plasticity which is at least partly determined by environmental factors (Dale & Elkington 1974), though a number of entities may nevertheless be recognisable (Allen 1981). This is a fiendishly complex group whose taxonomy has yet to be satisfactorily elucidated and it is best treated for the time-being as a single, polymorphic species.

The plant is self-incompatible and fruit set is often low. Salisbury (1965) gives an account of reproduction by plantlets which arise on the leaves.

The ± stalked lateral lobes of the rosette leaves are ± diagnostic to *Cardamine*, and the linear to ± oblong lobes of the upper stem leaves should separate this species from *C. amara* and *C. raphanifolia*.

Cardamine pratensis 55

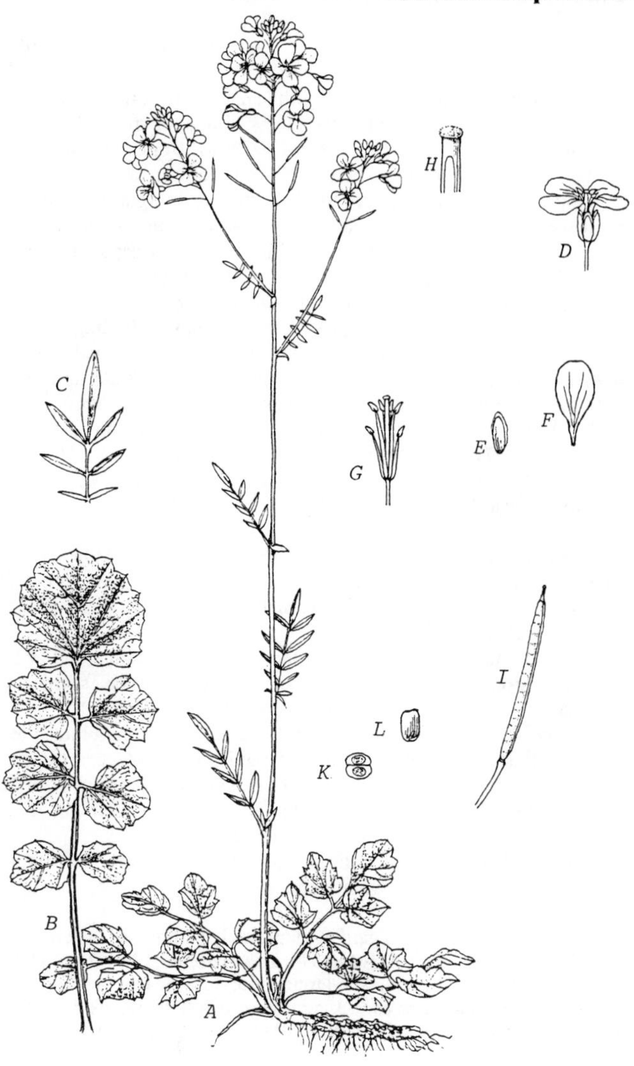

Ax0.5;B-Gx1;Hx10;Ix1;K,Lx3.

56. Cardamine raphanifolia Pourret

Map 56

Perennial 30-70 cm, glabrous or with simple hairs on the lower leaves. Stems decumbent at the base, rooting at the nodes, simple or sparingly branched. Basal rosette not persisting. Lower stem leaves to 30 cm, petiolate, pinnate; terminal lobe large, orbicular, obtuse at apex, cordate to rounded at base; lateral lobes (0-)1-6 pairs, smaller, stalked, ± ovate; margins sinuate. Upper stem leaves smaller, terminal lobe ovate to elliptic, cuneate to rounded at base, and 1-2 pairs of sessile, oblong to elliptic lateral lobes; margins sinuate to coarsely toothed. Inflorescence crowded. Sepals (?3-)4-5.5 mm, ovate, green (yellowing with age), erect. Petals 8-14 x 4-8.5 mm; limb elliptic to obovate, rounded to truncate at apex. Purple to pinkish (fading rapidly) or white; claw short, greenish. Petals *c.* 2-3 times as long as sepals. Stamens 6; anthers yellow. Stigma ± entire. Pedicels in fruit 18-35(-50) mm, slender, ascending to inclined. Fruits 15-30(-40) x 1.4-2 mm (but rarely fertile in Britain), linear, flattened (latiseptate), ascending to inclined, dehiscent. Valves 12-28(-37) mm, without veins. Persistent style 2-3.5(-?5) mm, slender. Seeds numerous, 1.2-2.1 mm, brown, oblong, uniseriate. 2n=44,46. Flowering May to June.

A garden escape occasionally established on lake shores, by streams, in ditches, fens, etc. Rare in England, Scotland and Wales, absent from Ireland. A native of streamsides and damp places in the mountains from Spain to Greece, and in Turkey.

The description above applies to the species as a whole and is based partly on European material. In Europe, it is a variable species with at least three subspecies (these have not been investigated in Britain).

In flower, the large purple-red (rarely white) petals and pinnate leaves with ± stalked lateral lobes should distinguish it. *C. pratensis* is similar but has upper leaves with ± linear lobes and all the leaves are in any case much smaller. *C. amara* without flowers may be similar but is again smaller. *C. bulbifera* has bulbils in its leaf axils.

It is probably self-incompatible and rarely sets fruit in Britain. It probably spreads mainly vegetatively.

57. Cardamine trifolia L. is a garden plant rarely naturalised in damp, shady woodlands and churchyards (Map 57). It is a perennial 10-30 cm, glabrous or with simple hairs; rhizomes 2-4 mm wide; rhizome leaves trifoliate with stalked leaflets; stem leaves 0-3, simple or trifoliate; petals 4.5-11 x 3-6 mm, white or pink; fruits 20-25 x 1.5-2 mm, rarely produced.

Cardamine raphanifolia 56

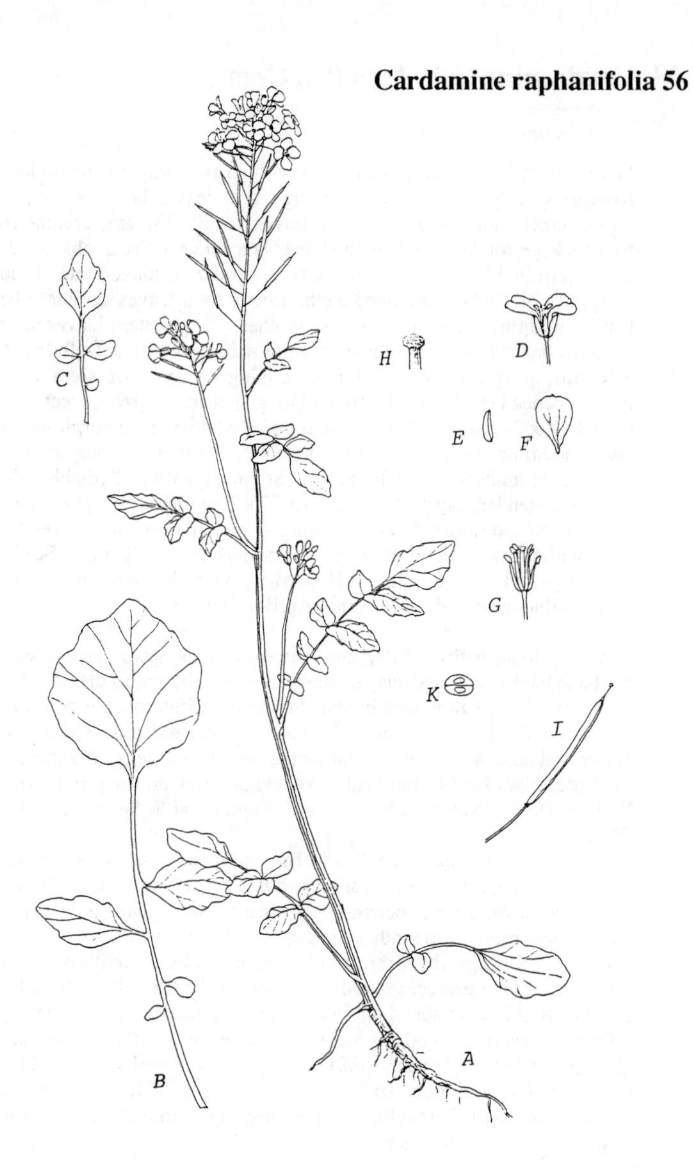

A-Cx0.4;D-Gx0.7;Hx5;Ix0.7;Kx2.

58. **Cardamine bulbifera** (L.) Crantz

Coralwort, Coralroot

Perennial 25-70(-92) cm, glabrous or with sparse, simple hairs on leaves. Rootstock scaly, with rhizomes giving rise to leaves between the stems. Stems erect, unbranched. Basal leaves to 20(-23) cm, arising from rootstock, petiolate, pinnate with a linear-lanceolate to linear-oblanceolate, acute, terminal lobe and 2-3(-4) pairs of similar, ± stalked lateral lobes; margins with shallow, rounded teeth. Lower stem leaves similar to basal leaves; margins obtusely to acutely toothed. Upper stem leaves simple, linear to linear-lanceolate, acute; margins acutely toothed. Bulbils 5-11 x 3-6 mm, purple-brown to black, arising in axils of stem leaves. Inflorescence lax. Sepals 4-6 mm, oblong, ± saccate, green, erect. Petals (9-)10-17 x 3-5 mm, oblanceolate, rounded at apex, pale purple-pinkish; claw indistinct, broad, greenish. Petals *c.* 3 times as long as sepals. Stamens 6; anthers greenish-yellow. Stigma ± entire. Pedicels in fruit 7-11 mm, slender, ascending. Fruits 15-37 x 2-3 mm but rarely developing, linear, flattened (latiseptate), ascending to erect, dehiscent. Valves 14-35 mm, without or with weak veins. Persistent style 1-3 mm. Seeds (if developing) 1.5-2 mm, ellipsoid, pale brown, uniseriate. 2n=96*(duodecaploid). Flowering April to June.

A very local native of dry and wet woodlands, shady riversides and roadside banks. On sand, clay or chalk soils, usually with a pH of (5.5-)6.0 or above. The plant is locally abundant in the Chilterns, but is local and rare in the Weald; it is possibly also native in Staffordshire. Occasionally cultivated and sometimes established in old gardens, parkland and roadsides elsewhere in England, Scotland and Ireland (map in Perring & Walters 1962). Native in Europe from France and S. Scandinavia to the Black Sea.

The description above applies to English material only, which varies little. Continental material varies particularly in leaf shape. It is also cultivated in Britain and occasionally escapes; such plants typically have more ovate, more deep toothed leaflets.

Instantly distinguished from all other crucifers by the axillary bulbils.

C. bulbifera reproduces mainly vegetatively by the axillary bulbils and is thus often locally abundant. The microdistribution in the Weald may reflect dispersal of bulbils by water (Rose 1966). Fruit-set is uncommon (Ferroussat 1982; Showler 1988) but is probably overlooked, and larger plants tend to set more fruit than small plants. The disjunct native distribution may reflect either a wider range in former times or a process of more recent colonization.

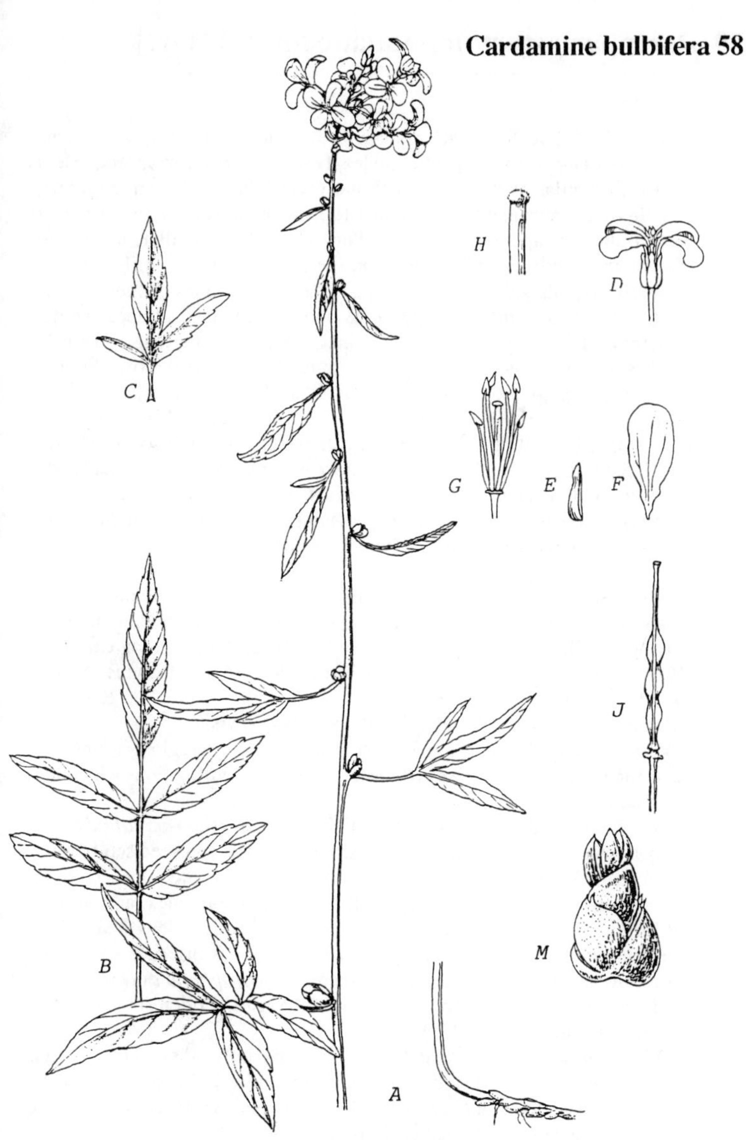

Ax0.5;B,Cx1;Dx1;E-Gx1.5;Hx10;Jx1;M=bulbil,x4.

59. Rorippa nasturtium-aquaticum (L.) Hayek

Watercress

Perennial 15-60(-200) cm, glabrous, dark green to purple. Stems decumbent below, rooting at the nodes, branched. Rosette absent. Stem leaves all similar or the upper with narrower lobes to 15 cm, petiolate, pinnate with a terminal lobe orbicular to lanceolate, cordate to cuneate at base, obtuse at apex, and (0-)1-5 pairs of slightly smaller, oblong to lanceolate, sessile, lateral lobes; margins entire, sinuate. Inflorescence crowded. Sepals 1.9-2.9 mm, oblong, green, erect to ascending. Petals 3.5-6.6 x 1.3-2.6 mm; limb elliptic to obovate, rounded at apex, white, sometimes flushed purple; claw distinct, about half as long, greenish. Petals about twice as long as sepals. Stamens 6; anthers yellow. Stigma entire to ± emarginate. Pedicels in fruit 6-20(-22) mm, slender, inclined to reflexed. Fruits (9-)11-19(-24) x 1.7-3.2 mm ((1.6-)1.9-2.7(-3) mm wide at middle), elliptic, ± terete, inclined to reflexed, dehiscent. Valves (9-)10.5-18(-23) mm, with or without a weak central vein. Persistent style 0.5-1.7(-2) mm, stout. Seeds numerous, 0.8-1.4 mm, ovoid, brown, coarsely reticulate with (6-)7-12 depressions across their width, biseriate. 2n=32*. Flowering May to October (-December).

A common plant of ditches, streams, fens, springs, marshes, wet meadows, ponds, canals, etc., throughout Britain and Ireland (map in Perring 1968). The ecology is described in detail in the Biological Flora (Howard & Lyon 1952). Throughout Europe to C. Asia, N. and E. Africa. Introduced to N. and S. America, southern Africa, the Far East and Australasia. Widely cultivated for at least 2000 years.

Very variable in size and leaf shape. Dwarf and giant forms are occasionally found but these appear to be habitat modifications which are not maintained in cultivation.

The best characters to distinguish *R. nasturtium-aquaticum, R. microphylla* and *R.* x *sterilis* are fruit shape and, size and seed sculpturing (Rich 1987b). Plants are plastic vegetatively and I am unconvinced of the value of characters such as foliage coloration, stomatal index, and anthers dehiscing inwards/outwards. Fruit curvature and pedicel length are of no use, and whilst *R. nasturtium-aquaticum* usually has biseriate seeds and "good" *R. microphylla* has uniseriate seeds, *R. microphylla* may also have seeds in two rows especially at the base of the fruit. Pollen size may be a useful character (Green 1955).

Other than *R. microphylla* and *R.* x *sterilis*, the only other white crucifer with pinnate leaves and stems rooting at the nodes is *Cardamine amara* which has flattened fruits and generally larger petals. *Apium nodiflorum* is similar vegetatively but has ± sheathing petioles.

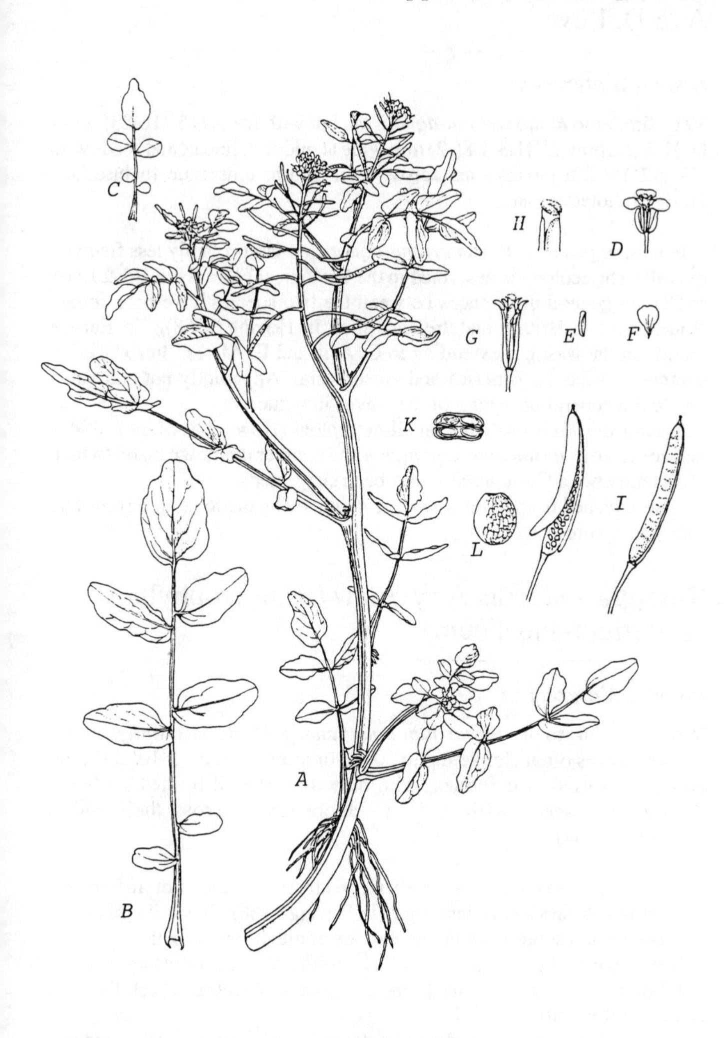

A-Cx0.5;D-Fx1;Gx2;Hx10;Ix1.5;Kx3;Lx8.

60. **Rorippa microphylla** (Boenn.) N. Hyl. ex A.& D. Löve

Brown Watercress

Very similar to *R. nasturtium-aquaticum* but with fruits (15-)16-23(-24) x (1-)1.5-2.2 mm ((1-)1.3-1.8(-2) mm wide at middle), linear, and seeds with (11-)12-18(-20) depressions across their width, uniseriate to biseriate. 2n=64*(allotetraploid).

In similar places to *R. nasturtium-aquaticum* but probably less frequent overall. The ecology is described in the Biological Flora (Howard & Lyon 1952); no general differences between the two species have been found. Widespread in Britain and Ireland (map in Perring 1968). In Europe mainly in the west but extending to C. Asia and E. Africa. Introduced to southern Africa, N. America and Australasia. Apparently not cultivated due to the general browning of the leaves in winter.

R. microphylla is probably an allotetraploid (Howard & Manton 1946) derived from *R. nasturtium-aquaticum* and another unknown taxon (which is certainly not a *Cardamine* as has been suggested).

Distinguished from *R. nasturtium-aquaticum* by the long, thin fruits and finely sculptured seeds.

61. **Rorippa x sterilis** Airy Shaw (R. microphylla x nasturtium-aquaticum)

Hybrid Watercress

Very similar to *R. nasturtium-aquaticum* and *R. microphylla* but inflorescences often elongating to 30 cm or more, fruits *c.* 5-10(-11) mm, aborted, dwarfed or deformed, with only 0-3(-4) well-formed seeds per loculus, the seeds with *c.* 10-14 depressions across their width. 2n=48*(triploid).

The commonest crucifer hybrid, widespread and not infrequent throughout Britain and Ireland (map in Perring 1968). In similar places to the parents and sometimes in the absence of either one or both.

Not every sterile watercress is *R.* x *sterilis* because occasional sterile individuals of the parents are found. It is thus as well to check the seed character if possible.

Also cultivated and spreading vegetatively. Widespread in Europe and occasionally introduced elsewhere (e.g. N. America, Australasia).

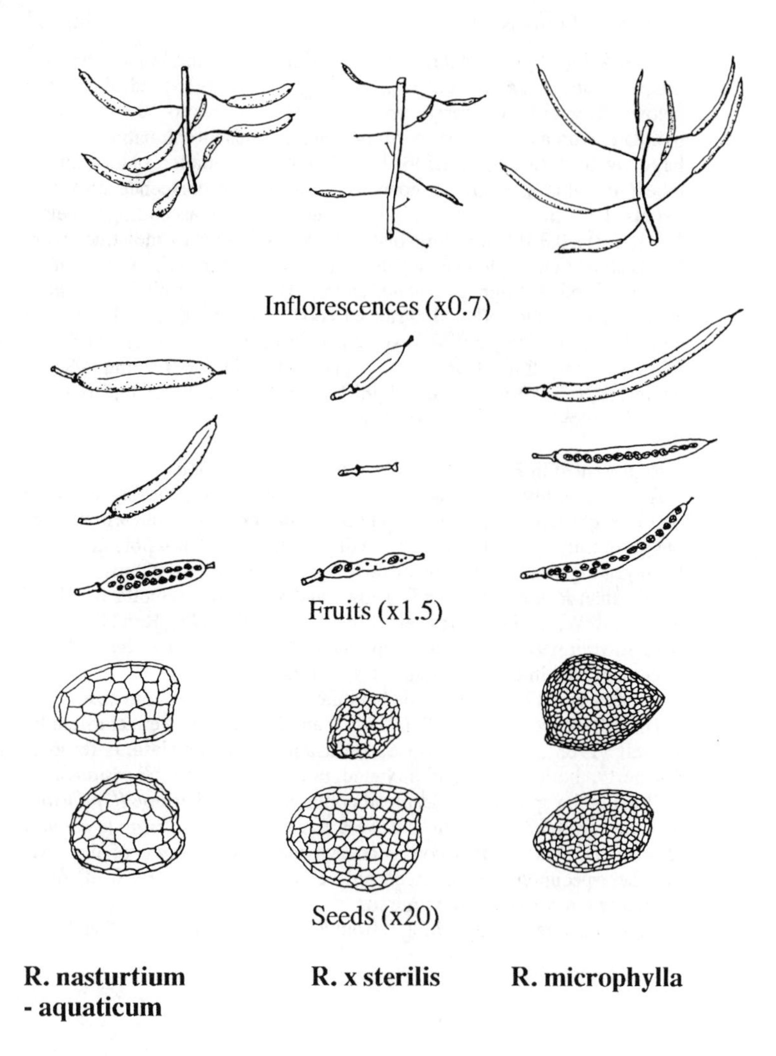
Inflorescences (x0.7)
Fruits (x1.5)
Seeds (x20)
R. nasturtium - aquaticum
R. x sterilis
R. microphylla

62. **Rorippa islandica** (Oeder ex Murray) Borbás

Northern Yellowcress Map 62

Annual 4-30(-40) cm, glabrous (rarely with sparse, simple hairs below). Stems prostrate, decumbent or ascending at base, branched above and below. Rosette leaves rarely persisting. Stem leaves to 10 cm, usually petiolate, with auricles very small or absent, pinnate to pinnatisect with a large, ovate, obtuse terminal lobe and 2-5 pairs of smaller, oblong lateral lobes; margins ± entire to coarsely toothed. Inflorescence crowded. Sepals 1.1-1.5(-2.4) mm, oblong, green, erect to ascending. Petals 1-1.5(-1.7) x 0.3-0.5 mm; limb obovate, pale yellow; claw indistinct, pale. Petals about twice as long as sepals. Stamens 6; anthers yellow. Immature ovaries elliptic. Stigma capitate, entire. Pedicels in fruit 2-4(-5) mm, slender, inclined to reflexed. Fruit set good. Fruits (5-)6-12 x 1.7-2.5(-3) mm, elliptic, ± terete, inclined to patent, dehiscent. Valves (5-)6-11.5 mm, without veins, thin, ± torulose. Persistent style 0.2-0.8 mm. Seeds numerous, 0.5-0.9 mm, ovoid, light brown, finely colliculate, biseriate. 2n=16*. Flowering July to October.

Represented in Britain and Ireland by subsp. *islandica*.

A native annual of small ponds and pools, locks, turloughs, damp pastures, etc, usually near the sea in open sites on damp mud which dries out in summer. Usually found in small populations and possibly dispersed by migrating birds. Rarely on disturbed ground or rubbish dumps, and less weedy than *R. palustris*. Only known and very rare in Scotland, Isle of Man and W. Ireland (Jonsell 1968, Scannell 1973, Randall 1974, MacGowran 1979), but probably overlooked. Norway, Iceland and Greenland and in the mountains of S. Europe.

Slightly variable in leaf shape but otherwise only varying in size.

The differences between *R. islandica* and *R. palustris* were clarified by Jonsell (1968), and still cause confusion for nomenclatural reasons. Formerly, both taxa were included under one taxon "*R. islandica*". Following the split, the widespread common species became *R. palustris* and the rare plant of north and west coasts became *R. islandica sensu stricto*. Because of this confusion and the critical nature of the taxa, voucher specimens (which need only consist of a few ripe fruit) of *R. islandica sensu stricto* are required.

For characters distinguishing it from *R. palustris*, see pages 48 and 156.

Rorippa islandica 62

A-Cx0.5;D-Fx3;Gx6;Hx10;Ix1;Kx2.

63. Rorippa palustris (L.) Besser

Marsh Yellowcress Map 63

Annual (rarely perennial) 5-60 cm, glabrous (rarely with a few sparse hairs). Stems erect, branched mainly above. Rosette leaves not persisting. Stem leaves to 18 cm, usually petiolate, with or without small auricles clasping stem, pinnate to pinnatisect with a large, ovate to lanceolate, obtuse to acute, terminal lobe and 2-6(-7) pairs of smaller, oblong to ovate, lateral lobes; margins ± entire to coarsely sinuate or toothed. Upper stem leaves with fewer lateral lobes than the lower. Inflorescence crowded. Sepals 1.6-3 mm, oblong, green, erect to ascending. Petals (1.4-)1.7-2.7(-2.8) x 0.5-1.1 mm; limb obovate, yellow to pale yellow; claw indistinct, pale. Petals about as long as sepals. Stamens 6; anthers yellow. Immature ovaries elliptic. Stigma capitate, entire. Pedicels in fruit 3-10 mm, slender, inclined to reflexed. Fruit set usually good. Fruits (4-)5-10(-12) x 1.3-3 mm, ellipsoid, ± terete, inclined to reflexed, dehiscent. Valves (4-)5-10(-12) mm, without veins. Persistent style 0.2-1(-1.2) mm. Seeds numerous, 0.5-0.9 mm, ovoid, brown, colliculate, biseriate. 2n=32*. Flowering May to October.

Represented in Britain and Ireland by subsp. *palustris*.

A native plant of riverbanks, pastures and watermeadows, pond and lake margins, particularly characteristic of damp mud. Also an infrequent weed of docks, railways, gardens, waste ground, rubbish dumps, etc. Common in England, frequent in Ireland, Wales and S. Scotland, rare in the far north (where it is possibly confused with *R. islandica s.s.*). Probably native throughout most of the world but introduced in S. America and Australasia.

Variable in size, and leaf and fruit shape. A number of other subspecies occur in N. America (see Jonsell 1968).

The tiny petals about as long as the sepals are virtually diagnostic and *R. palustris* is only likely to be confused with *R. islandica* from which it is most reliably distinguished by a chromosome count or microscopic (x25-50) examination of the seeds. The habit, pedicel:fruit length ratio, and petal and sepal lengths when taken together are reasonably reliable. *Descurainia sophia* also has petals equalling the sepals but has finely divided leaves and stellate hairs and looks completely different.

For details of the hybrid with *R. amphibia*, see **67**.

Rorippa palustris 63

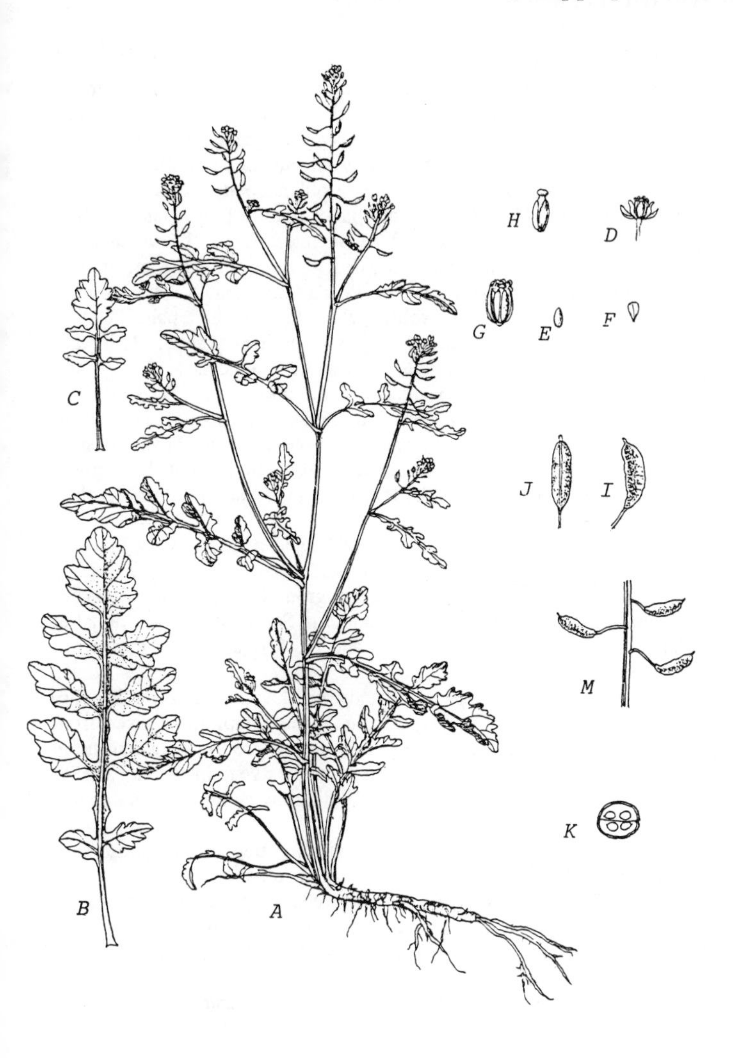

A-Cx0.5;D-Fx1;Gx6;Hx10;Ix1;Kx2.

64. Rorippa sylvestris (L.) Besser

Creeping Yellowcress Map 64

Perennial 15-60(-90) cm, glabrous or with a few sparse simple hairs, green, forming patches with spreading, white rhizomes. Stems decumbent to erect, branched mainly below. Rosette leaves not persisting. Stem leaves to 15 cm, petiolate (auricles rarely present but then small), pinnate to pinnatisect with an (ovate-)lanceolate to oblanceolate, acute to obtuse, terminal lobe and 3-6 pairs of (ovate-)lanceolate to linear, lateral lobes; margins entire to deeply toothed. Upper stem leaves with 1-3 lateral lobes or rarely simple. Inflorescence crowded. Sepals 1.8-3.5 mm, oblong, green to yellow, ascending to patent. Petals (2.2-)2.8-5.5 x 1.7-2.5 mm; limb obovate, rounded at apex, yellow; claw short, distinct, pale. Petals *c.* 1.5-2 times as long as sepals. Stamens 6; anthers yellow. Immature ovaries ± linear. Stigma capitate, emarginate. Pedicels 4-11(-12) mm, slender, ascending to patent. Fruit set usually poor. Fruits 7-23 x 0.9-1.8 mm, ± linear, ± terete to slightly compressed (latiseptate), inclined to patent, dehiscent. Valves 6.5-22 mm, without central vein. Persistent style 0.6-1.2(-1.3) mm, bilobed. Seeds numerous, 0.6-1 mm, ovoid, dark brown, uniseriate to biseriate. 2n=32,40,48*. Flowering (April-) June to October.

A native of riverbanks and shingle, lakes and pond margins, brackish swamps and damp ground, usually in barer, more open sites. Also a noxious weed of arable land, docks, waste ground and gardens. Frequent in England and Wales, mainly in the south in Scotland and occasional in Ireland. Native in Europe, W. Asia and N. Africa, widely introduced elsewhere in the world.

Very variable in leaf shape, habit and fruit shape. In Europe there are tetraploid and hexaploid cytotypes (plus rare pentaploids derived from tetraploid-hexaploid crosses), the latter being more vigorous and more aggressive weeds. Some of the variation may be due to introgression with *R. amphibia* and some populations in Britain may be impure. For details see Jonsell (1968).

Self-incompatible (but variably so) and often not setting fruit or seed thus giving problems with identification. Few other crucifers have small (less than 5.5 mm) yellow petals and more or less regularly pinnate leaves. *R. palustris* has petals only about as long as the sepals. *Barbarea* and *Sisymbrium* spp. usually have fruits more than 20 mm with uniseriate seeds.

For details of hybrids with *R. amphibia* and *R. austriaca*, see page 162.

Rorippa sylvestris 64

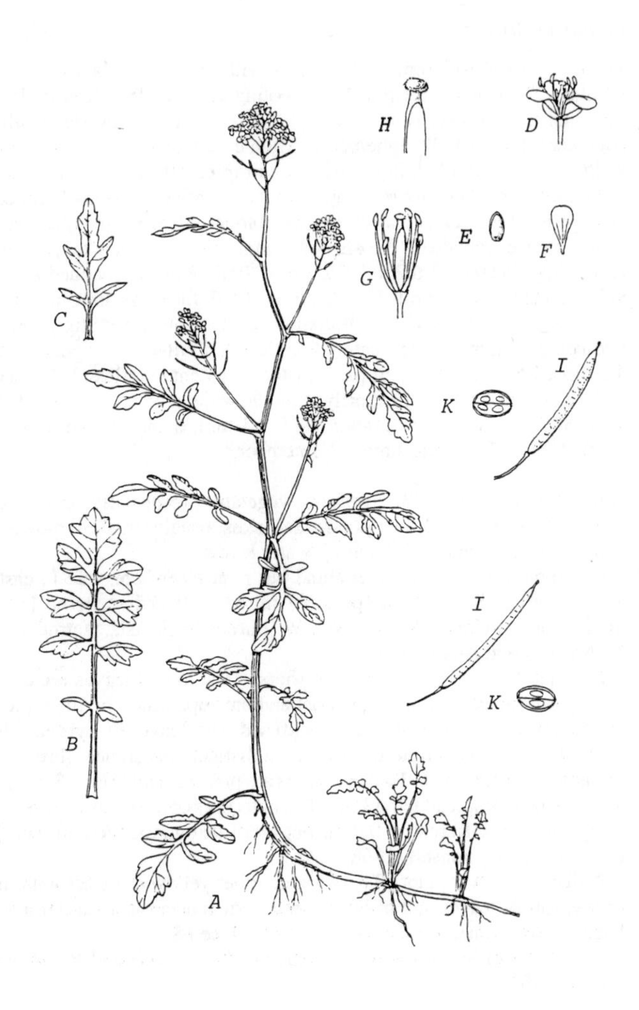

A-Cx0.5;D-Fx2;Gx6;Hx10;Ix1;Kx2.

65. Rorippa amphibia (L.) Besser

Great Yellowcress Map 65

Perennial (20-)40-120 cm, ± glabrous or with simple hairs below, forming clumps. Stems decumbent at base, rooting at the nodes. Rosette leaves not persisting. Lower stem leaves pinnatisect to pinnatifid (especially if submerged) and usually ephemeral. Middle stem leaves to 25 cm, sessile, with or without clasping auricles, simple, elliptic to obovate or oblong-oblanceolate, obtuse at apex; margins entire to shallowly lobed or acutely toothed. Upper stem leaves oblong-oblanceolate, often acute. Inflorescence crowded. Sepals 3-4.3 mm, oblong, green to yellowish, ascending. Petals 3.3-6.2 x 1.5-2.5 mm; limb obovate, rounded at apex, yellow; claw short, whitish. Petals *c.* 1.5-2 times as long as sepals. Stamens 6; anthers yellow. Immature ovaries oblong. Stigma entire. Pedicels 5-17 mm, slender, patent to reflexed. Fruit set often good. Fruits 4.5-7.5 x 1.5-2.7(-3) mm, ellipsoid, ± terete, patent to reflexed, dehiscent. Valves (2.5-)3-6 mm, very shortly stipitate, without veins. Persistent style 0.8-1.8(-2.5) mm. Seeds numerous, 0.6-1 mm, ovoid, brown, biseriate. 2n=16*,32*. Flowering June to September.

A native typical of tall riverbank vegetation (especially on the outer margins), swamps, ditches, ponds and lakes, usually in slow-moving or still water. Tolerant of fluctuating water levels.

Common in C. and S. E. England but rarer elsewhere, rare in eastern Wales, rare in S. Scotland (possibly introduced) and scattered but not infrequent in Ireland. Native from W. Europe to E. Asia. Introduced to N. America and Australasia.

A variable plant in leaf and fruit shape. The lowest leaves are usually deeply pinnatisect but do not persist and are superseded by the typical ± simple leaves; these must not be confused with leaves of hybrids. Irish plants have deeply serrate leaves with a distinct appearance. Fruit set is variable but often good. Two cytotypes are present, a diploid in S. England and a more widespread tetraploid. The cytotypes correlate to a degree with morphological differences but do not merit taxonomic recognition. For more details, see Jonsell (1968).

A distinct, robust plant. There are no other yellow crucifers with more or less simple leaves and biseriate seeds which occur in a similar aquatic habitat. Sometimes similar to *R. austriaca* (see **68**).

For details of hybrids with *R. austriaca, R. palustris* and *R. sylvestris*, see page 162.

A-Cx0.5;D-Fx2;Gx6;Hx10;I,Kx1.7.

66. **Rorippa x anceps** (Wahlenb.) Reichenb. (R. amphibia x sylvestris)

Perennial, ± intermediate between the parents. Rhizome creeping. Stem leaves pinnatifid to pinnatisect, usually lobed at least 1/2 way to midrib at least at the base of the leaf, auricles usually small or absent. Sepals 2-3.8 mm. Petals 3.5-5 mm. Immature ovaries oblong. Pedicels patent to deflexed. Fruit set variable. Fruits 3-10 x 1.2-2.5 mm. Persistent style (0.8-)1.2-2.5(-3) mm. 2n=32*,40*.

A hybrid probably widespread throughout S. Britain and Ireland (Jonsell 1968) and probably widely overlooked (Map 66). In similar places to the parents, sometimes without one or the other. Widespread in Europe.

The deeply lobed upper stem leaves distinguish it from *R. amphibia*. *R. sylvestris* has leaves regularly pinnatifid along the length of the leaf but *R. x anceps* has them much less divided at the apex. It is fertile, and variable due to back-crossing with the parents.

67. **Rorippa x erythrocaulis** Borbás (R. amphibia x palustris)

Perennial, ± intermediate between the parents. Leaves pinnatifid with 4-6 pairs of lateral lobes; auricles conspicuous, clasping stem. Sepals (1.8-)2.5-2.7 mm. Petals (2-)2.5-3.6 mm. Immature ovaries elliptic. Pedicels patent to deflexed. Fruit set usually failing, sometimes irregular. 2n=24*,32*.

A rare sterile hybrid so far only recorded from England and Sweden (Britten 1909, Jonsell 1968); Map 67.

Rorippa x armoracioides (Tausch) Fuss (R. austriaca x sylvestris)

Perennial with spreading rhizomes, ± intermediate between the parents. Leaves acutely toothed or lobed, auricles conspicuous, clasping stem. Sepals 1.8-3 mm. Petals 3-4.5 mm. Immature ovaries elliptic. Pedicels in fruit ascending to patent. Fruit set variable. Fruits (3-)3.5-9 x 1.5-2 mm. Persistent style (0.8-)1.2-1.5 mm. 2n=32*. A rare hybrid only confirmed from London, Glasgow and Oban. Widespread in Europe (Jonsell 1968).

Rorippa x hungarica Borbás (R. amphibia x austriaca)

Perennial with shortly creeping rhizomes, ± intermediate between the parents. Leaves acutely toothed with conspicuous auricles clasping stem. Sepals 3.2-3.6 mm. Petals 4.8-5.7 mm. Immature ovaries elliptic. Pedicels patent to deflexed. Fruit set failing. A rare hybrid only confirmed for Essex (Rich & Wurzell 1988). Reported rarely in Europe.

Rorippa x barbaraeoides (Tausch) Celak. pro parte (*R. palustris x sylvestris*) has been reported but not confirmed for Britain and is rare on the Continent. It could occur but would be difficult to determine without a chromosome count.

Rorippa x anceps 66

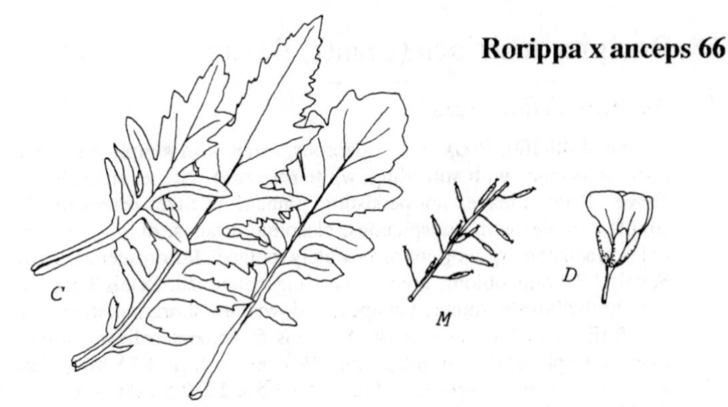

Cx0.5;Dx2.5;M=inflorescence x1.

Rorippa x erythrocaulis 67

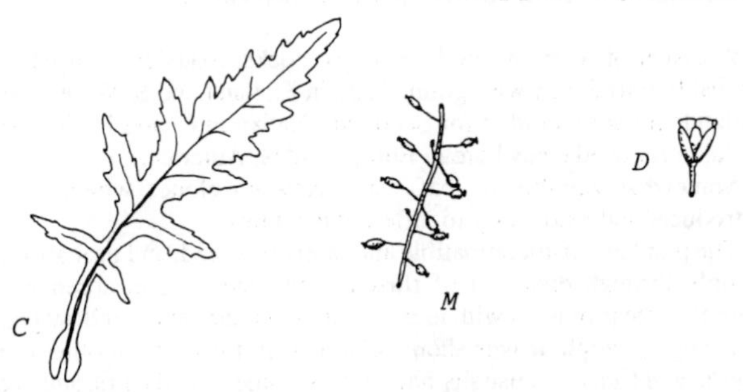

Cx0.5;Dx2.5;M=inflorescence x1.

Rorippa x armoracioides

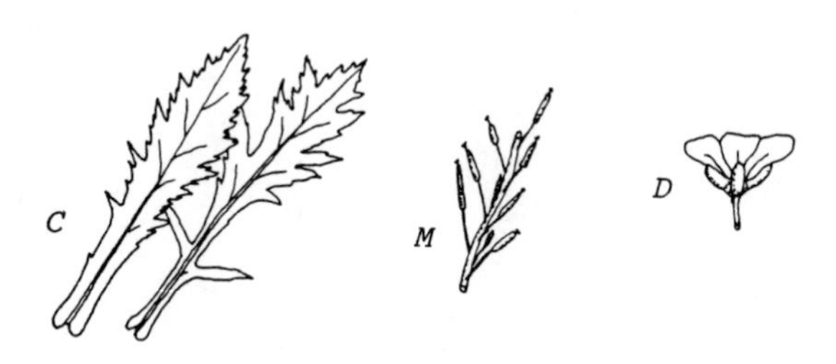

Cx0.5;Dx2.5;M=inflorescence x1.

68. **Rorippa austriaca** (Crantz) Besser

Austrian Yellowcress Map 68

Perennial 40-100(-200) cm, ± glabrous, often with simple hairs below, forming patches with spreading, white rhizomes. Stems erect, branched above. Rosette leaves not persisting, pinnatifid. Stem leaves to 15 cm, simple, sessile, with conspicuous, clasping auricles, oblong-obovate to oblong, acute; margins ± entire to acutely toothed. Inflorescence crowded. Sepals 2-3.1 mm, oblong, green, ascending to inclined. Petals 3-5 x 1.7-2.5 mm; limb obovate, rounded at apex, yellow; claw short, indistinct. Petals *c.* 1.5 times as long as sepals. Stamens 6; anthers yellow. Immature ovaries ± spherical. Stigma entire. Pedicels in fruit 4-15 mm, slender, ascending. Fruits rarely set. Fruits 2.5-3.5 x 1.5-2.5 mm, ± globose to ovoid, ascending, dehiscent. Valves 1.5-2.5 mm, veins absent. Persistent style 1-1.4(-2) mm. Seeds 2-8 in each loculus, 0.7-0.9 mm, ovoid, brown, ± biseriate. 2n=16*. Flowering May to September.

A casual of waste ground, railways, docks, roadsides and near rivers, probably introduced with grain. Rare in England and S. Wales, extinct in Ireland and not recorded for Scotland. Native in C. and E. Europe to W. Asia, introduced elsewhere in Europe and N. America.

Somewhat variable in leaf shape between clones which have been introduced independently to different localities.

The plant is self-incompatible and rarely sets seed, and probably spreads mainly through dispersal of rhizome fragments. The absence of fruit causes some problems with identification, but the small, yellow petals and clasping, ± simple leaves should distinguish it from most other crucifers. *Neslia* and *Camelina* usually have at least some forked hairs, and *Lepidium draba* can be similar vegetatively but has white flowers. The more or less globose valves, ascending pedicels and creeping rhizomes distinguish it from *R. amphibia*.

For hybrids with *R. amphibia* and *R. sylvestris*, see p.162.

Rorippa austriaca 68

A-Cx0.5;D-Fx2;Gx6;Hx10;I,Kx3.

69. **Barbarea stricta** Andrz.

Upright Wintercress, Small-flowered Wintercress

Biennial or perennial 20-100 cm, ± glabrous sometimes with simple hairs below, bright yellow-green. Stems erect, branched above. Rosette leaves to 15 cm, usually with clasping auricles, pinnate with a large ovate to oblong, obtuse terminal lobe and (0-)1-3(-4) pairs of small ± oblong lateral lobes; margins ± entire to sinuately lobed. Lower stem leaves similar but smaller, with 0-2 pairs of lateral lobes. Uppermost stem leaves sessile, with clasping auricles, simple, obovate to ovate, obtuse; margins ± entire to coarsely sinuately lobed. Inflorescence crowded. Sepals 2.6-3.2 mm, oblong, ± awned, at least sparsely hairy, green, erect. Petals 3.2-6 x 0.7-1.3 mm; limb obovate, rounded at apex, deep yellow; claw indistinct. Petals *c.* 1.5 times as long as sepals. Stamens 6; anthers yellow. Stigma entire to ± emarginate. Pedicels in fruit 3-7 mm, slender, erect to ascending. Fruits 13-28(-35) x 1.2-2.2 mm, linear, ± terete to 4-angled, erect to ascending, dehiscent. Valves 12-28(-34) mm, with a strong central vein and 0-2 weak or inconspicuous lateral veins. Persistent style 0.5-1.8(-2.3) mm, stout. Seeds numerous, (1-)1.1-1.7(-1.8) mm, ovoid, brown, uniseriate. 2n=(14-)16(-18). Flowering May to September.

Probably native (Rich 1987d). A very local lowland plant of canal and riverbanks, ditches, quiet backwaters, marshes and also a casual of waste land. Scattered in England, rare in Wales and Scotland (map in Perring & Walters 1962), and a rare casual in Ireland. Native from Europe to W. Asia, introduced to N. America and Australasia.

Varying slightly in leaf shape and lobing.

The ± simple upper leaves, hairy flower buds and small petals distinguish it from other *Barbarea* species. Similar to *B. vulgaris* from which it may be distinguished in fruit by the short persistent style (see also Sprague & Hutchinson 1908).

Few other genera have yellow flowers and clasping upper leaves. *Camelina, Neslia* and *Arabis* have at least some stellate hairs, *Isatis* has pendulous fruits, *Rorippa* has fruits less than 20(-22) mm, and *Brassica* and *Conringia* have generally, larger petals and fruits.

The history of *B. stricta* in Britain is described by Jackson (1908) and Rich (1987d).

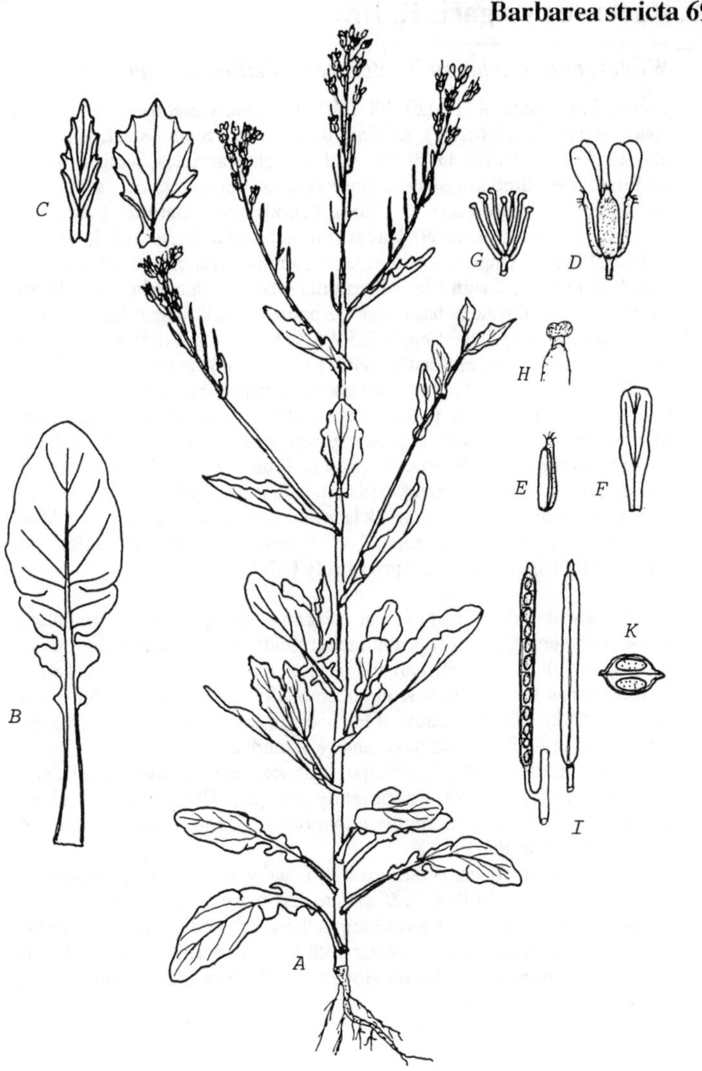

Ax0.2;B,Cx0.5;D-Gx4;Hx10;Ix1.5;Kx10.

70. **Barbarea vulgaris** R. Br.

Wintercress, Common Wintercress, Yellow Rocket

Biennial or perennial (10-)20-90(-130) cm, usually glabrous (rarely with sparse simple hairs below), shining, dark green. Stems erect, branched above. Rosette leaves to 20 cm, with clasping auricles, pinnate with a terminal lobe elliptic to ovate, obtuse at apex, cuneate to cordate at base and with (0-)2-5(-6) pairs of oblong lateral lobes; margins ± entire to coarsely sinuate. Lower stem leaves similar but smaller, with 1-4(-5) pairs of lateral lobes. Uppermost stem leaves sessile with clasping auricles, ± simple or pinnatifid with a broad terminal lobe, obovate to broadly elliptic (rarely ovate), cuneate at base, and 1-2 pairs of small, linear, lateral lobes. Inflorescence crowded. Sepals 2.5-4.5 mm, oblong, slightly saccate, ± awned, glabrous (exceptionally with 1 or 2 hairs), green, erect. Petals 4.9-7(-8) x 1.1-3.5(-4.1) mm; limb obovate, rounded at apex, pale to deep yellow; claw indistinct, paler. Petals about twice as long as sepals. Stamens 6; anthers yellow. Stigma capitate, entire to emarginate. Pedicels in fruit 2-6 mm, slender, erect to patent. Fruits (7-)15-32 x 1.1-1.8 mm, linear, terete to 4-angled, erect to patent, dehiscent. Valves (5-)13-30 mm, with a strong central vein and weak laterals. Persistent style (1.7-)2-3.5(-4) mm, slender. Seeds numerous, 1.1-1.8 mm, ovoid, brown, uniseriate. 2n=(14-)16(18). Flowering April to July (-October).

A locally abundant plant of river banks and shingle, ditches, woodland rides, road verges, farmyards and arable fields, waste ground, docks, etc. Widespread throughout Britain and Ireland, rare in the far north and west, in most areas the commonest member of the genus (map in Perring & Walters 1962). Native in most of Europe and eastwards to the Himalayas, introduced to Africa, Australasia and N. America.

A very variable plant in leaf shape, fruit size, angle of the pedicels, etc., but no infraspecific taxa are currently accepted (Rich 1987d) and the species is best regarded as highly polymorphic. More variable in the rest of Europe than in Britain.

In fruit, the long, linear persistent style usually separates it from the other *Barbarea* species. In flower, *B. stricta* has hairy flower buds and deep yellow petals only *c.* 1.5 times as long as the sepals and usually more entire leaves. *B. intermedia* and *B. verna* both have pinnatifid uppermost stem leaves with narrower terminal lobes. For characters distinguishing *Barbarea*, see under **69**.

Barbarea vulgaris 70

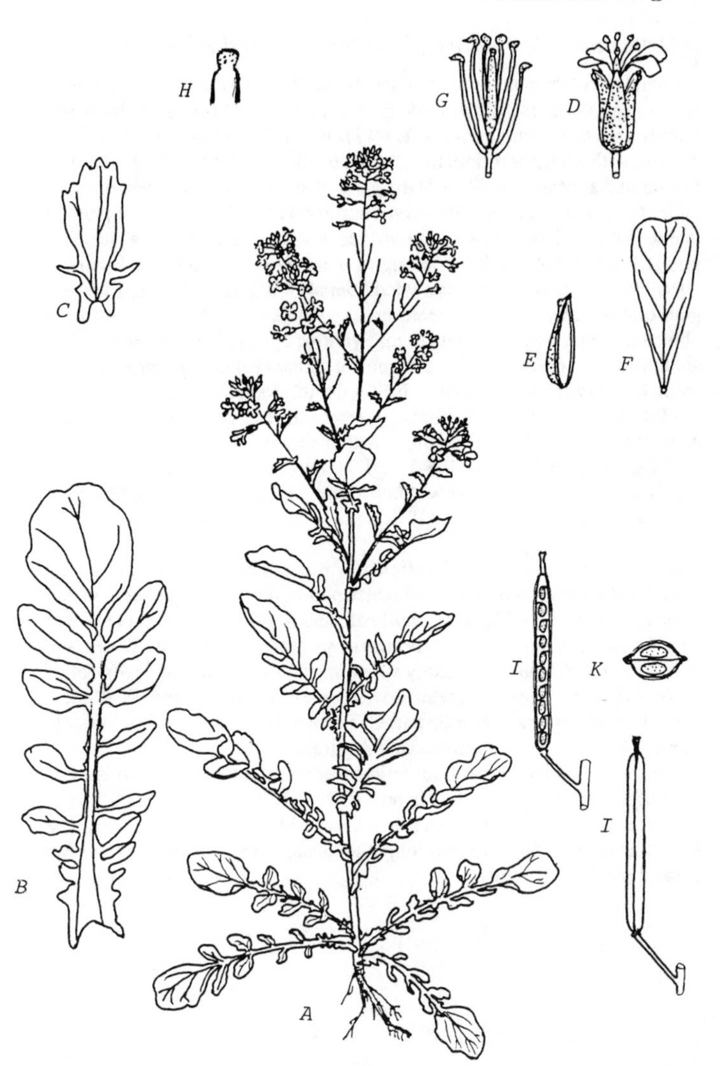

Ax0.2;B,Cx0.5;D-Gx4;Hx10;Ix1.5;Kx10.

71. **Barbarea intermedia** Boreau

Intermediate Wintercress, Medium-flowered Wintercress

Biennial (rarely annual or perennial) (10-)15-60(-70) cm, glabrous or with sparse, simple hairs below, dark green, shining. Stems erect, branched mainly above. Rosette leaves to 12(-17) cm, with clasping auricles at base, pinnate with a large terminal lobe, ovate to elliptic, obtuse at apex, cordate to rounded at base, and (0-)2-5(-6) pairs of smaller lateral lobes; margins entire to sinuate. Lower stem leaves similar but smaller, with 1-4 pairs of lateral lobes. Uppermost stem leaves sessile, with clasping auricles, pinnatifid, the terminal lobe oblong to oblanceolate, with (1-)2-3 pairs of lateral lobes. Inflorescence crowded. Sepals 2-3.5 mm, oblong, ± awned, glabrous, slightly saccate, green, erect. Petals 4-6.3 x 0.9-2 mm; limb obovate, rounded at apex, pale to mid yellow; claw indistinct, pale. Petals about twice as long as sepals. Stamens 6; anthers yellow. Stigma capitate, entire to emarginate. Pedicels in fruit 3-6 mm, stout, erect to patent. Fruits 15-35(-40) x 1.3-2.3 mm, linear, ± terete to 4-angled, erect to patent, dehiscent. Valves 14-34(-39) mm, with a strong central vein and weaker laterals. Persistent style 0.6-1.6(-1.7) mm, stout. Seeds numerous, 1.7-2.3(-2.4) mm, ovoid, brown, uniseriate. 2n=16. Flowering March to July (-August).

A casual of roadsides, arable fields, waste and disturbed ground, often associated with building and road construction sites. Rarely abundant and often as isolated individuals or small patches. Rarely persistent but in some sites known for many years. Frequent in most of England, Wales and Ireland and S. Scotland, probably spreading and in some cases overlooked as *B. vulgaris*. More frequent now than shown in Perring & Walters (1962). Probably native in S. and C. Europe, N. and E. Africa. Widely introduced elsewhere in Europe and Australasia.

Not very variable, except sometimes in size. Intermediate in general appearance between *B. vulgaris* and *B. verna*. The former has a longer slender style, smaller seeds and much less divided upper stem leaves. The latter has fruits at least 40 mm long and usually larger petals. For further details, see Rich (1987d).

Barbarea intermedia 71

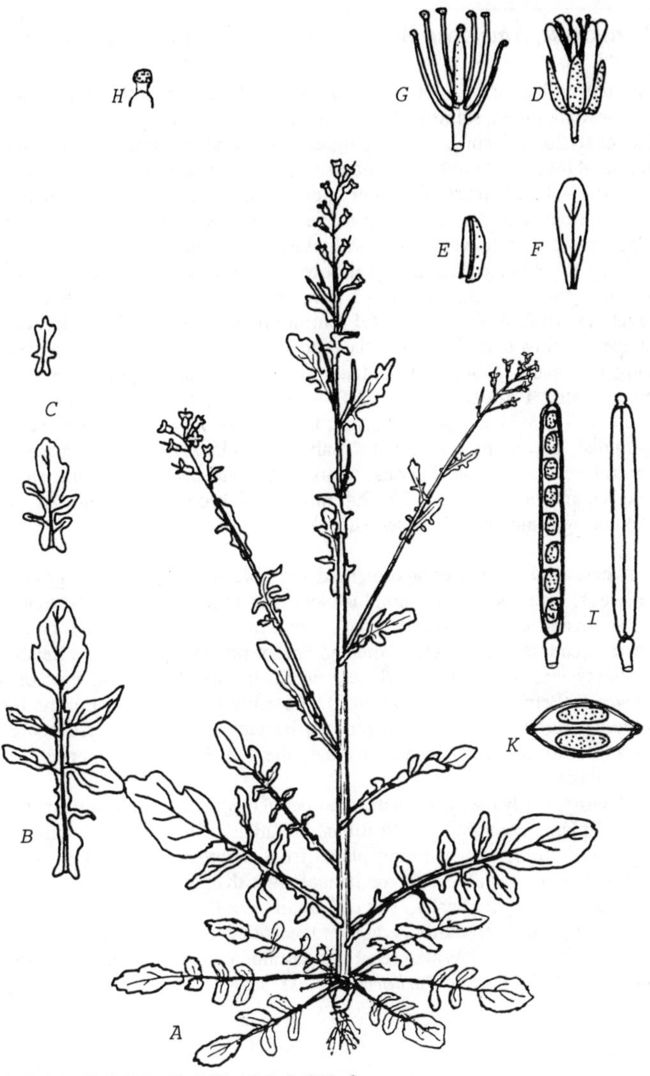

Ax0.2;B,Cx0.5;D-Gx4;Hx10;Ix1.5;Kx8.

72. **Barbarea verna** (Miller) Ascherson

Landcress, American Wintercress, Early-flowering Wintercress

Annual or biennial (5-)30-90(-130) cm, glabrous or with simple hairs below, dark green, shining. Stems erect, branched mainly above. Rosette leaves to 20(-25) cm, with clasping auricles at base, pinnate with a large terminal lobe, ovate to elliptic, obtuse at apex, cordate to rounded at base, and (0-)4-10(-11) pairs of smaller lateral lobes; margins entire to sinuate. Lower stem leaves similar to basal leaves but with fewer lateral lobes. Uppermost stem leaves sessile with clasping auricles, pinnatifid with an oblong to oblanceolate terminal lobe and 2-4 pairs of lateral lobes. Inflorescence lax. Sepals 2.8-5.1 mm, oblong, ± awned, saccate, green, erect. Petals (5.6-)6-9.6 x 1.5-4.1 mm; limb obovate, rounded to truncate at apex, yellow to dark yellow; claw indistinct, pale. Petals about twice as long as sepals. Stamens 6; anthers yellow. Stigma capitate, entire to emarginate. Pedicels in fruit (2-)3-8 mm, stout, ascending to inclined. Fruits (28-)35-71 x 1.8-3(-3.4) mm, linear, upwardly curved, ± terete to 4-angled, ascending, dehiscent. Valves (27-)35-70 mm, with a strong central vein and weak laterals. Persistent style 0.6-2(-2.3) mm, stout. Seeds numerous, (1.6-)1.7-2.3(-2.4) mm, ovoid, brown, uniseriate. 2n=16. Flowering March to July (-December).

A casual or garden escape of roadsides, waste and disturbed ground, railways, bare and stony ground usually near to habitation. Persistent in the south and west and near the coast. Seeds may remain dormant for many years. Scattered through England and Wales, rare in Scotland and Ireland, probably less frequent than *B. intermedia* but more frequent now than shown in Perring & Walters (1962). Possibly native in W. Europe but widely naturalized in Europe in general; the native range has been obscured by cultivation. Introduced to Asia, the Far East, N. America and Australasia.

Phenotypically the most plastic member of the genus, small, depauperate annual plants contrasting with robust biennials. This range of variation may give problems separating plants from *B. intermedia*, but *B. verna* almost always has at least some fruits longer than 40 mm. Plants on the coast are often quite fleshy. For full details, see Rich (1987d).

B. verna has been cultivated since the 16th century as a salad crop (a winter substitute for Watercress) and commercial seed is still available today. Other species of *Barbarea* are less tasty.

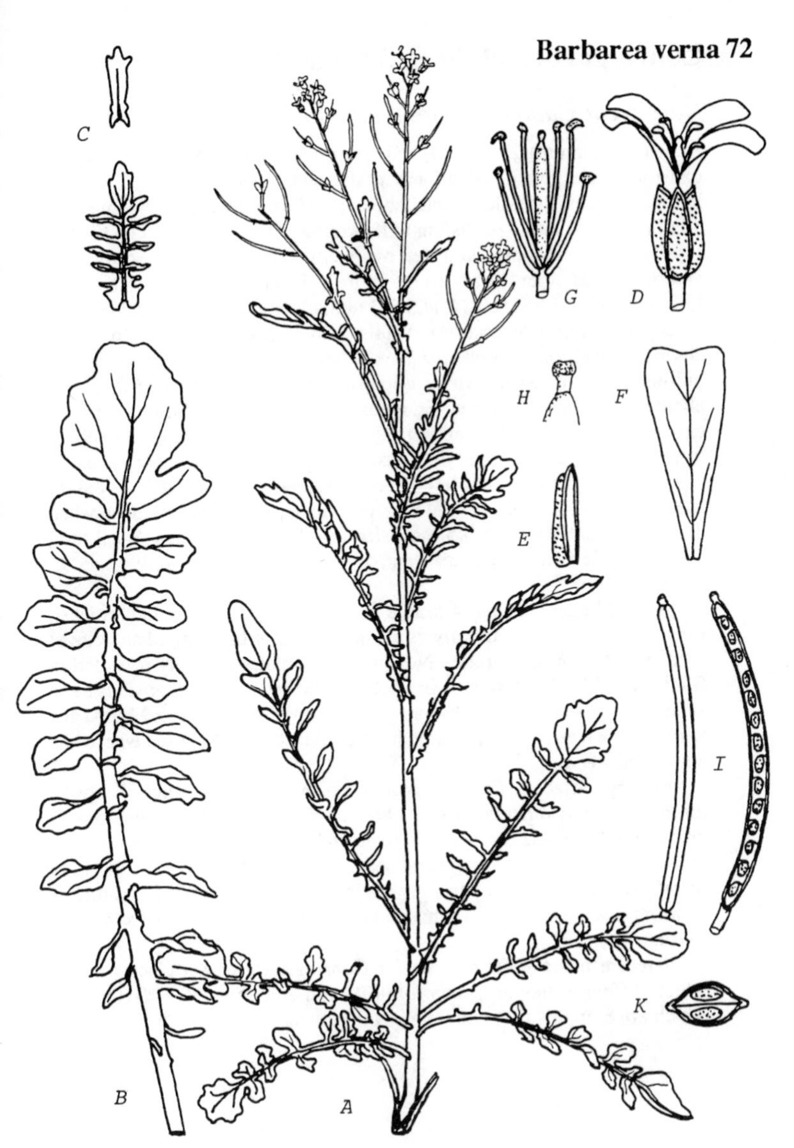

Ax0.2;B,Cx0.5;D-Gx4;Hx10;Ix1.5;Kx6.

73. Conringia orientalis (L.) Dumort.

Hare's-ear Cabbage, Hare's-ear Mustard Map 73

Annual (?biennial) 20-60(-80) cm, glabrous (?rarely glandular pilose), glaucous, pale green. Stems erect, sparsely branched above. Rosette leaves to 10 cm, ± sessile, simple, obovate, obtuse, cuneate at base; margins ± entire; leaves rarely persisting to flowering. Stem leaves to 11(-13) cm, sessile and perfoliate, simple, ovate, elliptic or oblong, obtuse to emarginate at apex; margins ± entire. Upper stem leaves smaller. Inflorescence lax. Sepals (4-)4.5-7.3 mm, oblong, saccate, awned or not, green, erect (rarely ascending). Petals (7-)8-14 x 1.5-3(-?4) mm or rarely absent; limb elliptic, rounded at apex, white to yellow or greenish (?rarely purplish); claw indistinct, about as long as the limb, pale or greenish. Petals about twice as long as sepals. Stamens 6; anthers yellow. Stigma capitate, entire to emarginate. Pedicels in fruit 6-20 mm, slender to stout, ascending to inclined. Fruits (45-)60-140(-150) x (?1-)1.5-3 mm, linear, strongly 4-angled (± mediseptate-latiseptate), ascending, dehiscent. Valves (43-)60-138(-149) mm, central vein prominent, somewhat torulose when dry. Persistent style 0.8-3.5 mm, slender. Seeds numerous, 2-2.9 mm, ovoid, dark brown, uniseriate. 2n=14. Flowering May to September.

Once a frequent casual of arable fields, docks, waste ground, chicken runs, rubbish tips, etc., usually as a contaminant of bird, cereal and clover seed from C. and E. Europe. Now a rare alien, occasionally recorded in England and Wales, rarely so in Scotland and Ireland and there is little to suggest it persists. Probably native in Europe to C. Asia and N. Africa where it is a weedy summer annual. Introduced to N. America where it is a frequent weed.

Not very variable. Once known, the plant is distinct and will not be mistaken for any other crucifers. Often confused (principally due to poor keys) with *Brassica rapa* and *B. napus* which also have entire clasping upper stem leaves but have round seeds, a conical beak to the fruit more than 5 mm and much less strongly angled valves. A diagnostic feature of *Conringia* is the thin (*c.* 0.1 mm) translucent margin to the leaves (opaque in *Brassica*).

There are 6 pre-1950 records for *Conringia austriaca* (Jacq.) Sweet which differs in having petals 6-8(-10) mm and fruits 50-80(-100) mm which are 8-angled (3 ± equal veins on each valve).

Conringia orientalis 73

Ax0.2;B,Cx2;D-Gx3;Hx20;Ix1;Kx8;Lx10;M=inflorescence x0.5.

74. Erysimum cheiranthoides L.

Treacle Mustard

Annual (2-)15-90 cm, roughly hairy with appressed medifixed and stellate hairs, green. Leaves to 10 cm, sessile to shortly stalked, elliptic to narrowly lanceolate or linear-oblanceolate, acute to obtuse at apex, cuneate at base; margins entire to distinctly toothed. Inflorescence crowded. Sepals 1.8-3 mm, oblong, awned, green, erect. Petals 2.9-5 x 0.9-1.3 mm; limb obovate, rounded at apex, yellow; claw indistinct, pale. Petals *c.* 1.5 times as long as sepals. Stamens 6; anthers yellow. Stigma emarginate to 2-lobed. Pedicels in fruit 4-12(-15) mm, slender, ascending to patent. Fruits (12-)15-25(-27) x 1-2 mm, linear, ± terete to strongly 4-angled, erect to inclined, dehiscent. Valves (12-)14-24(-26) mm, with a prominent central vein. Persistent style 0.5-1.4 mm, linear, emarginate. Seeds numerous, 0.9-1.4(-1.7) mm, oblong, sometimes partially winged, pale brown, uniseriate (sometimes partially biseriate). 2n=16. Flowering May to November.

A characteristic arable weed or on waste ground, walls, docks, railways, roadsides, gardens, etc. Locally abundant especially on sandy ground, but often in only small populations. Roberts & Boddrell (1983) found it could form a seed bank and thus become a persistent weed, but it is susceptible to herbicides and is probably less frequent now than 50 years ago. Probably introduced many centuries ago.

Frequent in S. E. England, occasionally in the Midlands, S. W. England and Wales, rare or casual elsewhere (map in Perring & Walters 1962). In Ireland, mainly found in the centre. Probably native from Europe to C. Asia and in N. America but now a cosmopolitan weed.

Varying little except in size. Ahti (1961) distinguished two subspecies. British material is largely refereable to subsp. *cheiranthoides* but some Scottish material may be subsp. *altum* Ahti. Further work is required.

The appressed medifixed and stellate hairs (the former may appear simple under a lens) give it a characteristic rough feel to the fingertips. These hairs, the more or less entire, elliptic to lanceolate leaves and tiny yellow flowers should separate it from other crucifers.

Erysimum cheiranthoides 74

Ax0.2;B,Cx0.3;D-Fx4;G,Hx12;Ix1;Kx6;Lx10.

75. Erysimum repandum L.

Map 75

Annual 8-40 cm, with appressed, medifixed hairs. All leaves similar, petiolate, linear-lanceolate to linear-oblong; margins entire to sinuate-dentate. Petals 6-10 mm, pale yellow, about twice as long as sepals. Fruits 45-100 x 0.7-1.5 mm, linear, ± terete, ± torulose, dehiscent. Persistent style 4-7 mm, linear. Seeds numerous, cylindrical, uniseriate.

A casual of docks, waste ground, grain silos, fields, etc, but now much rarer than formerly. It is probably native in the Mediterranean but is also widespread as a weed (e.g. N. America).

Distinguished from the other *Erysimum* species by the combination of long, linear fruits and small petals.

76. Erysimum allionii hort.

Siberian Wallflower

Similar to the next species **77** *E. cheiri*: Annual or biennial 20-60 cm, with appressed medifixed hairs. Leaves petiolate to sessile, narrowly elliptic, acute; margins sparsely toothed. Petals 15-25 mm, clear bright orange (less commonly yellow), about twice as long as sepals. Fruits *c.* 40-70 x *c.* 2 mm, linear, 4-angled, dehiscent. Persistent style *c.* 2-3.5 mm, 2-lobed. Seeds numerous, cylindrical, uniseriate.

A plant grown in gardens and rarely naturalised on waste ground, rubbish tips, etc. and unknown in the wild. There are 6 post-1950 records but it could well be under-recorded.

The clear bright orange petals are distinctive once known, and will separate it from all other crucifers. Otherwise it is very similar to *E. cheiri* which has flattened fruits and woody stems.

There appears to be no validly published name for this plant. The specimens reported as *E. perofskianum* Fischer & C.A. Meyer (Clapham *et al.* 1962) may be this taxon.

Erysimum repandum 75

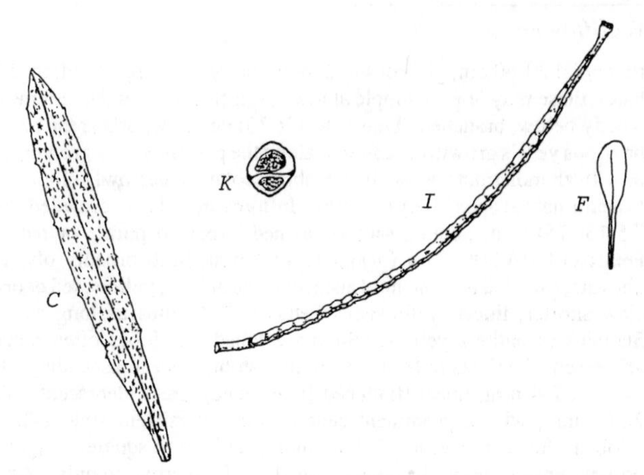

Cx2;Fx2;Ix1;Kx10.

Erysimum allionii 76

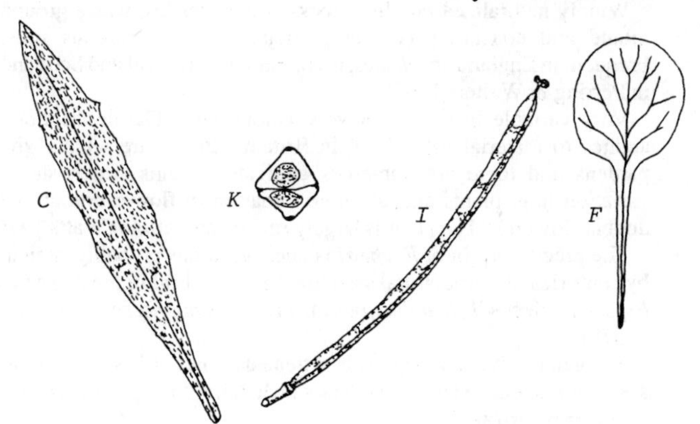

Cx1;Fx2;Ix1;Kx4.

77. **Erysimum cheiri** (L.) Crantz

Wallflower

Perennial 20-60 cm, glabrous to densely hairy with appressed, medifixed hairs (these may appear simple at low magnification), green. Stems erect, woody below, branched. Leaves to 12(-15) cm, in whorls at the top of the previous year's growth or scattered along the present year's growth, sessile or with an indistinct petiole, linear-oblanceolate to narrowly elliptic, acute to acuminate at apex; margins entire. Inflorescence lax to crowded. Sepals 7.5-13(-15) mm, oblong, saccate, awned, green to purple or red, erect. Petals (14-)16-30(-33) x (5-)6.5-16(-20) mm; limb broadly obovate to obovate, rounded to truncate at apex, cuneate at base, yellow, red or orange; claw shorter, linear, yellowish. Petals *c.* 2-3 times as long as sepals. Stamens 6; anthers yellow. Stigma 2-lobed, the lobes often erect and appressed. Pedicels in fruit 6.5-16 mm, stout, erect to ascending. Fruits 24-70 x 2-4 mm, linear, flattened (latiseptate), erect, dehiscent. Valves 22-67 mm, with a prominent central vein. Persistent style 2-3.5 mm, 2-lobed. Seeds numerous, 2.2-3.7 mm, ± oblong to square, winged, pale brown, uniseriate to biseriate. 2n=12. Flowering mainly March to September, and sometimes in the winter.

Widely naturalized on cliffs, rocks, walls, castles, waste ground, etc., inland and coastal, persistent particularly on calcareous substrates. Frequent in England and Wales, less common in Ireland and Scotland (map in Perring & Walters 1962).

Very variable in size of flowers, colour, etc. The description above applies to material naturalized in Britain. Plants are widely grown in gardens and there are numerous cultivated strains which show more variation (e.g. plants annual, more variation in flower colour and size, double flowers). The plant is largely self-incompatible (Watts 1976).

The precise origin of *E. cheiri* is unclear. It has probably been derived by centuries of horticultural selection from a hybrid in the Aegean within *Erysimum* Sect. *Cheiranthus*, possibly *E. corinthium* x *senoneri* (Snogerup 1967b).

The genus *Cheiranthus* L. is best treated as part of *Erysimum* L. as there are no good, consistent characters which delimit the genera satisfactorily (Snogerup 1967a).

The large, yellow flowers combined with the appressed, medifixed hairs are distinct and it is only likely to be confused with other cultivated *Erysimum* species.

Erysimum cheiri 77

Ax0.3;D,Ex1;F,Gx0.6;Hx20;Ix0.7;Kx1.5;Lx2.

78. Matthiola incana (L.) R. Br.

Stock, Hoary Stock, Gilliflower, Queen Stock

Perennial (?biennial) 30-80 cm, densely hairy with stellate hairs (minute glandular hairs also present), grey-green. Stems erect, woody, branched above. Stem leaves to 15 cm, in whorls at the top of the previous year's growth or scattered along the present year's growth, broadly petiolate, simple, narrowly oblanceolate to narrowly oblong, obtuse at apex; margins entire. Inflorescence lax. Flowers often imperfect. Sepals 11-14 mm, oblong, saccate, green to purple, erect (often partially fused). Petals (19-)21-31 x (6-)7-16 mm; limb obovate, truncate to rounded at apex, cuneate at base, white to pink or purple; claw about as long as limb, greenish, thickened, linear. Petals about twice as long as sepals. Stamens 6; anthers yellow. Stigma ± emarginate. Pedicels in fruit 8-19 mm, stout, ascending to inclined. Fruits (27-)45-140 x 3-5 mm, linear, curved or straight, ± terete to flattened (latiseptate), erect to patent, dehiscent. Valves (27-)44-140 mm, with a distinct central vein. Persistent style 0.2-4 mm, distinct and with or without two lateral horns. Seeds numerous, 2.5-3.5 mm, ± square, flattened, broadly winged, brown to blackish, uniseriate. 2n=14,16. Flowering (March-) April to June (-October).

Represented in Britain by subsp. *incana*.

A conspicuous plant of sea cliffs (especially of chalk or limestone), shingle, walls and banks near the sea and occasionally in towns or cities. Well naturalized in S. and S. W. England and Wales, rarely in Scotland and Ireland (map in Perring & Walters 1962), often appearing a native member of the vegetation but undoubtedly introduced. Native in the Mediterranean, N. W. Africa and W. France. Introduced to Australasia.

Very variable, most populations showing characters of the plants grown in gardens as Queen, Brompton or Bedding Stocks, such as white to purple or variegated, large, often double-width petals, ± inflated fruits and partial "double" flowers, though the annual habit is usually lost. The description above applies to material naturalized in Britain; wild-type material is less variable. Saunders (1928) discusses the genetics.

Distinguished from *M. sinuata* by all leaves entire and by the absence of large, conspicuous glands. *M. longipetala* has two pronounced horns to the fruits and is a slender annual. *Malcolmia maritima* has thinner fruits tapering to a linear style. *Erysimum cheiri* is superficially similar out of flower but has orange or deep red petals.

Widely cultivated in gardens for at least 400 years. The flowers are exquisitely fragrant.

Matthiola incana 78

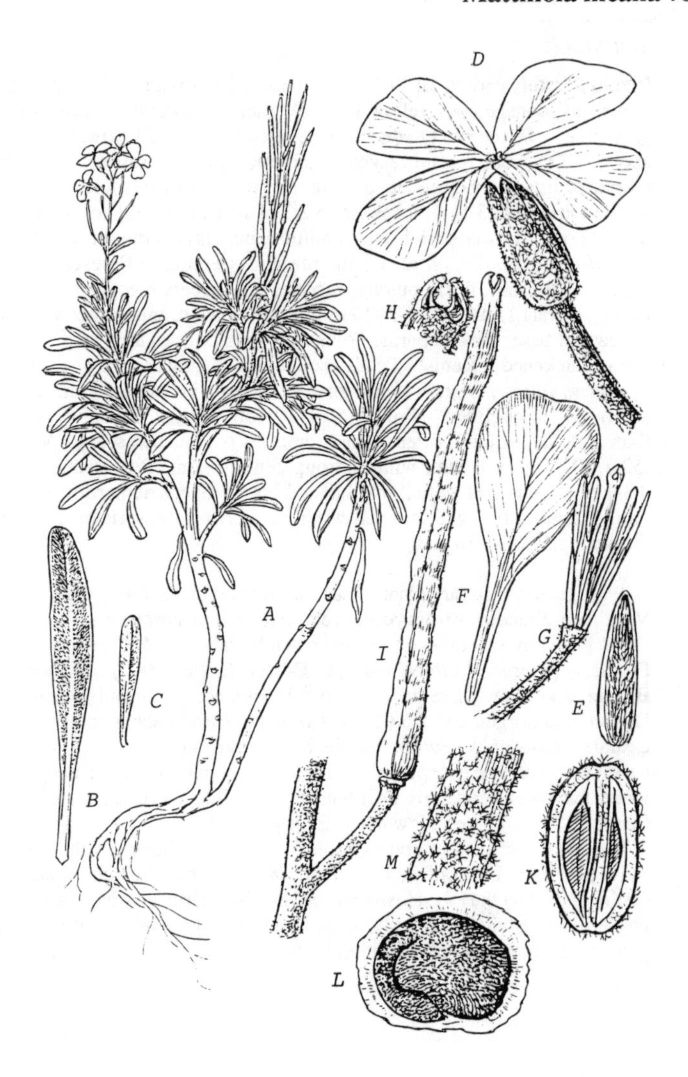

Ax0.2;B,Cx0.3;D-Gx2;Hx10;Ix0.7;Kx6;Lx10;M=hairs on stem.

79. Matthiola sinuata (L.) R. Br.

Sea Stock

Biennial (rarely perennial) 40-100 cm, densely hairy with stellate hairs and numerous, conspicuous yellow to black glands (especially when fresh), grey-hoary. Stems erect to decumbent, often toughened below, branched above. First year rosette leaves to 35 cm, petiolate, simple, narrowly elliptic to linear-oblanceolate, obtuse; margins shallowly sinuate-toothed to pinnatifid with 3-5(-6) lobes cut to 2/3 of way to midrib; not persistent to flowering. Lower stem leaves ± similar, but less deeply lobed. Upper stem leaves ± sessile, linear-oblong; margins ± entire. Inflorescence lax. Sepals 12-14 mm, linear-oblong, saccate, grey-green, erect (often partly fused). Petals 18-26 x (5.5-)7-10 mm; limb obovate, emarginate at apex, cuneate at base, pink to purple (?white); claw about as long as the limb, linear, thickened, greenish. Petals about twice as long as sepals. Stamens 6; anthers yellow. Stigma ± emarginate. Pedicels in fruit 6-13 mm, stout, ascending to inclined. Fruits (50-)110-150(-160) x 3-5 mm, linear, flattened (latiseptate), often curled, erect to patent, dehiscent. Valves (50-)110-150(-160) mm, with 1 strong central vein. Persistent style *c.* 0.3-1 mm, indistinct, with 2 small lateral horns. Seeds numerous, 2.5-4.2 mm, ± square, flat, with a broad wing, pale brown, uniseriate. 2n=14*. Flowering June to August.

A gorgeous, very rare plant of sand dunes and cliffs, rarely on shingle. Varying in abundance from year to year, establishing best in open disturbed sites. Probably sensitive to frost and eaten by rabbits (McClintock 1955). Formerly recorded from N. Wales to Dorset (Britten 1900) in S. and W. Britain, and from Clare and Wexford in Ireland. Now only known in Devon, Glamorgan and the Channel Islands. Rarely also occurring as a casual. Coasts of W. Europe and the Mediterranean.

Not very variable except in size. The upper stem leaves are usually little lobed, the lower stem or rosette leaves show the characteristic lobing better. Saunders (1928) gives a few notes on its genetics.

The large, conspicuous, pink-purple flowers are striking and the plant is only likely to be confused with *M. incana*, from which it can be distinguished by the lobed lower leaves. Fresh material also has distinctive large yellow to black glands (*M. incana* has minute glandular hairs); similar glands are otherwise only found in *Bunias*.

Matthiola sinuata 79

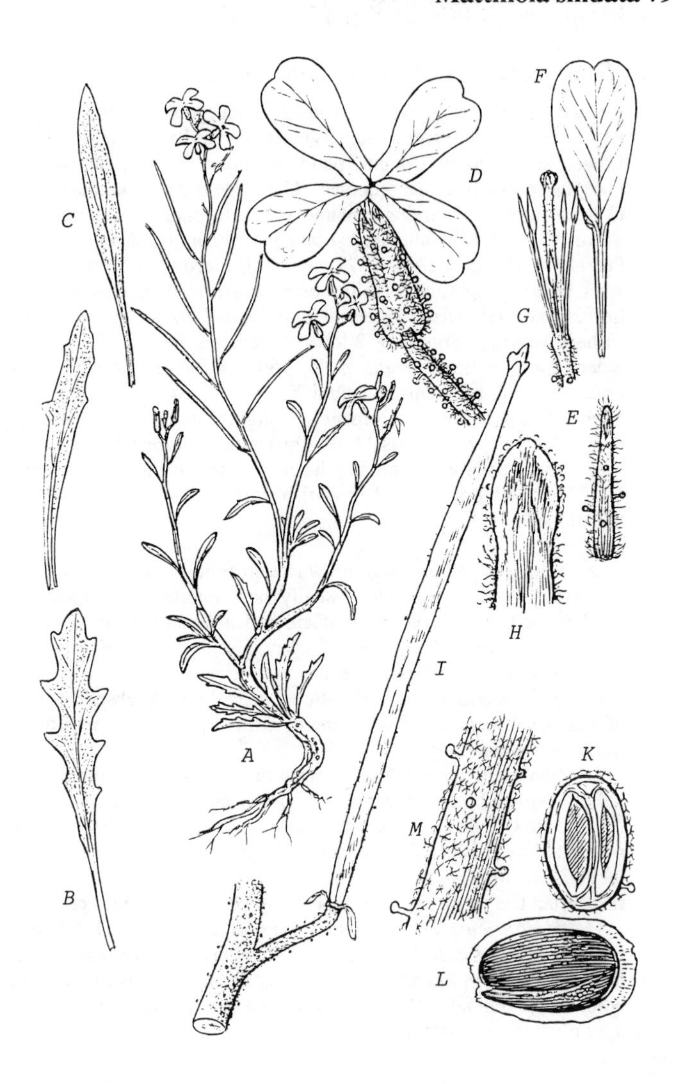

Ax0.2;B,Cx0.3;D-Gx2;Hx10;Ix0.7;Kx6;Lx10;M=glandular hairs on stem.

80. **Matthiola longipetala** (Vent.) DC. subsp. **bicornis** (Sibth. & Sm.) P. W. Ball

Night-scented Stock, Night-flowering Stock

Annual 15-50 cm, sparsely to densely hairy with stellate and small glandular hairs, dull green. Stems erect to decumbent, sparingly branched. Lower leaves to 9(-10) cm, petiolate, linear-oblong, obtuse at apex; margins sparingly sinuate-toothed. Upper stem leaves similar but obtuse to acute at apex; margins sinuate-toothed to entire. Inflorescence lax. Sepals 8-10.5(-12) mm, linear-oblong, saccate, ± awned, green, erect. Petals (15-)16-25 x (3-)4-6(-7) mm; limb narrowly elliptic, rounded at apex, pink to purple above, yellow or brownish beneath; claw ± longer than limb, linear, yellowish. Petals about twice as long as sepals. Stamens 6; anthers yellow. Stigma ± 2-lobed. Pedicels in fruit 1-3 mm, stout, ascending to inclined. Fruits 35-95(-150) x 1-2 mm (excluding horns), ± linear, with 2 pronounced lateral horns 2-7(-10) mm at apex (like a pitch-fork), ± terete (mediseptate) or slightly compressed, ascending to inclined, dehiscent. Valves 33-89(-140) mm, torulose, with a weak central vein. Persistent style (excluding horns) 2-7 mm, stout. Seeds numerous, 1.8-2.7 mm, oblong, winged, brownish, uniseriate. 2n=14. Flowering July to October.

Occasionally naturalised in waste ground, roadsides, abandoned cultivated ground, tips, etc., usually near habitation. Uncommon in England and Wales, rare in Scotland and not recorded in Ireland. The species is native in the Aegean, Turkey, Ukraine, etc. Widely grown in gardens, particularly for the scent.

A variable plant in the wild with between 3 and 7 subspecies only one of which, subsp. *bicornis*, is apparently cultivated. This varies little except in size and branching.

The two pronounced lateral horns at the fruit apex are distinctive. The yellow-brown backs to the petals are usually noticeable because the flowers are shut during the day and open in the evening. They are probably pollinated by moths. The slender habit and horns distinguish it from *M. incana* and *M. sinuata*. *Malcolmia maritima* appears similar when in flower but has relatively smaller, broader petals and longer pedicels.

Matthiola tricuspidata (L.) R. Br. and *M. fruticulosa* (L.) Maire have been recorded as casuals and are also illustrated. The former differs from M. *longipetala* in having 3 equally long horns at the fruit apex, and the latter is a perennial with stigma horns 0-3 mm and narrow petals. There are only a few records of each.

Matthiola longipetala 80
***M. tricuspidata**
#M. fruticulosa

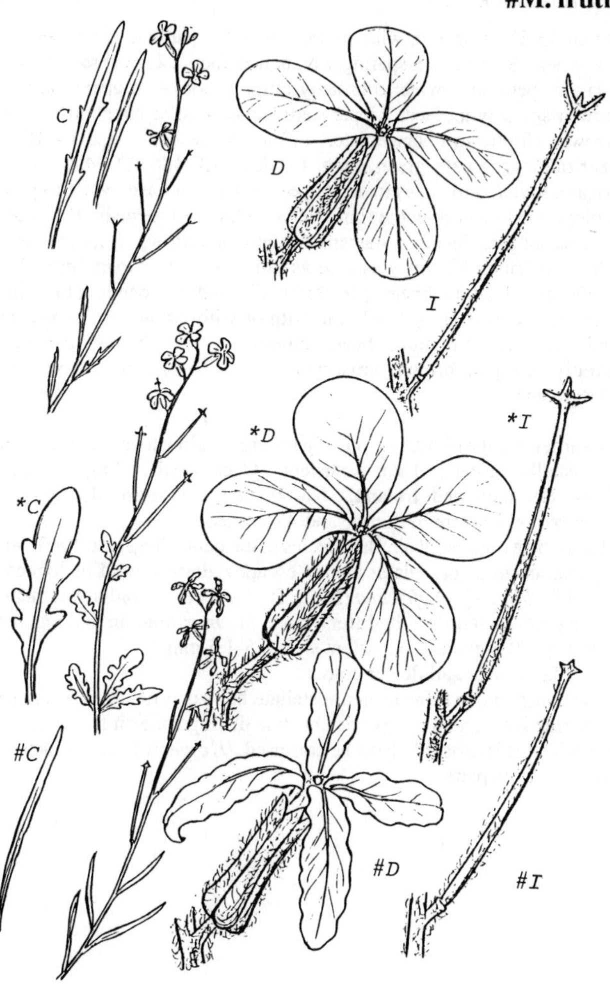

Cx0.3;Dx2;Ix1(all taxa).

81. **Malcolmia maritima** (L.) R. Br.

Virginia Stock

Annual 15-50(-100) cm, with simple, forked or stellate appressed hairs, grey-green. Stems erect, sparingly branched above. Lower stem leaves to 5(-7) cm, petiolate, ovate to broadly elliptic, obtuse at apex; margins ± entire or sparsely toothed. Upper stem leaves similar but shortly stalked, narrowly elliptic to oblanceolate. Inflorescence lax. Sepals 4-10 mm, linear-oblong, saccate, green, erect. Petals (9-)10-18(-21) x 4-9 mm; limb obovate, rounded to emarginate at apex, cuneate at base, white to pink or purple; claw longer than the limb, linear, white or greenish. Petals about twice as long as sepals. Stamens 6; anthers yellow. Stigma ± entire. Pedicels in fruit 3-12 mm, ± as wide as fruits, ascending to inclined. Fruits (15-)20-70 x 1-2 mm, linear, ± terete (mediseptate), ascending to inclined, dehiscent. Valves (13-)20-67 mm, with or without inconspicuous veins. Persistent style 2-6 mm. Seeds numerous, 1.5-2.3 mm, cylindrical, partially winged, brown, uniseriate. Flowering May to November (-December).

A garden escape of waste ground, paths, tips, rubbish dumps, pavements, etc., usually casual and not persistent. Occasional in England, rare in Wales, Scotland and Ireland. Native in the eastern Mediterranean, naturalized elsewhere in Europe and Australasia.

There are a number of similar species in the genus (e.g. Ball 1963) which may also occur as casuals or garden escapes; these should be looked for critically. *Malcolmia africana* (L.) R. Br. has been recorded but there are no post-1950 records. It differs from *M. maritima* in having petals (5-)8-10(-12) mm and a persistent style 0.5-1.5 mm.

Variable in size and flower colour.

The conspicuous flowers, dense stellate hairs and linear, tapering fruits are distinctive. The linear persistent style distinguishes it from *Matthiola* with which it is most likely to be confused. *Hesperis* is very much more robust in all its parts.

Malcolmia maritima 81

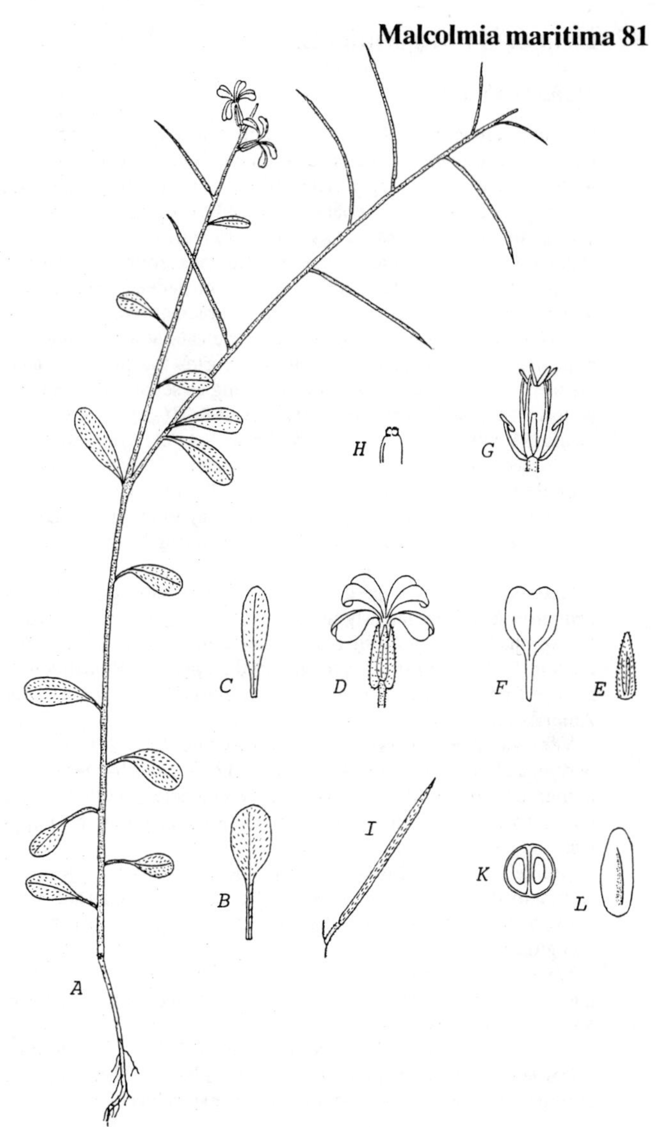

A-Cx0.4;D-Gx1;Hx10;Ix0.7;Kx8;Lx7.

82. Hesperis matronalis L.

Dame's Violet

Perennial (?biennial) 30-100(-150) cm, nearly glabrous or with simple, branched and glandular hairs, dark green. Stems erect, branched above. Rosette leaves to 30 cm, petiolate, elliptic, acute to obtuse at apex, cuneate at base; margins finely to coarsely toothed. Lower stem leaves to 15 cm, similar to rosette leaves. Upper stem leaves smaller, sessile or shortly stalked, lanceolate, acute to acuminate at apex, rounded at base; margins usually finely toothed. Inflorescence crowded. Sepals 7-10 mm, oblong-ovate, saccate, ± awned, white to green or red, erect. Petals 17-29 x 7-14 mm; limb obovate, rounded to truncate at apex, cuneate at base, pink, purple or white; claw about as long as the limb, distinct, linear, greenish-white. Petals *c.* 2-3 times as long as sepals. Stamens 6; anthers yellow. Stigma 2-lobed. Pedicels in fruit 8-21(-30) mm, stout, inclined. Fruits (35-)50-115 x *c.* 1.5-2.5 mm, linear, ± terete to slightly compressed (± latiseptate), ascending to patent, dehiscent. Valves (34-)48-114 mm, irregularly torulose, central vein strong. Persistent style 1-2 mm, often bilobed. Seeds numerous, 2.5-3.9 mm, cylindrical, partially winged, brown, uniseriate. 2n=24(?26,28,32). Flowering May to August.

Extensively naturalized on roadsides, waste ground, woodland, hedgerows, and particularly in damp places beside streams and rivers, in sun or shade. Throughout Britain and Ireland (commoner now than in Perring & Walters 1962), rare only in the uplands. Probably native in S. Europe from Spain to Italy but widely introduced elsewhere in Europe, N. America and Australasia.

Very variable in flower colour from white through to deep pink and purple, and in hair quantity and quality. The species has been divided into numerous infraspecific taxa using these characters, but the variability of the cultivated forms in Britain has made it not possible to apply these classifications with any confidence.

Unlikely to be confused with any other crucifers though in fruit looks similar to *Sisymbrium strictissimum*. The tall, leafy stems and conspicuous flowers later replaced by the long, thin, torulose fruits are readily recognisable.

Widely grown in gardens since at least the 16th century. Double forms are also known (Hodgkin 1971). The vernacular name is a corruption of Violette de Damas (Damascus).

Hesperis laciniata All. can be distinguished from *H. matronalis* by the lower leaves which are pinnatifid near their bases. There are 2 post-1950 records but it can scarcely be regarded as naturalised.

Hesperis matronalis 82

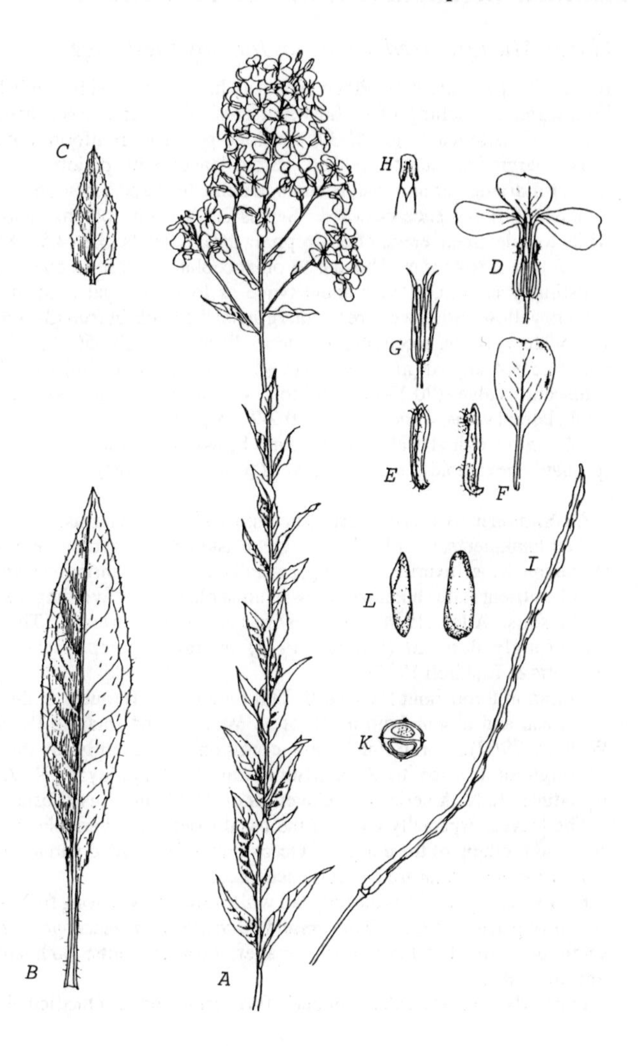

Ax0.2;B,Cx0.3;D-Gx1;Hx10;Ix1;Kx3;Lx4.

83. **Alliaria petiolata** (Bieb.) Cavara & Grande

Garlic Mustard, Hedge Garlic, Jack-by-the-hedge

Biennial (? perennial) 30-120 cm, sparsely hairy below with simple hairs, bright green, smelling of garlic when crushed. Stems erect, branched above. Rosette leaves to 25(-30) cm, long petiolate, reniform, obtuse at apex; margins sinuate to toothed. Stem leaves with petioles *c.* 1 cm, reniform, triangular or ovate, acute to acuminate at apex; margins acutely toothed. Inflorescence crowded. Sepals 2.5-4.7 mm, oblong, ± awned, white to pale green, erect, often dropping early. Petals (3.8-)4.5-7.8(-8.3) x (1.7-)1.9-3.8(-4) mm, obovate to oblanceolate, obtuse at apex, white, indistinctly clawed. Petals about twice as long as sepals. Stamens 6; anthers yellow. Stigma entire to emarginate. Pedicels in fruit (3-)4-9 mm, ± as wide as fruits, ascending to patent. Fruits (20-)30-65(-75) x 1.2-2.2 mm, linear, 4-angled (mediseptate to ± latiseptate), ascending to inclined, dehiscent. Valves (20-)30-64(-74) mm, with a strong central vein and 0-2 weak, lateral veins. Persistent style 0.2-2 mm, indistinct. Seeds numerous, 2.4-4.5 mm, cylindrical, pale to dark brown, uniseriate. 2n=36*,42* (probably hexaploid). Flowering April to June (-August).

A common native spring herb of hedgerows, wood margins, road verges and embankments, river banks, shingle, waste ground, gardens, etc., often abundant. Most luxuriant in damp, partially shaded localities on soils with a high nutrient status but it will grow in most places only avoiding the more acidic soils. Allen (1984) describes it as calcicole in Ireland. The seeds are strongly dormant (Lhotská 1975) though apparently short-lived (Roberts & Boddrell 1983).

Common throughout lowland Britain though less frequent in the north and west, and absent from much of N. W. Scotland (map in Perring & Walters 1962). In Ireland, less common and mainly in the east. Throughout Europe to Asia Minor and the Himalayas, N. Africa. Introduced to N. America (see Cavers *et al.* 1979) and Australasia.

The leaves, especially those of the basal rosette, are variable in shape, size and toothing of the margins. Occasional short-fruited variants occur. No infraspecific taxa are currently accepted.

Easily distinguished as the only tall white crucifer with long fruits which smells of garlic. *Thlaspi alliaceum* and *Pachyphragma macrophyllum* also smell of garlic but have much smaller, broader fruits; both are rare introductions.

Formerly used as a salad, pot herb, flavouring and as a medicinal plant.

Alliaria petiolata 83

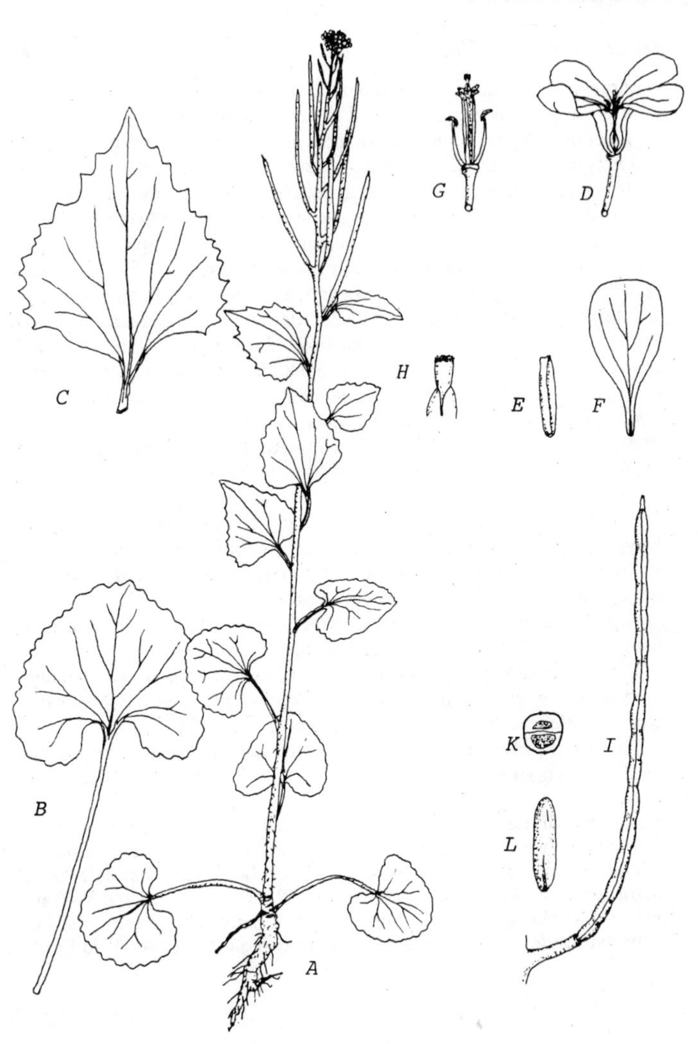

Ax0.2;B,Cx0.3;D-Gx4;Hx10;Ix1.2;Kx3;Lx4.

84. Lunaria annua L.

Honesty, Money Flower Map 84

Biennial (rarely annual or perennial) 20-150 cm, with dense, simple hairs, dark green. Stems erect, branched above. Rosette leaves to 25 cm, with long petioles, ovate, acuminate at apex, cordate at base; margins coarsely and irregularly toothed. Lower stem leaves similar, often opposite. Upper stem leaves smaller, sessile or with petiole to 7(-10) mm. Inflorescence crowded. Sepals 7-10 mm, oblong, saccate, green to purple or white, erect. Petals 15-24 x 5.5-12 mm; limb obovate to broadly elliptic, rounded at apex, reddish-purple, rarely white; claw long, distinct, greenish. Petals about twice as long as sepals. Stamens 6; anthers yellow to purple. Stigma ± 2-lobed. Pedicels in fruit (5-)9-25 mm, slender, ascending to patent. Fruits 25-50(-75) x 15-32(-35) mm (excluding stipe), elliptic to broadly ovate, rounded to truncate at base and apex, flattened (latiseptate), stipitate (the stipe (4-)5-17 mm), ascending to patent, dehiscent. Valves 21-47(-65) mm, without veins. Persistent style (3-)4-10(-12) mm, linear. Seeds (0-)1-3(-4) per loculus, 5-10 mm, reniform, flattened, broadly winged, dark brown, biseriate. 2n=28. Flowering April to June (-August).

A garden escape on waste ground, banks, roadsides, hedgerows and waysides, etc., frequent in Britain near habitation, rarer in Ireland. Probably native in S. E. Europe but widely cultivated and naturalized throughout Europe, N. America and Australasia.

Slightly variable in fruit size and shape, and in flower colour. The plants naturalized in Britain and Ireland are subsp. *annua*.

The large, flat fruits are distinctive and *Lunaria* species are unlikely to be confused with any other genus. *L. rediviva* differs in the longer fruits with cuneate to subacute ends of the valves.

Cultivated in Britain for over 400 years, the silvery septa being much used for decoration.

85. Lunaria rediviva L. has been rarely recorded as an escape from cultivation in England and Wales (there are 2 post-1950 records). It is similar to *L. annua*: Perennial to 150 cm; leaves lanceolate, with spinulate-dentate margins; upper leaves with petioles (5-)10 mm or more; stipe 7-45 mm; valves (40-)45-80 x 15-35 mm, cuneate to subacute at apex and base; persistent style 1-7 mm. It is native in calcareous beechwoods in Europe.

Lunaria annua 84
***L. rediviva 85**

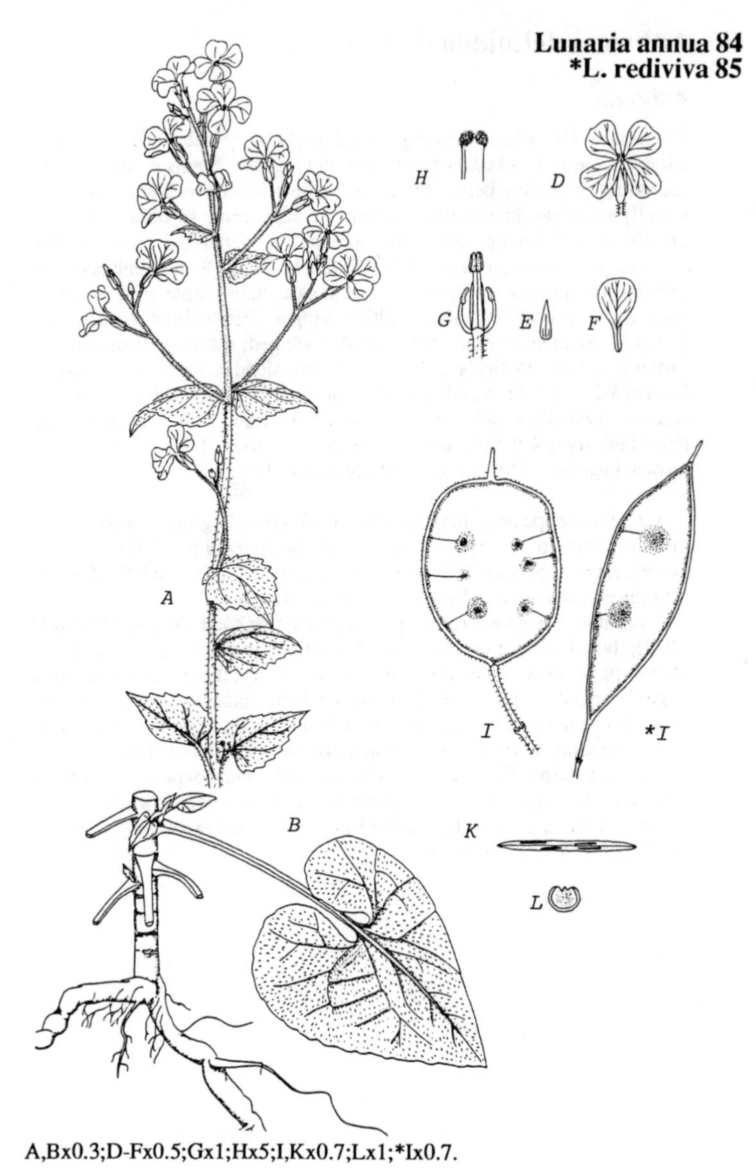

A,Bx0.3;D-Fx0.5;Gx1;Hx5;I,Kx0.7;Lx1;*Ix0.7.

86. Aubrieta deltoidea (L.) DC.

Aubretia

Perennial 7-30(-50) cm, forming mats, densely hairy with stellate hairs and a few simple and forked hairs, green to grey-green. Stems procumbent to ascending, branched below the leaves. Stem leaves to 3 cm, sessile or broadly petiolate, kite-shaped, cuneate at base, acute to obtuse at apex; margins with 1-3 acute teeth. Inflorescence lax. Sepals 6-11 mm, oblong, saccate, green to purple or red, erect. Petals 12-28 x 4-8 mm; limb obovate, truncate to emarginate at apex, cuneate at base, blue-purple, pink or white; claw long, linear-oblanceolate, whitish, winged. Petals about twice as long as sepals. Stamens 6; filaments toothed or winged; anthers yellow. Stigma capitate, entire. Pedicels in fruit 6-15 mm, slender, erect to ascending. Fruits 13-23 x 2.5-4 mm, ellipsoid (excluding style), ± terete (mediseptate), erect to ascending, dehiscent. Valves 5-15 mm, with 1 central vein. Persistent style 4-9 mm, linear. Seeds numerous, 1.2-1.6 mm, ovoid, brown, biseriate. Flowering mainly March to July.

A garden escape occasionally naturalized on sandy ground, walls, waste ground, paths, etc. always near habitation. Scattered through England and Wales, rarer in Scotland. Native in C. Europe, Balkans and W. Turkey, introduced elsewhere in Europe and N. America.

A variable species with many geographical races and varieties (Mattfeld 1939); these have not been examined in British material. The description above applies only to material seen naturalized, and there is more variation in garden plants and in the wild. It is possible that other species of *Aubrieta* may also be naturalized; it is distinguished from them by the presence of long, spreading hairs on the fruits in addition to the stellate hairs.

The blue-purple flowers are distinctive and should separate it from all other crucifers. Even when the flowers are pink or white, or when they are absent, there are no other sprawling, mat-forming crucifers with a persistent style 4-9 mm in fruit.

Aubrieta deltoidea 86

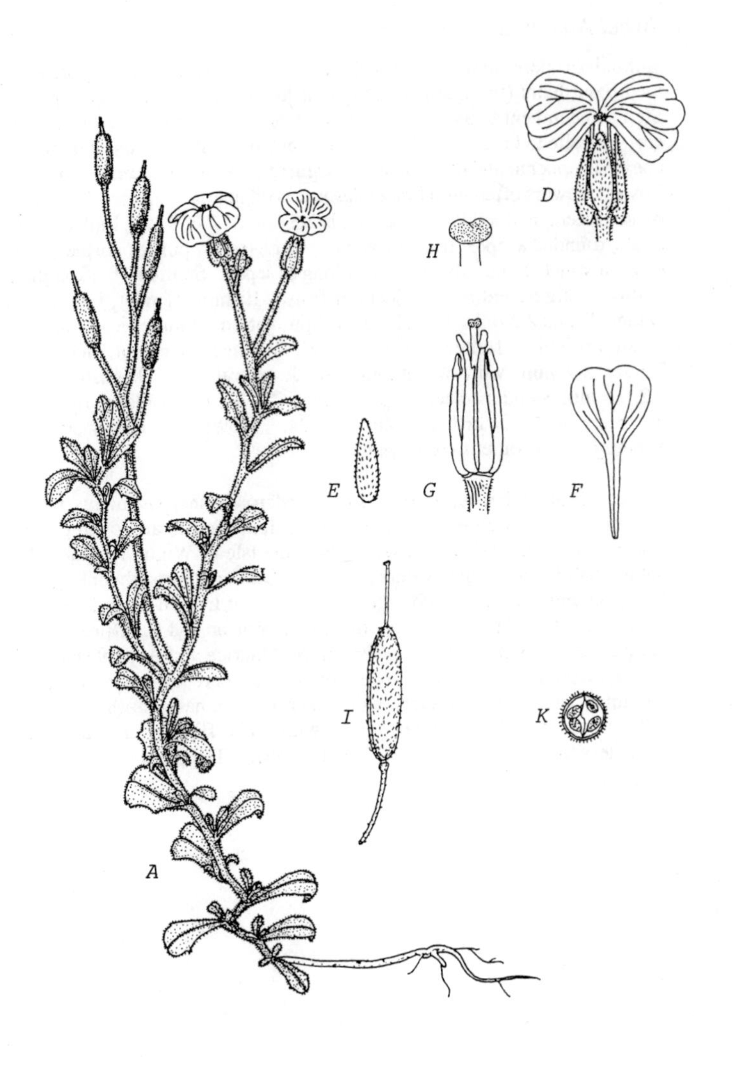

Ax0.5;D-Gx1.5;Hx10;I,Kx2.

87. Lobularia maritima (L.) Desv.

Sweet Alison, Sea Alyssum

Annual or perennial 10-30(-40) cm, densely hairy with appressed, medifixed hairs (these appear simple at low magnification), grey-green. Stems decumbent to ascending at base, branched below. Leaves to 4(-5) cm, ± sessile to broadly petiolate, linear-oblanceolate, the lower obtuse at apex, the upper acute, often fleshy; margins entire. Inflorescence crowded. Lowest flowers often with bracteoles and no stamens. Sepals 1.4-2.5 mm, ovate, green, inclined to patent. Petals 2.3-4.4 x 1.7-3.4 mm; limb broadly ovate, rounded at apex, truncate at base, white (rarely pink or purple); claw short, distinct. Petals about twice as long as sepals. Stamens (0-)6; anthers yellow. Stigma entire. Pedicels in fruit 4-10 mm, slender, inclined to patent. Fruits 2.2-4.2 x 1.4-2.3 mm, elliptic (often asymmetrical), obovate or suborbicular, flattened (latiseptate), inclined to patent, dehiscent. Valves 2-4 mm, with or without a weak central vein. Persistent style 0.2-0.6 mm, slender. Seeds 1 per loculus, 1.2-1.8 mm, oblong, flattened, narrowly winged, brown. 2n=(?22)24. Flowering mainly May to September but sometimes all year.

A casual or garden escape of waste ground, roadsides, pavements, banks, etc., in towns and cities and also on sand dunes, cliffs and open ground near the sea. Naturalized on sea cliffs in the Isle of Wight, typical of the natural habitat in which it occurs in Europe. Scattered throughout England, Scotland and Wales, less commonly recorded in Ireland (map in Perring & Walters 1962). Native around the Mediterranean and N. Africa, widely naturalized elsewhere in Europe and in N. America and Australasia.

Very variable in size and flower colour, and there are many cultivated variants. These revert towards wild material when naturalized.

There are no other crucifer genera with white flowers and latiseptate fruits less than 5 mm long with 1-seeded loculi. The plant should, in any case, be familiar from gardens.

Cultivated widely for many years.

Lobularia maritima 87

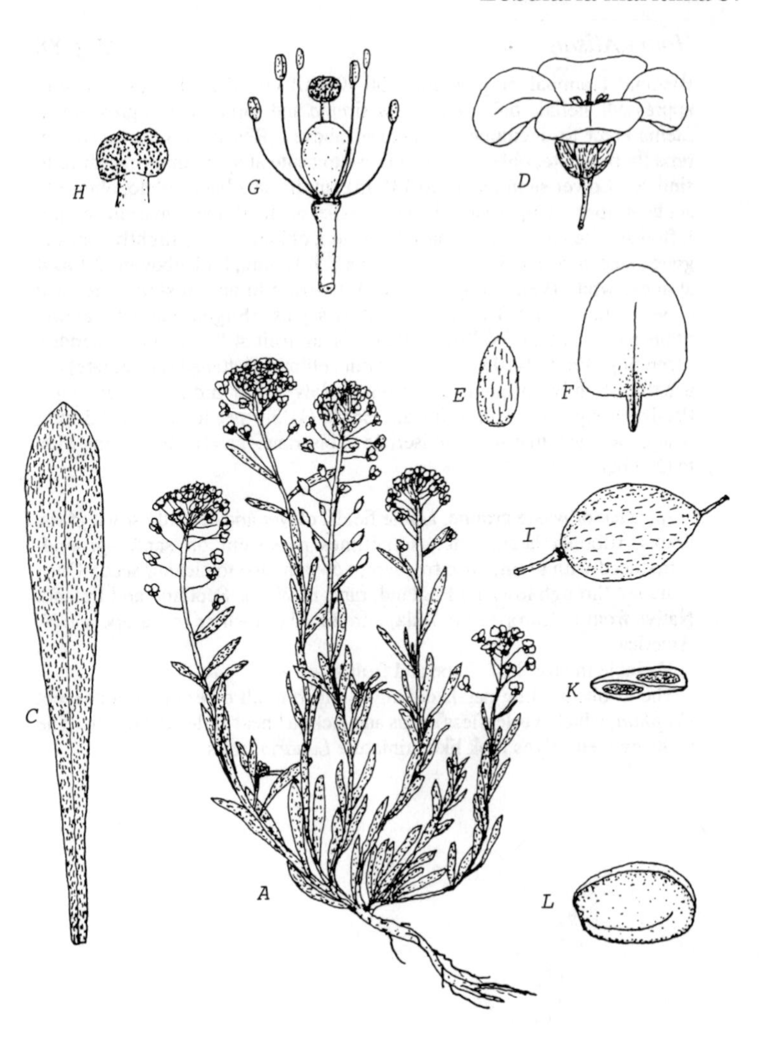

Ax0.5;Cx3;D-Gx8;Hx16;I,Kx10;Lx15.

88. **Berteroa incana** (L.) DC.

Hoary Alison Map 88

Biennial (?annual or perennial) 30-70(-90) cm, densely covered with appressed, stellate hairs and a few simple and forked hairs, grey-green. Stems erect to decumbent, branched above. Rosette leaves to 10 cm, broadly petiolate, oblanceolate, obtuse to acute at apex; margins entire to sinuate. Lower stem leaves to 13(-15) cm, linear-oblanceolate, obtuse to acute at apex. Upper stem leaves sessile, acute at apex; margins entire. Inflorescence crowded. Sepals 2.2-4 mm, oblong-ovate, slightly saccate, green, erect to ascending. Petals 5-8.5 x 2.5-3.5 mm; limb obovate, 2-lobed at apex, white (yellowing when dry), tapering to an indistinct, greenish claw. Petals *c.* 1.5-3 times as long as sepals. Stigma capitate, entire. Stamens 6; anthers yellow. Pedicels in fruit 4-9(-11) mm, slender, ascending. Fruits 4-10 x (1.7-)2.5-4 mm, elliptic, flattened (latiseptate) but usually slightly inflated, dehiscent. Valves 3-8 mm, without veins. Persistent style 1-3.5 mm, linear. Seeds (1-)3-6 per loculus, 1-2.3 mm, ovate, flattened, margined, uniseriate to biseriate. 2n=16. Flowering May to October.

A casual of waste ground, arable fields, clover and recently sown grass, docks, rubbish heaps, etc., sometimes persistent on sandy ground. Introduced with grain, once frequent, now rare due to cleaner seed. Once scattered through lowland England, rarer in Wales, Scotland and Ireland. Native from E. Europe to E. Asia, introduced elsewhere in Europe and N. America.

Variable in size, leaf shape and fruit size.

The 2-lobed, white petals distinguish it from all other crucifers except *Erophila*, which has leafless stems and lacks a linear style on the fruit. The fruits by themselves look like miniature *Lunaria* fruits.

Berteroa incana 88

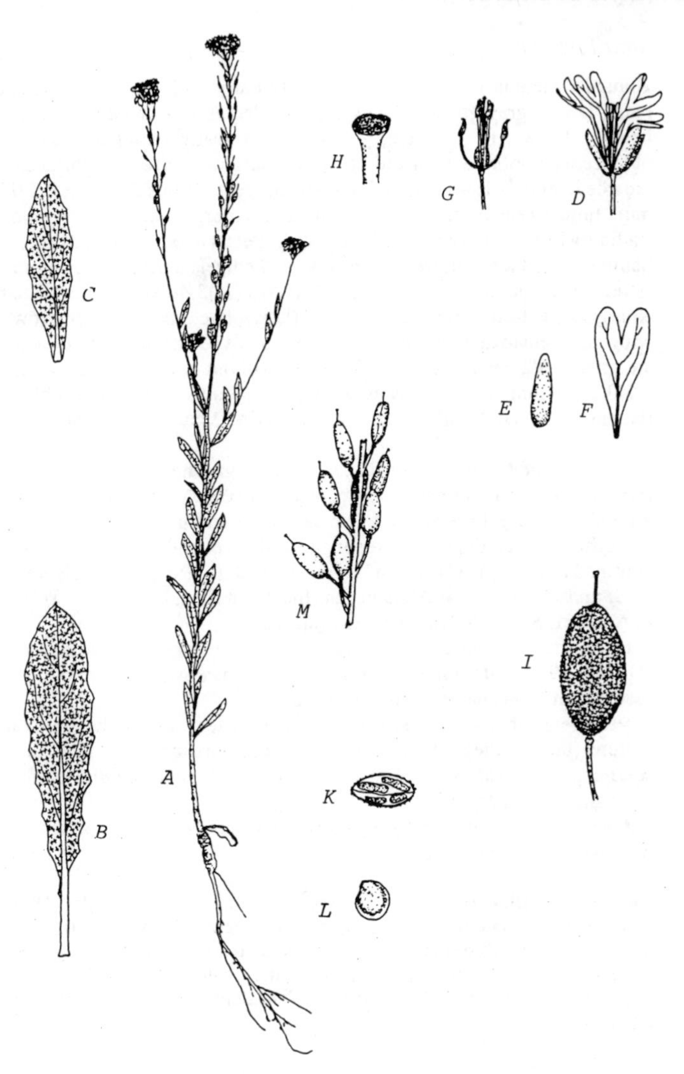

Ax0.2;B,Cx0.5;D-Gx3;Hx10;I,Kx3;Lx3.

89. Alyssum alyssoides (L.) L.

Small Alison

Annual (?biennial) 5-30(-50) cm, densely hairy with appressed, stellate hairs, hoary, grey-green. Stems erect to decumbent at base, branched below. Leaves to 3(-4) cm, ± sessile to broadly petiolate, narrowly oblanceolate, obtuse to acute at apex; margins entire. Inflorescence crowded. Sepals 1.4-3 mm, oblong, green, erect. Petals 2-3(-4) x 0.3-0.7 mm; limb linear-oblanceolate, rounded to emarginate at apex, yellow (fading white), not or indistinctly clawed. Petals about as long as sepals. Stamens 6; anthers yellow. Stigma entire. Pedicels in fruit 2.5-5(-6) mm, slender, inclined to patent. Fruits (2-)3-4(-5) x (2-)3-4(-5) mm, ± orbicular, emarginate to truncate at apex, flattened (latiseptate), the margins narrowly winged, patent to inclined, dehiscent, with sepals persistent whilst the fruits are green. Valves (2-)2.5-3.5(-4.2) mm, without veins. Persistent style 0.2-0.5(-1) mm, linear. Seeds 2 in each loculus, 1.5-2 mm, oblong, flattened, winged, biseriate. 2n=32. Flowering May to September.

A casual of fields and arable land, waste ground, docks and tracks. Introduced with grain and rarely persistent (except sometimes on sandy ground), now rarely seen due to cleaner seeds and herbicides. Formerly not infrequent in England, rarer in Ireland, Scotland and Wales. Now confined to E. England (map in Perring & Walters 1962). Probably native in S. and C. Europe to Afghanistan, the Near East, N. Africa. Widely introduced, N. and S. America, Australasia.

British material varies little except in size. Elsewhere a variable species (Dudley 1965) with a number of infraspecific taxa which have not been assessed in plants naturalized in Britain.

Very few other crucifers have sepals persistent whilst the fruits are mature (though they do drop once the fruits turn brown). *Coronopus squamatus* has leaf-opposed racemes, and rare casual *Lepidium* species have angustiseptate fruits.

Alyssum is a taxonomically very difficult genus. About 12 species have been recorded as very rare casuals.

90. Aurinia saxatilis (L.) Desv. (*Alyssum saxatile* L.; see Dudley 1964), *Golden Alison*, occurs rarely as a garden escape. Perennial 20-50 cm, densely grey-green stellate hairy. Leaves to 10 cm, oblanceolate; margins ± entire. Petals 2.5-6(-8) mm, yellow, emarginate. Fruits (3-)4-8(-12) x (2-)3-8(-15) mm, flattened, latispetate. Persistent style 0.5-2.5 mm. Seeds 2 per loculus.

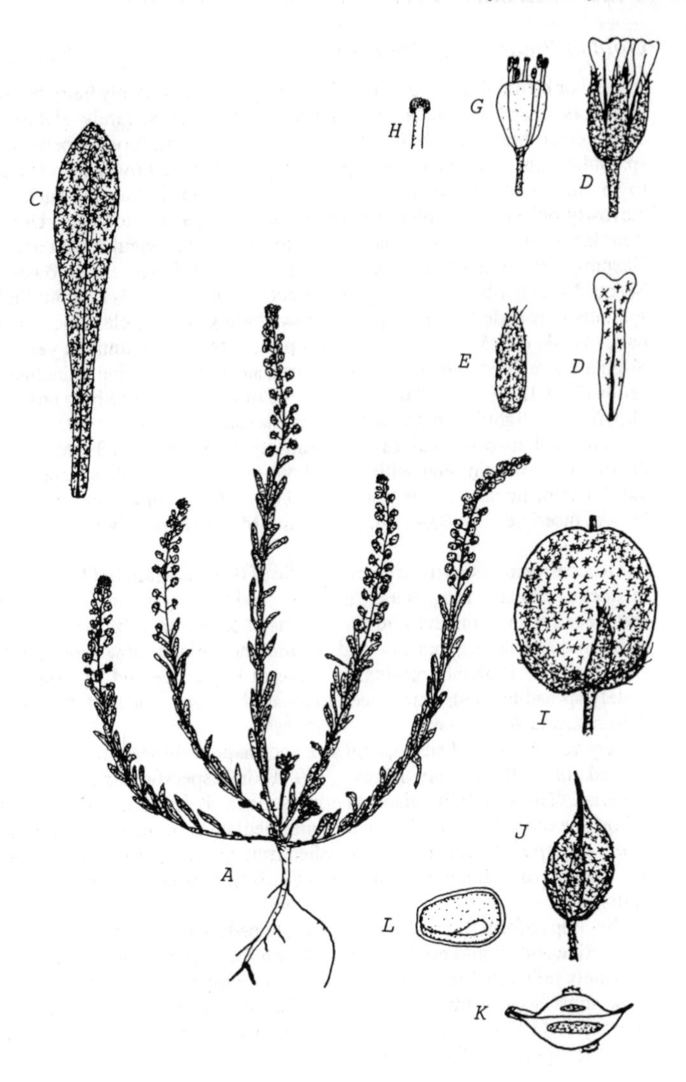

Ax0.3;Cx2;D-Gx10;Hx20;I-Lx8.

91. Camelina sativa (L.) Crantz *sensu stricto*

Gold of Pleasure, False Flax Map 91

Annual or biennial (20-)30-100(-130) cm, sparsely to densely hairy below with forked and stellate hairs and fewer simple hairs, rarely glabrous. Stems erect, branched above. Rosette leaves to 9 cm, broadly petiolate, spathulate, obtuse at apex, rarely persisting to anthesis. Lower stem leaves to 15 cm, broadly petiolate often with small, rounded, clasping auricles, narrowly oblanceolate, obtuse at apex; margins sparsely toothed. Upper stem leaves smaller, sessile, lanceolate to narrowly triangular with acute, clasping auricles, acute at apex; margins ± entire. Inflorescence crowded. Sepals 2-4 mm, oblong-ovate, green, erect. Petals 3.4-6 x 1-1.3 mm; limb spathulate, rounded at apex, pale yellow (fading white); claw indistinct, pale. Petals *c.* 1.5 times as long as sepals. Stamens 6; anthers yellow. Stigma entire. Pedicels in fruit 10-25 mm, slender, ascending to inclined. Fruits (7-)9-13 x (3.6-)4-8 mm, (3-)3.2-4.5 mm thick, narrowly to broadly obpyriform, slightly compressed (latiseptate) and margined, inflated, erect to inclined, dehiscent. Valves (excluding style) (5.5-)6.5-10.3 mm, with a distinct central vein, and with or without lateral veins. Persistent style 1.3-2.5 mm, linear. Seeds numerous, (1.5-)1.6-2.1 mm, ellipsoid, pale brown, biseriate. 2n=40,42,48. Flowering May to November.

An occasional casual most characteristic of flax and grain fields, and also in docks, chicken runs, waste ground, roadsides, gardens, cities, etc. Introduced with grain and bird seed (Hanson & Mason 1985). Once said to be frequent throughout lowland Britain and Ireland, now rarer due to cleaner seed. Probably originally native in E. Europe and W. Asia but widely spread by man. Introduced to most of Europe, Far East, N. and S. America, and Australasia.

Very variable in leaf shape, fruit size and shape, hairiness, etc, and it has proved difficult to apply Mirek's (1981) infraspecific taxa to British material. The variability also sometimes makes it difficult to distinguish it from the other species. *C. sativa* is intermediate between *C. alyssum* and *C. microcarpa*. It is readily distinguished from the latter by the larger seeds and thicker fruits, but less readily from *C. alyssum* which is larger in all its parts.

The obpyriform (i.e. inverted pear-shaped) fruits of *Camelina* are distinctive and should not be confused with fruits of any other genera.

Widely cultivated in prehistoric times as a crop for oil, food and fibre. Now most significant as a weed of flax and grain fields in Europe. Hjelmqvist (1950) gives a fascinating account of the evolution of *Camelina* and other species in flax fields.

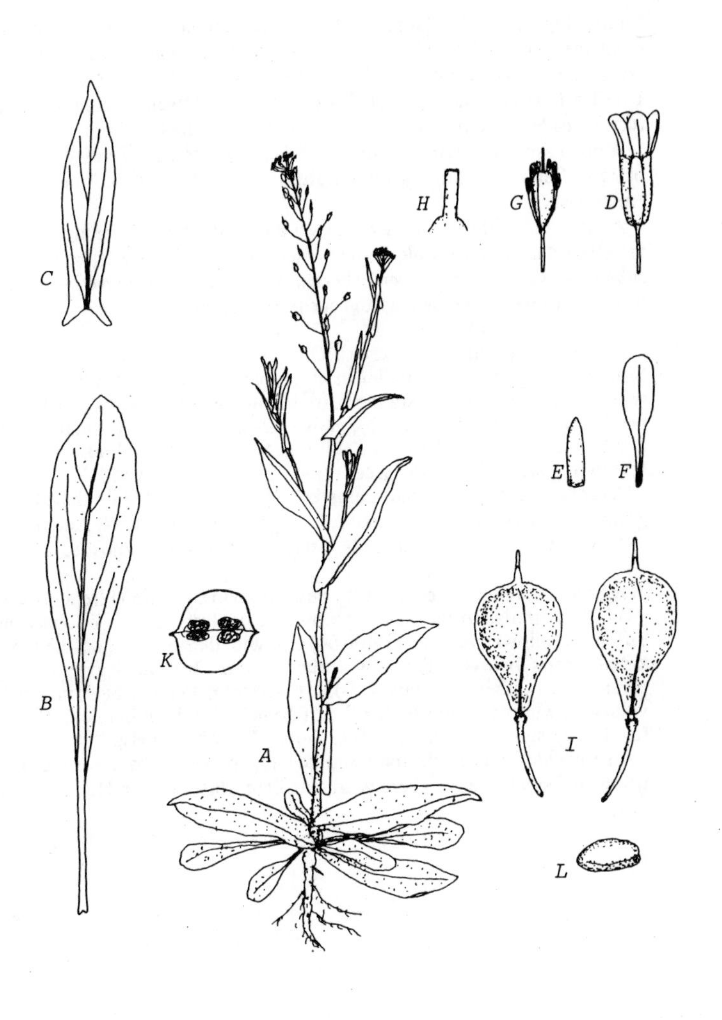

Ax0.2;B,Cx1;D-Gx4;Hx10;I,Kx2.5;Lx6.

92. Camelina alyssum (Miller) Thell. subsp. **integerrima** (Celak) Smejkal

Similar to *C. sativa* but larger in all its parts: Annual 30-100 cm, glabrous or sparsely covered with branched hairs below. Rosette leaves usually not persisting. Sepals 2.5-4.1(-4.2) mm. Petals 4.3-6.7 x 1.3-2.1 mm. Fruits 11-13 x (4.6-)5-7 mm, (3.5-)4-5.5 mm thick. Valves (excluding style) (6.5-)7-10.5(-12) mm. Seeds (1.8-)2.1-2.9 mm. 2n=40.

A rare casual in Britain and Ireland (Map 92), introduced with grain. An obligate flax weed in E. Europe and Russia. Introduced to S. America and Australasia.

A strict view of this species is taken and only plants agreeing with all the characters above should be named as such. Only a few specimens of subsp. *alyssum* have been seen in herbaria; it is distinctive in having deeply toothed or lobed lower leaves and seeds 1.8-2 mm.

93. Camelina microcarpa Andrz. ex DC.

Similar to *C. sativa* but smaller in all its parts. Annual 15-60(-100) cm, with many branched and simple hairs below. Rosette leaves usually not persisting. Sepals 1.9-3.1 mm. Petals (2.2-)2.6-4.6 x 0.9-1.1 mm. Fruits 6-9.5 x (2-)2.5-4.5(-4.8) mm, (1.7-)1.9-3(-3.2) mm thick. Valves (excluding style) 4-7.5 mm. Seeds 0.9-1.3(-1.4) mm. 2n=40.

A rare casual in Britain and Ireland (Map 93), probably introduced with grain. Probably native in C. and E. Europe. Introduced to N. America.

The combination of small seeds, petals and fruits is diagnostic.

94. Camelina rumelica Velen. is a very rare casual in Britain and Ireland (Map 94). It is similar to the other species (particularly *C. microcarpa*) but differs as follows: Annual 15-40(-60) cm, with dense, long (1.5-3.5 mm), simple hairs below and a few forked hairs. Rosette leaves usually persisting to anthesis. Sepals (2.7-)3-4(-4.5) mm. Petals 5-8(-9) mm, pale yellow to whitish. Fruits 6-10 x 3-4.5(-5) mm, 1.7-3.2 mm thick. Valves (excluding style) 5-8 mm. Seeds (1.2-)1.3-1.5 mm. 2n=12.

Useful characters are the long, simple hairs, persistent basal rosette, long petals and small seeds. There is an excellent illustration in Holland *et al.* (1986).

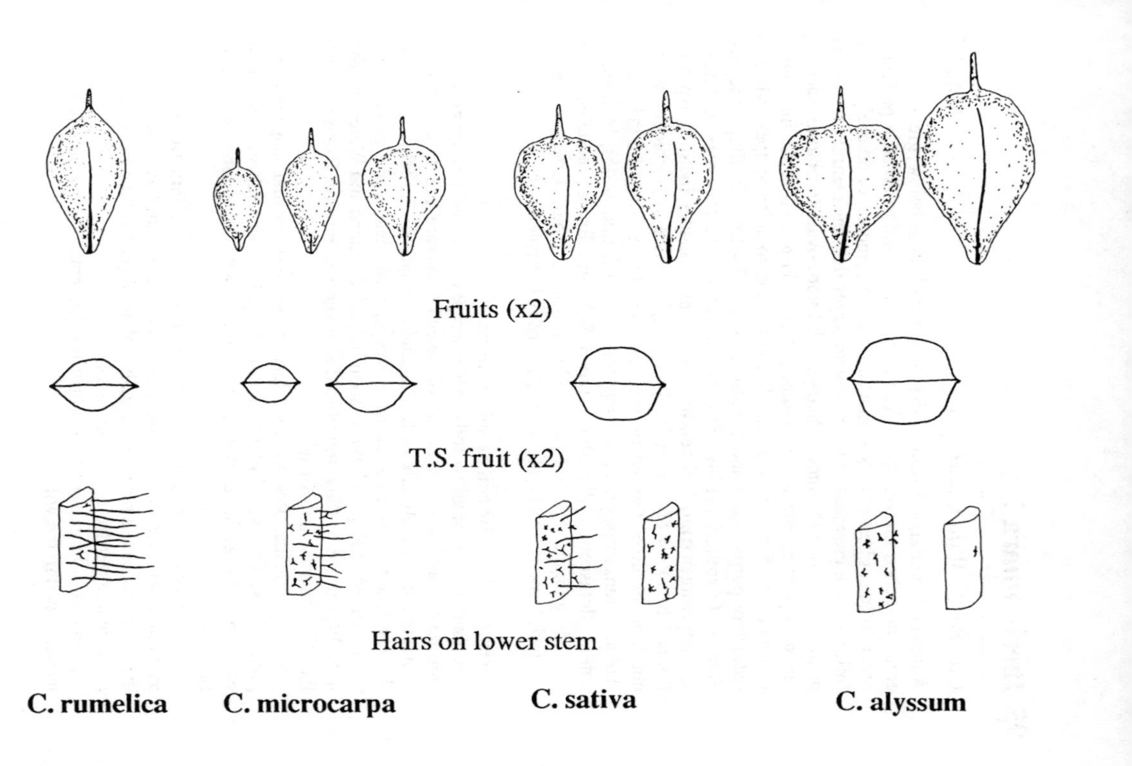
Fruits (x2)
T.S. fruit (x2)
Hairs on lower stem
C. rumelica
C. microcarpa
C. sativa
C. alyssum

95. Iberis amara L.

Candytuft, Wild Candytuft Map 95

Annual (?biennial) 7-35 cm, with sparse, simple hairs below, green. Stems erect, branched above. Leaves to 5(-6) cm, sessile to broadly petiolate, oblanceolate, obtuse at apex; margins ± entire, sinuate, or coarsely lobed. Inflorescence crowded. Outer flowers larger than the central ones, outer petals larger than the inner. Sepals 1-3.2 mm, ovate to oblong, saccate, green to purple, ascending. Outer petals (3-)4-8 x (1-)2-4.5 mm, inner petals (1.5-)2-5 x (0.5-)1-3 mm; limb obovate, rounded at apex, white to reddish or purple; claw short, distinct. Petals 2-4 times as long as sepals. Stamens 6; anthers yellow. Stigma capitate, entire. Racemes contracted or elongating in fruit. Pedicels in fruit 2-10 mm, slender, inclined to patent. Fruits 4-6(-7) x 4-6(-7) mm (excluding style), ovate to ± orbicular with a notch at the apex formed by the acute to obtuse wings of the valves, flattened (angustiseptate, the septum *c.* 0.9-1.2 mm wide), inclined to patent, dehiscent. Persistent style 0.8-2 mm, linear, included in or exceeding notch. Seeds 1 per loculus, 2-3 mm, ovoid, flattened, slightly winged, brown. 2n=14,16. Flowering May to September.

Represented in Britain by subsp. *amara*.

A plant of open, shallow soils, rabbit scrapes and scree in chalk grassland and scrub, and also as an arable weed. A rare plant, particularly characteristic of the chalk of S. England as a native. Probably declining. Widely reported as a casual of railways, roadsides, waste ground, etc, through most of England, rarely so in Ireland, Scotland and Wales. Native in N. W. Europe to Italy, introduced elsewhere in Europe, Russia, the Far East, S. America and Australasia.

Variable in size, flower colour, leaf shape, fruit size and shape between populations. Cultivated plants (*I. coronaria* hort. and cultivars) are more variable and may have fruits to 8 mm long with acute lobes and larger flowers.

Iberis species are distinguished from other crucifers by the large, asymmetrical flowers. *I. amara* is often confused with *I. umbellata* which has very contracted racemes, usually coloured (rather than usually white) petals and larger fruits with acute lobes.

I. amara is largely self-incompatible (Bateman 1954) but fruit set is usually good in the wild.

Iberis amara 95

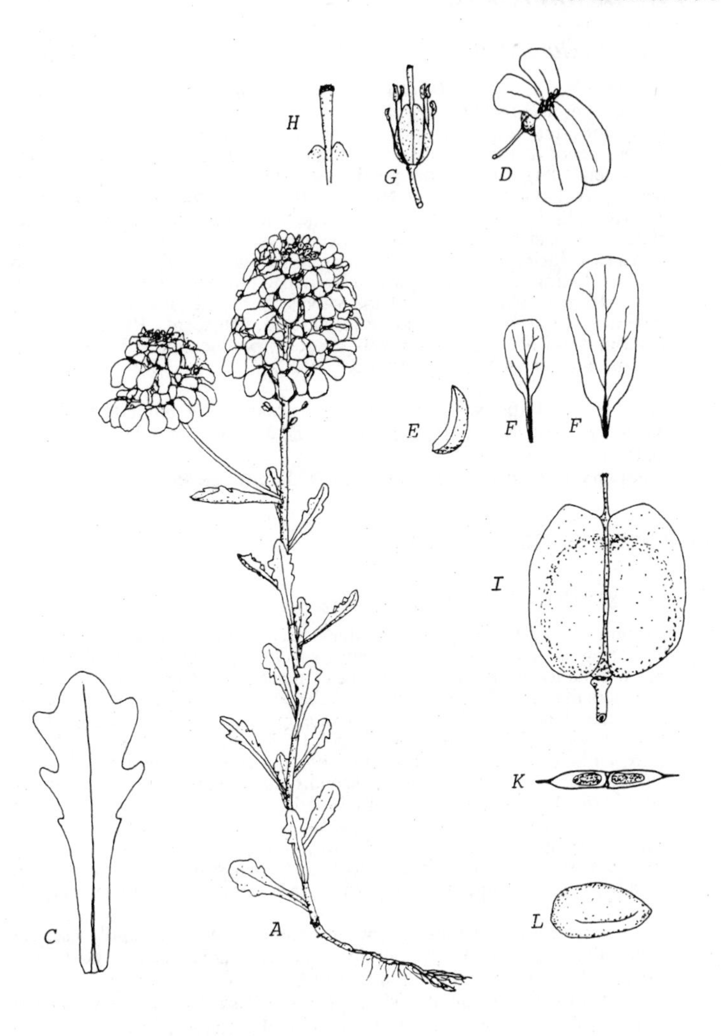

Ax0.3;Cx1;Dx2.5;E-Gx4;Hx10;I,Kx6;Lx8.

96. Iberis umbellata L.

Garden Candytuft Map 96

Annual (?biennial) 10-30(-60) cm, glabrous, dark green. Stems erect, without non-flowering rosettes. Leaves to 5 cm, sessile, linear-oblanceolate, acute at apex; margins entire or coarsely toothed. Inflorescence dense. Outer flowers larger than the central flowers and the outer petals larger than inner petals. Sepals 3-5.5 mm, obovate, green to purple, erect to ascending. Outer petals 10-16 x 5-7 mm, inner petals 6-10 x 2-5 mm; limb obovate, truncate to rounded at apex, white, pink or purple; claw distinct, short, linear, paler. Racemes remaining densely contracted in fruit. Pedicels 4-12 mm, ascending to inclined. Fruits (6-)7-10 x 5-8 mm (excluding style), ovate, with a V-shaped notch at the apex formed by the prominent, acute wings of the valves, flattened (angustiseptate, the septum *c.* 1mm wide), ascending to inclined, dehiscent. Persistent style 2-4.5 mm, linear, exceeding notch. Seeds 1 per loculus, 2-3 mm, ovoid, brown. Flowering May to October.

A casual of tips, waste ground, etc., usually near habitation but not persisting. British material is a cultivar and differs from wild S. European material.

For characters distinguishing it from *I. amara*, see **95**.

97. Iberis sempervirens L.

Evergreen Candytuft

Perennial 10-20 cm, glabrous, dark green, bushy, with decumbent flowering stems and non-flowering rosettes. Stem leaves to 5 cm, linear-oblanceolate, obtuse at apex; margins entire. Inflorescence crowded, the outer flowers larger than the central ones and the outer petals larger than inner. Sepals 2-4.5 mm, ovate, green, erect to ascending. Outer petals 5-13 x 2-7 mm, inner petals 3-6.1 x 1-4 mm, obovate, rounded at apex, white to purplish, with a short, distinct claw. Racemes elongating little in fruit. Pedicels 5-11 mm, ascending to patent. Fruits often poorly set, 6-8 (excluding style) x 5-6 mm, broadly ovate, with a V-shaped notch at apex formed by the prominent, acute wings of the valves, flattened (angustiseptate, the septum *c.* 1 mm), inclined to patent, dehiscent. Persistent style 1-3 mm, linear, prominent beyond the wing. Seeds 1 in each loculus, *c.* 3 mm, ovoid, brown. 2n=22. Flowering April to July.

I. sempervirens is widely grown in gardens and occasionally escapes onto walls, banks, waste ground, tips, etc. There are at least 10 post-1950 records. It is native in the mountains of S. Europe and Turkey.

It is distinguished from the other species of *Iberis* by the perennial habit with non-flowering rosettes. Other perennial species of *Iberis* are grown in gardens and may also escape.

Iberis umbellata 96

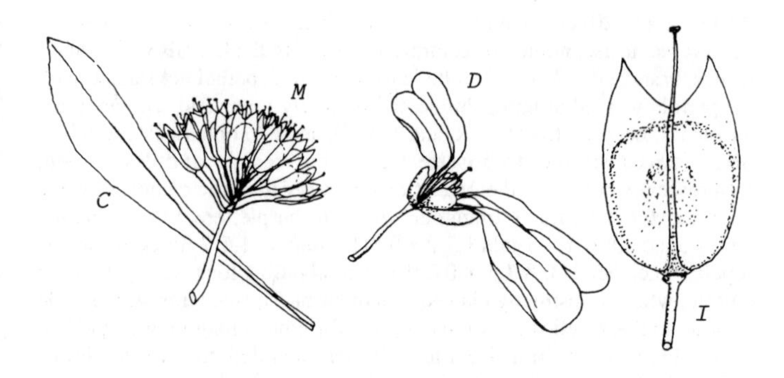

Cx2;Dx3;Ix4;M=inflorescence in fruit x0.7.

Iberis sempervirens 97

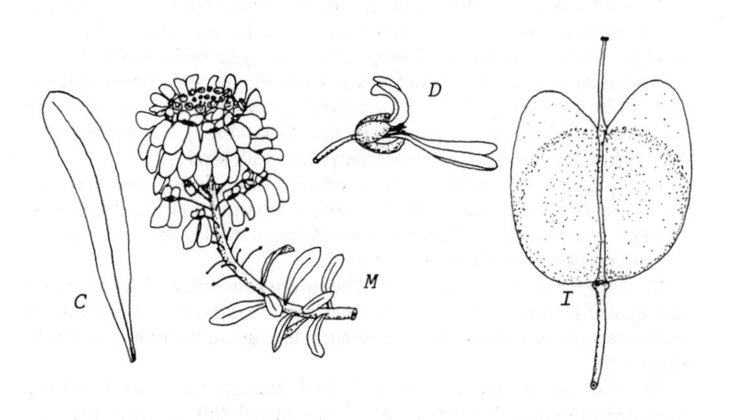

Cx2;Dx3;Ix4;M=inflorescence, x0.7.

98. Teesdalia nudicaulis (L.) R.Br.

Shepherd's Cress

Annual 2-15(-20) cm, glabrous or sparsely hairy below with simple hairs. Stems erect to ascending, the central stem leafless, the laterals with 1-4(-5) leaves, branched at base. Rosette leaves to 5 cm, spathulate, stalked with the petiole winged at base, simple or irregularly pinnatifid with an ovate terminal lobe, rounded at apex, and 1-3(-4), acute to obtuse, lateral lobes. Stem leaves similar to rosette leaves but sessile and less lobed. Inflorescence crowded, the outer flowers larger than the central flowers. Sepals *c.* 0.6-1.1 mm, ± triangular, green to purple, erect to ascending. Petals unequal: outer petals 1.5-2 x 0.8-1.1 mm, *c.* 1.5-2 times as long as sepals; inner petals 0.8-1.5 x 0.6-0.7 mm, about as long as sepals; limb elliptic, white, indistinctly clawed. Stamens 6(-5); filaments with a scale at base; anthers yellow. Stigma entire. Racemes elongating slightly in fruit. Pedicels in fruit 2-5 mm, slender, inclined to patent. Fruits (2.5-)3-4(-4.5) x 2.2-3.4(-4) mm; ± obovate to ± orbicular, notched to emarginate at apex, ± concave, flattened (angustiseptate, the septum *c.* 0.7-1.2 mm wide), narrowly winged, inclined to patent, dehiscent. Valves (2.5-)3-4(-4.5) mm, with a strong marginal vein. Persistent style 0-0.3 mm, ± equalling or shorter than notch. Seeds usually 2 per loculus, 0.9-1.7 mm, ovoid, yellow-brown, uniseriate. 2n=36*. Flowering March to June (-July).

A very local native of heathland, sandy ground, banks, dunes, shingle and gravel, mountain screes and gravels, coal bings and cinder tips. Populations can fluctuate markedly from year to year as the plant has little in the way of a seed bank. About 90% of the seeds germinate in the autumn and flower and fruit production the following spring is very dependent on favourable weather conditions (Newman 1964, 1965). The plant also occurs rarely as a casual, perhaps transported with sand.

Very locally abundant but widely scattered through England, Scotland and Wales (map in Perring & Walters 1962). Very rare in Ireland (Lambert 1971). Mainly in W. Europe from Portugal to Poland, rare in the Mediterranean. Introduced to N. America.

Not very variable except in size. The tiny asymmetrical flowers should distinguish it from other crucifers. *Iberis* also has asymmetrical flowers but these are much larger. The tiny scale at the base of the filaments is also diagnostic.

Teesdalia coronopifolia (Bergeret) Thell. was reported once from Eigg (specimen in E), and is best regarded as a casual; this is one of the very doubtful Heslop-Harrison records of the 1940s.

Teesdalia nudicaulis 98

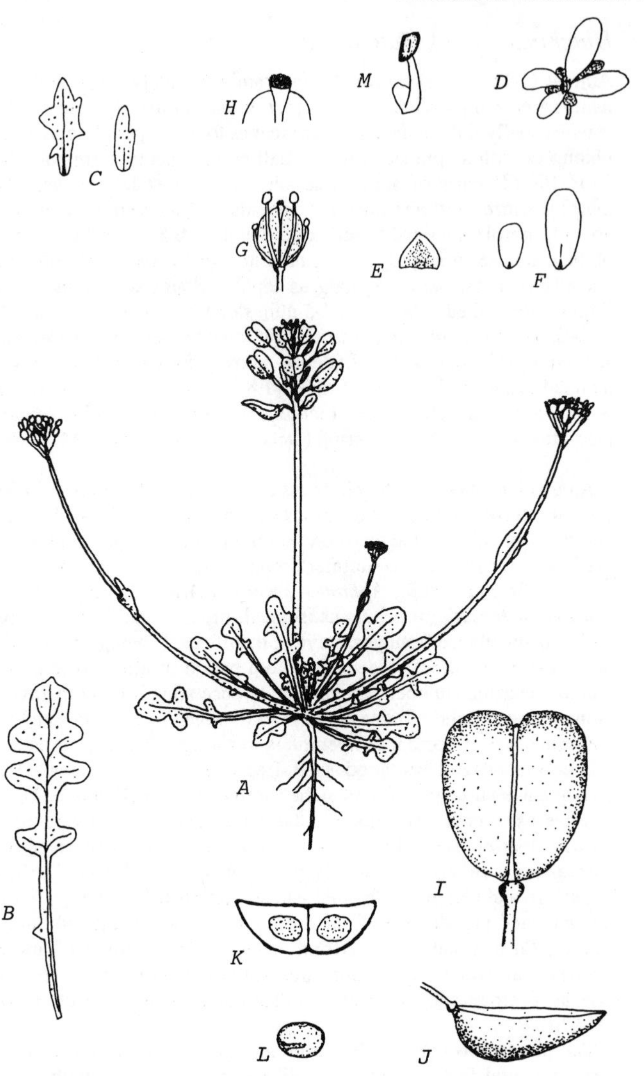

Ax0.5;B,Cx1.3;D-Gx7;Hx10;I-Kx0.8;Lx5.

99. Hornungia petraea (L.) Reichenb.

Hutchinsia, Rock Hutchinsia

Annual (1-)2-10(-15) cm, glabrous (rarely minutely hairy with stellate hairs), green to purplish. Stems erect, branched mainly below. Rosette leaves usually not persisting. Stem leaves to 3 cm, petiolate to ± sessile, oblong in outline, pinnate with a small elliptic, acute, terminal lobe and (2-)3-10(-12) pairs of similar, sessile (rarely ± stalked), lateral lobes; margins entire. Inflorescence lax. Sepals *c.* 0.5-1 mm, oblong, hooded, green to purple, inclined to reflexed. Petals *c.* 0.5-1 x 0.2-0.3 mm; limb oblong, obtuse to slightly emarginate at apex, white or slightly purple, unclawed. Petals about as long as sepals. Stamens 6; anthers yellow. Stigma entire. Pedicels in fruit 2-5 mm, slender, inclined to patent. Fruits (1.8-)2-3 x 1-1.7 mm, ovate to elliptic, compressed (angustiseptate, the septum *c.* 0.7 mm wide), flattened above, convex below, unwinged, inclined to patent, dehiscent. Valves (1.8-)2-3 mm, with veins on margins only. Seeds usually 2 per loculus, 0.6-0.8 mm, ovoid, orange- to pale-brown. 2n=12*. Flowering (December-) February to May (-June).

A rare plant of open, sunny habitats on shallow, carboniferous limestone soils and screes and open calcareous dunes, rarely on old walls and mine debris. A member of the therophyte communities on open soils with high pH (above 6.8) subject to summer drought, associated with other annuals such as *Erophila* spp., *Aphanes arvensis, Arenaria serpyllifolia* and *Cardamine hirsuta.* It occurs as isolated, usually small, populations and varies in abundance from year to year, the life cycle being very dependent on climatic conditions. Seeds require a period of after-ripening for 2-3 months over the summer and germinate once soils become permanently moist in the autumn. Plants overwinter as vegetative rosettes and flower with the onset of warmer temperatures in spring. They are automatically self-pollinated and set good seed freely in most localities. Further ecological details can be found in the Biological Flora (Ratcliffe, D. 1959).

It has a scattered distribution in the British Isles, probably a relict from a wider distribution earlier in the interglacial and it has survived only in specialised habitats. S. W. England, N. and S. Wales, Derbyshire, Yorkshire and Cumbria, Channel Islands (Ratcliffe, D. 1959). Extinct in Scotland and introduced to Ireland but now extinct. Scattered through S. and W. Europe and N. W. Africa, extending eastwards to Turkey and Crimea, and with a disjunct area around the southern Baltic. The general habitat, ecology and distribution is similar in many respects to *Draba muralis*.

The plant shows very little variation other than in size in response to environmental factors, and it is unlikely to be confused with any other British species.

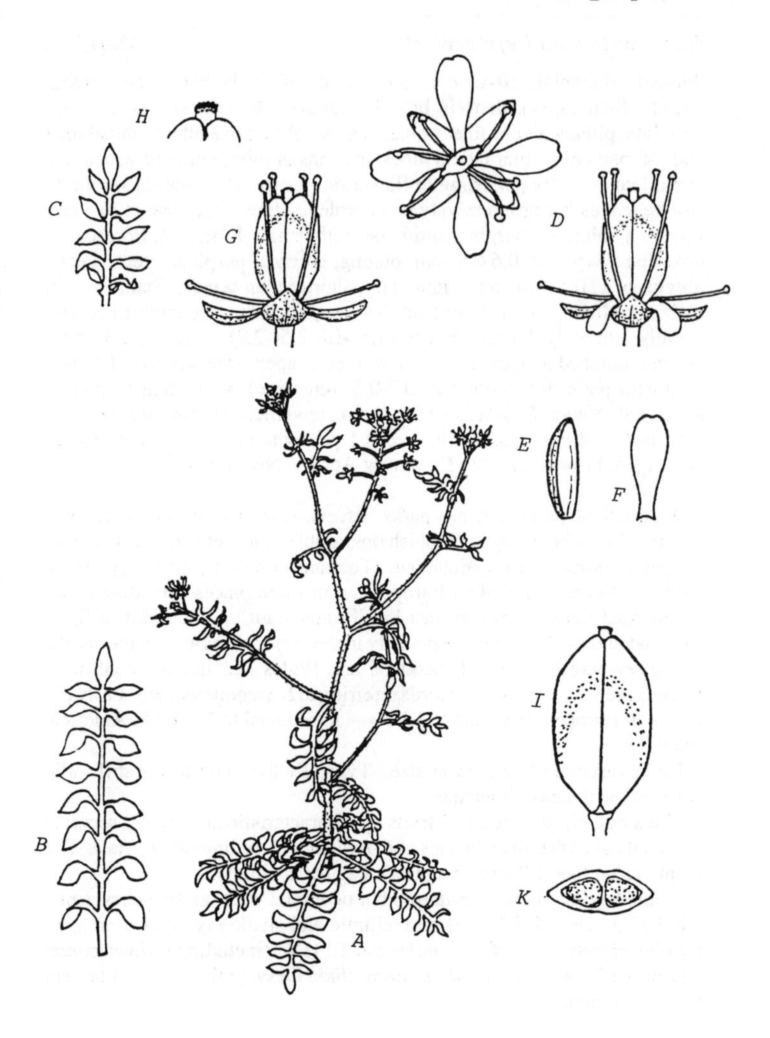

A-Cx2;D-Gx20;Hx30;I,Kx10.

100. Lepidium ruderale L.

Narrow-leaved Pepperwort Map 100

Annual (?biennial) 10-40 cm, glabrous or minutely hairy, dark green, foetid. Stems erect, densely branched above. Rosette leaves to 8 cm, petiolate, pinnate with a linear-oblanceolate, obtuse to acute, terminal lobe and 3-6 pairs of similar lateral lobes; margins of lobes entire or with a few teeth; leaves rarely persisting to flowering. Lower stem leaves similar to rosette leaves but progressively less divided. Upper stem leaves to 3 cm, linear, ± obtuse; margins entire or with small lobes. Inflorescence crowded. Sepals *c.* 0.6-0.9 mm, oblong, green to purplish, erect. Petals absent, or *c.* 0.5 mm and rudimentary, shorter than sepals. Stamens *c.* 2; anthers yellow. Pedicels in fruit 1-4 mm, slender, ascending to patent, usually minutely hairy. Fruits (1.5-)1.8-2.5(-2.7) x 1.5-2(-2.3) mm, elliptic, notched at apex, narrowly winged at apex or unwinged, flattened (angustiseptate, the septum *c.* 0.3-0.6 mm wide), ascending to patent, dehiscent. Valves (1.5-)1.8-2.5(-2.7) mm, unveined. Persistent style *c.* 0.1 mm, included in apical notch. Seeds 1 per loculus, 1-1.6 mm, ellipsoid, unwinged, brown. 2n=32. Flowering April to November.

A casual of roadsides, car parks, streets in towns and cities, rubbish dumps, flower beds, docks, rubbish tips, arable fields, etc., often persistent if open conditions are maintained. Tolerant of salinity (it is apparently native in the salt pans of C. Europe) and in some places spreading along salted road verges. Sometimes locally abundant. Occasional in S. E. England and the Midlands (especially in the larger cities and on the coast), rare elsewhere in England, Scotland and Wales. In Ireland, confirmed records are very rare, most records referring to *L. virginicum*, etc. Probably native in Europe and C. and S. W. Asia, introduced to N. and S. America, India.

Little variable other than in size. Typically the plant forms stiff, little "bushes" with dense branches.

The small elliptic, notched fruits are characteristic and should separate it from most of the other species of *Lepidium* with rudimentary petals. The plant is also foetid (leave in a plastic bag for a while!).

Plants keying out to *L. ruderale* but perennial with lax branches, fruits 1.8-3.2(-3.7) x 1.4-2.1(-2.3) mm, elliptic and shallowly notched at apex may be referable to *L. africanum* (Burm. fil.) DC. (including *L. divaricatum* Aiton, see Jonsell 1975a (=*L. hyssopifolium* Desv.)); these should be sent for determination.

Lepidium ruderale 100

Ax0.2;B,Cx0.3;D-Gx12;Hx20;Ix8;Lx9;M=inflorescence x0.5.

101. **Lepidium densiflorum** Schrader

Least Pepperwort Map 101

Annual or biennial 10-50 cm, glabrous or sparsely hairy. Stems erect, branched above. Basal leaves to 5 cm, petiolate, oblanceolate, acutely lobed to pinnatifid with a large, acute terminal lobe and 2-4 pairs of lateral lobes; margins acutely toothed. Lower stem leaves similar, but less lobed. Upper stem leaves sessile or shortly stalked, linear to linear-oblanceolate; margins entire or acutely toothed. Inflorescence crowded. Sepals 0.6-0.9 mm, oblong, green, erect. Petals 0.3-0.9 mm, absent or reduced, inconspicuous, equalling or shorter than sepals. Stamens 2; anthers yellow. Pedicels in fruit 1.5-3.5(-4) mm, slender, ascending to patent, hairy. Fruits 2-3.5 x 1.7-3 mm, obovate to oblong-obovate, narrowly winged above and notched, flattened (angustiseptate, the septum 0.5-0.6 mm wide), ascending to patent, dehiscent. Valves 2-3.5 mm, without veins. Persistent style *c.* 0.2 mm, included in notch. Seeds 1.2-1.5 mm, obovate, flattened, partially winged, brown. Flowering June-October.

A rare casual of fields, railways, waste ground, tips, docks and roadsides. Probably introduced with grain and bird seed and rarely persistent. Rare in England, Wales, Scotland and Ireland. Native in N. America, introduced widely in Europe and Australasia. Possibly more frequent than *L. virginicum* but less frequent than *L. ruderale*.

A variable plant in fruit shape and size (4 or 5 varieties are recognised in N. America but have not been investigated here). The species is difficult to identify convincingly because each specimen seems different possibly due to predominant self-pollination and the establishment of inbreeding lines. As described here, *L. densiflorum* includes *L. neglectum* Thell. (but see Duvigneaud & Lambinon 1975).

Difficult to distinguish from specimens of *L. virginicum* which have petals reduced or absent. *L. densiflorum* usually has smaller, obovate fruits (usually larger and ± orbicular in *L. virginicum*). The seeds of *L. densiflorum* are typically only partially winged (usually narrowly winged down at least one side in *L. virginicum*). *L. ruderale* has narrower, elliptic fruits with subacute apices to the wings and seeds entirely unwinged.

Very few crucifers (other than *Lepidium* species) lack petals completely. *Coronopus didymus* has leaf-opposed racemes. A number of other genera (*Capsella, Descurainia, Hornungia, Rorippa* and *Subularia*) have petals about equalling sepals but only dwarfed *Capsella* is likely to be confused with *Lepidium*; it has stem leaves with clasping auricles and many-seeded fruits.

Lepidium densiflorum 101

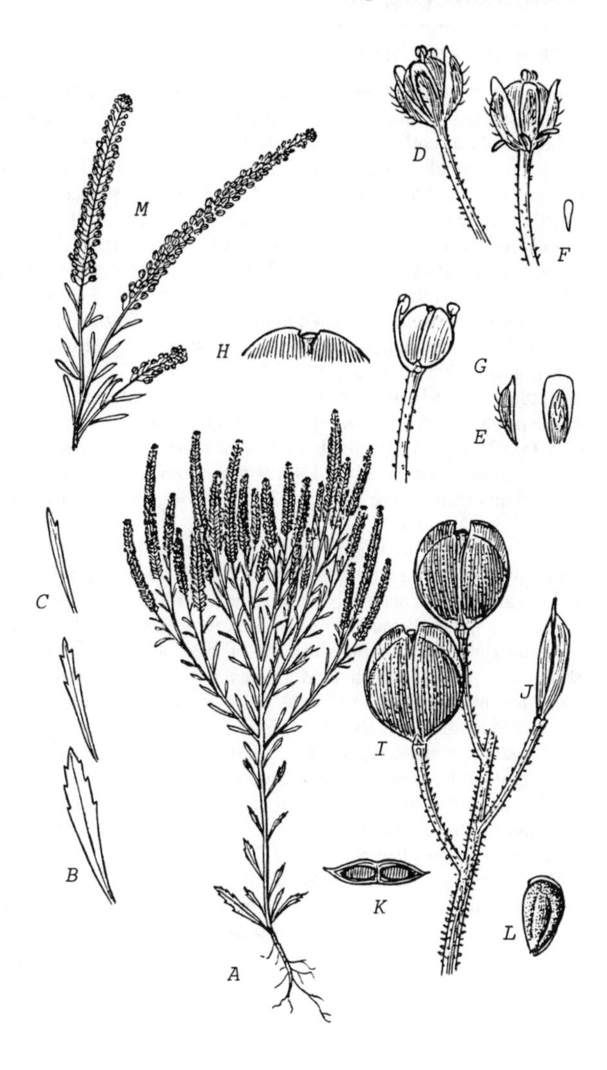

Ax0.2;B,Cx0.3;D-Gx12;Hx20;Ix8;Lx9;M=inflorescence x0.5.

102. Lepidium virginicum L.

Virginian Pepperwort Map 102

Annual or biennial 20-40(-60) cm, glabrous or minutely hairy, green. Stems erect, branched above. Rosette leaves to 7(-9) cm, petiolate, oblanceolate in outline, pinnatifid with a large, acute terminal lobe, and 2-5 pairs of smaller lateral lobes; margins acutely toothed. Lower stem leaves smaller with shallower lateral lobes. Upper stem leaves sessile or shortly stalked, linear to linear-oblanceolate, acute at apex; margins acutely toothed. Inflorescence crowded. Sepals 0.5-1.1 mm, oblong, green, erect. Petals 1.1-1.5 x 0.2-0.6 mm or sometimes absent; limb oblong, white, rounded at apex, indistinctly clawed. Petals 1-2 times as long as sepals. Stamens 2(-?4); anthers yellow. Stigma entire. Pedicels in fruit 2.5-5 mm, slender, hairy on upper side, ascending to patent, often curving outwards. Fruits (2.5-)3-3.7(-4) x (2.5-)2.7-3.2(-4) mm, ± orbicular, narrowly winged above and notched, flattened (angustiseptate, the septum *c.* 0.5 mm wide), ascending to patent, dehiscent. Valves (2.5-)3-3.7(-4) mm, unveined. Persistent style *c.* 0.1-0.2 mm, included in notch. Seeds 1 per loculus, 1.4-1.9 mm, obovate and flattened, usually winged on one side, brown. 2n=32. Flowering May to November.

A rare casual of docks, railways, roadsides, arable fields, waste ground, probably introduced with grain and bird seed and not persistent. Rare in Britain and Ireland and probably confused with *L. densiflorum*. Native in N. America, introduced to Europe, S. America, Africa and Australasia.

Fortunately varying little except in size, and in that the petals are sometimes absent which then causes difficulty separating it from *L. densiflorum* (see **101**) and from *L. ruderale* which has elliptic fruits, unwinged seeds and is foetid.

The small, angustiseptate fruits with 1 seed in each loculus should distinguish it from all other crucifers except *Lepidium* species and *Iberis* (which has large, asymmetrical petals). Not likely to be confused with *Lepidium* species other than *L. densiflorum, L. ruderale* and a number of rare casual species (see Ryves 1977).

Plants keying out as *L. virginicum* but with upper stem leaves pinnatifid to pinnate (not ± deeply serrate to entire) may be referable to *Lepidium bonariense* L., a rare casual with 4 post-1950 records.

Lepidium virginicum 102

Ax0.2;B,Cx0.3;D-Gx12;Hx20;Ix8;Lx9.

103. Lepidium perfoliatum L.

Map 103

Annual (?biennial) 7-35(-40) cm, glabrous or with sparse, simple hairs below, green. Stems erect, branched above. Leaves strongly dimorphic. Rosette and lower stem leaves to 10 cm, petiolate, pinnate, bipinnate or tripinnate with acute, linear lobes. Upper stem leaves simple, ovate to suborbicular, perfoliate, acute at apex; margins entire. Inflorescence crowded. Sepals *c.* 0.7-1.3 mm, ovate, green, erect to inclined. Petals *c.* 0.9-1.8 x 0.2-0.5 mm; limb spathulate, rounded at apex, pale yellow; claw long, distinct. Petals *c.* 1-1.5 times as long as sepals. Stamens 6; anthers yellow. Stigma entire. Pedicels in fruit 3-5(-7) mm, slender, glabrous, ascending. Fruits *c.* 3-4.5(-4.8) x 3-4 mm, ovate to rhombic, narrowly winged above and notched at apex, flattened (angustiseptate, the septum *c.* 1 mm wide), ascending, dehiscent. Valves 3-4.5(-4.8) mm. Persistent style 0.1-0.5 mm, about as long as notch. Seeds 1 per loculus, 1.7-2.3 mm, ovate, flattened, narrowly winged, dark brown. 2n=16. Flowering ?May to August.

A casual of arable fields, docks, railways, roadsides, waste ground, mills, etc., introduced with foreign grain and seed. Usually in small numbers and not persisting. Once frequent in England and occasional in Ireland, Scotland and Wales, now rare due to cleaner seed. Native probably in E. Europe and W. Asia where it is a characteristic grain weed, introduced through Europe, N. Africa, N. America, the Far East and Australasia.

Probably little variable other than in size.

The yellow petals should distinguish it from the other species of *Lepidium,* but when the petals are absent, the strikingly dimorphic lower and upper leaves are very distinctive. It is rarely misidentified as *Thlaspi perfoliatum* (which has entire lower leaves, white petals and 3-5 seeds per loculus), but the similarity between the names is a more frequent source of confusion.

Ax0.2;B,Cx0.3;D-Gx12;Hx20;Ix8;Lx9.

104. Lepidium graminifolium L.

Tall Pepperwort Map 104

Perennial 25-50(-60) cm, glabrous or minutely hairy below, dark green. Stems erect (sometimes ascending) with widely spreading branches above. Rosette leaves to 7(-10) cm, petiolate, the blade elliptic to oblanceolate, obtuse at apex, coarsely toothed to pinnately lobed; leaves withering by flowering. Lower stem leaves to 5 cm, sessile, simple, oblanceolate, acute at apex; margins entire. Upper stem leaves ± linear, acute at apex; margins entire (rarely with 1-2 teeth). Inflorescence crowded. Sepals 0.9-1.4 mm, ovate, green to purple, erect to ascending. Petals 1.6-2.2 x 0.9-1.2 mm; limb obovate, rounded at apex, white, indistinctly clawed. Petals *c.* 1.5 times as long as sepals. Stamens (?4-)6; anthers yellow (?purple). Stigma capitate. Pedicels in fruits 2.5-4 mm, slender, glabrous or minutely hairy, ascending to inclined. Fruits (2.5-)3-4.2 x 1.5-2.5(-3) mm, ovate, unwinged, not notched, compressed (angustiseptate, the septum *c.* 0.5 mm wide), ascending to inclined, dehiscent. Valves (2.5-)2.8-4 mm, glabrous. Persistent style 0-0.2 mm, ± prominent, capitate. Seeds 1 per loculus, 1.5-1.8 mm, elliptic, unwinged, brown. 2n=16. Flowering July to November.

A rare casual of docks, railways, waste ground, rubbish tips, grain siftings, car parks, etc., probably introduced with grain. Mostly casual but sometimes persistent. Rarely recorded in England and S. Wales, absent from Scotland and Ireland. Native in S. Europe around the Mediterranean, and in N. W. Africa. Introduced to Australasia and N. America.

Somewhat variable in size. Two subspecies are recognised in Europe but it has been difficult to place British material; some is definitely referable to subsp. *graminifolium* but the presence of subsp. *suffruticosum* (L.) P. Monts. remains to be clarified. *L. graminifolium* is probably an aggregate of taxa, closely related to the C. Asian and Iranian *L. lyratum* complex and the African *L. armoracia* (Jonsell 1975a).

Distinguished from *L. latifolium* by the ± linear upper stem leaves, and larger, ovate fruit. The small, ovate fruit with a prominent style should distinguish it from other *Lepidium* species with obvious petals and stalked upper leaves.

Lepidium graminifolium 104

Ax0.2;B,Cx0.3;D-Gx12;Hx20;Ix8;Lx9.

105. Lepidium latifolium L.

Dittander

Perennial 40-120(-130) cm, glabrous or with sparse, simple hairs below, glaucous especially below, forming patches with creeping rhizomes. Stems erect, branched above. Basal leaves to 30 cm, long-petiolate, the blade elliptic, ovate or lanceolate, rounded to acuminate at apex, cuneate to truncate at base; margins entire or toothed. Lower stem leaves similar but with shorter petioles (?rarely pinnately lobed). Upper stem leaves smaller, sessile to ± stalked, lanceolate, acute at apex, cuneate to truncate at base; margins entire or toothed. Inflorescence crowded. Sepals 0.9-1.5 mm, ovate, green-purple to whitish, erect to inclined. Petals 1.8-2.5 x 0.8-1.2 mm; limb spathulate, rounded at apex, white (to pinkish); claw distinct, long. Petals about twice as long as sepals. Stamens 6; anthers yellow. Stigma capitate to emarginate. Pedicels in fruit 2-6 mm, slender, ascending to inclined, glabrous. Fruits 1.6-2.7 x 1.3-1.7 mm, elliptic to orbicular, flattened (angustiseptate, the septum *c.* 0.5 mm wide), unwinged (margined when dry), apical notch absent, ascending to inclined, dehiscent. Valves 1.6-2.5 mm, glabrous or hairy, unveined. Persistent style 0.2-0.3 mm, capitate, prominent. Seeds 1 per loculus, 0.8-1.3 mm, oblong to elliptic, unwinged, brown. 2n=24. Flowering July to September.

A native plant of saltmarshes, creeks, brackish ditches and marshes, wet sand, etc, near the coast and inland in S. E. England, the Severn Estuary and N. E. England. Introduced or casual elsewhere in England and Wales, S. Scotland and S. and E. Ireland, on canals, riverbanks, railways, docks, waste ground, etc. Often forming large patches and persistent. In some localities a relict of cultivation. Native in Europe, S. W. Asia and N. Africa, introduced to N. America and Australasia.

Somewhat variable in leaf and fruit shape. Plants from the Baltic have dark purple sepals and narrower, more toothed upper stem leaves, and some from the E. Mediterranean have leaves with dense, stellate hairs.

A robust, distinct plant unlikely to be confused with any other crucifers.

Once cultivated and used for flavouring, the roots and leaves being "extremely hot and burning bitter" (Grigson 1958), but it was apparently ousted by pepper and horse-radish.

Lepidium latifolium 105

Ax0.2;B,Cx0.3;D-Gx12;Hx20;Ix8;Lx9.

106. Lepidium draba L.

Hoary Cress, Hoary Pepperwort, Thanet Cress Map 106

Perennial 20-60(-90) cm, glabrous to densely hairy with simple hairs, green to greyish, with creeping rhizomes, usually forming large patches. Stems erect to decumbent at base, branched above. Leaves variable. Basal and rosette leaves to 15 cm, with a short, broad petiole, obovate to elliptic, obtuse at apex; margins entire to irregularly toothed. Upper stem leaves sessile, ovate, obovate, lanceolate, oblanceolate or elliptic with rounded to acute auricles clasping stem, obtuse to acuminate at apex; margins entire to coarsely toothed. Inflorescence crowded. Sepals 1.5-2.6 mm, oblong, green, erect to patent. Petals 2.5-4.5 x 1-2.2 mm; limb spathulate, rounded at apex, white; claw long, distinct. Petals 1.5-2 times as long as sepals. Stamens 6; anthers yellow. Stigma capitate to emarginate. Pedicels 4-15 mm, slender, ascending to patent. Fruit set variable. Fruits 4-6(-6.5) x (3.5-)3.8-5.5 mm, cordate to broadly triangular, emarginate to truncate at base, cuneate to obtuse at apex, flattened but somewhat inflated (angustiseptate), margins ± keeled, ascending to patent, ± indehiscent. Valves 2.5-4 mm, often reticulate. Persistent style 0.7-1.8 mm, linear, prominent. Seeds 1 per loculus, *c.* 1.5-2.5 mm, ovoid, unwinged, brown. 2n=64*. Flowering April to July (-November).

A plant of roadsides, railways, docks, waste ground, saltmarshes, arable and pasture land, etc. It reproduces by seed or from root fragments and readily colonises disturbed soil often forming large patches. A noxious, persistent weed which can be controlled with selective herbicides. The ecology is described in detail by Scurfield (1962) and Mulligan & Findlay (1974).

First introduced in 1802 at Swansea and independently to a number of other ports (including Thanet) about this time (Scurfield 1962). It spread rapidly and is now common in England (especially in the south east), scattered in Wales and uncommon in Scotland and Ireland. Probably native in S. Europe and S. W. Asia, and widely introduced around the world (Willis 1953).

Very variable, probably represented by many morphologically different clones. Many infraspecific taxa have been described but none are currently accepted. The characters separating *Cardaria* from *Lepidium* are unworkable and *Cardaria* is therefore here included in *Lepidium.* Mulligan & Frankton (1962) maintain it is a separate genus.

Distinguished from the other *Lepidium* species with white flowers and clasping upper leaves by the cordate, unwinged fruit.

107. Lepidium chalepense (L.) L. is very similar to *L. draba* but differs in having flatter fruits, cuneate to rounded at base, and the often oblong leaves. It is a rare casual; Map 107 (see Rich 1988a).

Lepidium draba 106
*L. chalepense 107

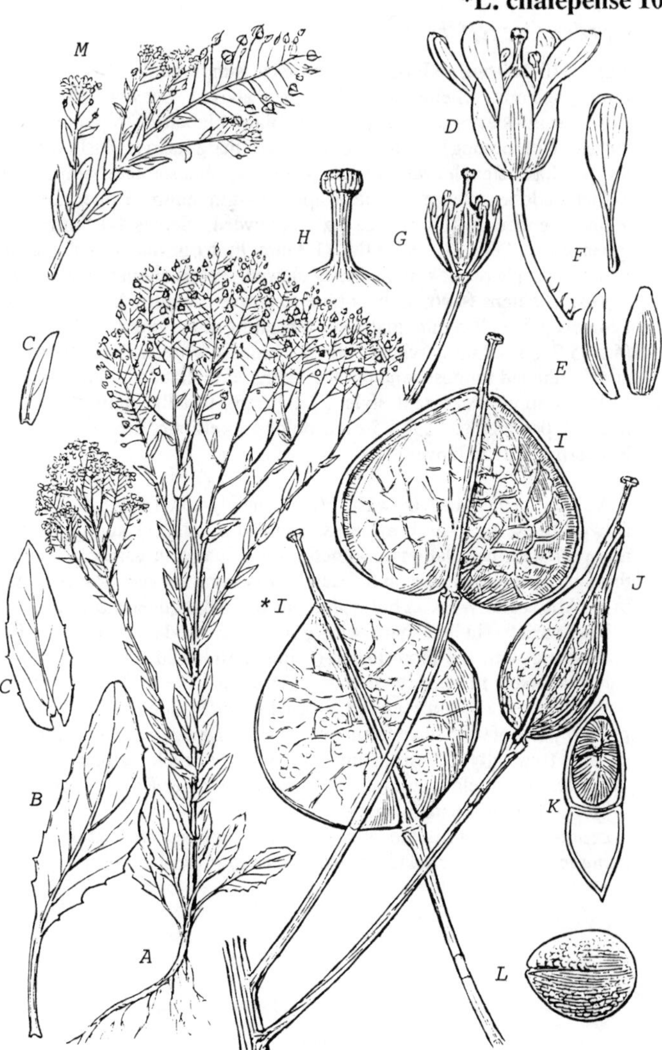

Ax0.2;B,Cx0.3;D-Gx12;Hx20;Ix8;Lx9;M=inflorescence x0.5;*Ix8.

108. Lepidium sativum L.

Garden Cress, Cress Map 108

Annual 10-100 cm, glabrous or with sparse, simple hairs below, pale green, often glaucous. Stems erect, branched above. Leaves very variable. Lower stem leaves to 8(-10) cm, petiolate, pinnate to bipinnate (rarely simple) with oblong to ovate, obtuse lobes; margins irregularly toothed or lobed. Upper stem leaves ± sessile, ± linear to oblanceolate, simple or with 1-4 lateral lobes, obtuse to acute at apex; margins entire. Uppermost leaves ± linear, entire. Inflorescence lax or crowded. Sepals 1-1.8 mm, ovate, green, erect. Petals 2.3-3.8 x 0.7 - 1.3 mm; limb obovate, rounded at apex, white to purple or pink, indistinctly clawed. Petals about twice as long as sepals. Stamens (4-)6; anthers blue to purple (?-yellow). Stigma entire. Pedicels in fruit 2-6 mm, glabrous, ascending or curving outwards. Fruits (4.5-)5-7 x 3-5.8 mm, ovate to elliptic, broadly winged and notched at the apex, flattened (angustiseptate), ascending to inclined, dehiscent. Valves (4.5-)5-7 mm. Persistent style 0.2-0.8 mm, usually included in notch. Seeds 1 per loculus, 2.5-3.5 mm, oblong to elliptic, brown. 2n=16,24. Flowering May to November.

A casual in bird seed, on waste ground, newly-sown grass, docks, towns, flower beds, tips, arable fields, roadsides, etc. Usually in small, non-persistent populations. Sometimes cultivated in gardens. Scattered throughout England, rare in Scotland, Wales and Ireland. Possibly native in N. Africa and W. Asia but widely introduced all over the world.

Very variable in leaf shape with ± simple, crisped cultivated leaf forms contrasting with the more usual ± bipinnatifid wild type. Much of the variation is due to selective breeding for salad.

A distinct plant, but widely confused possibly due to poor keys. It is probably one of the 5 most frequent species of *Lepidium*. The combination of large (typically 5-7 mm) fruits with 1 seed in each loculus and stalked upper leaves should distinguish it from all the other British species. The 3-fid cotyledons are also unique.

Lepidium sativum is the "cress" of "mustard and cress" but is largely being replaced in our salads by *Brassica napus* (Rich 1988b).

Lepidium sativum 108

Ax0.2;B,Cx0.3;D-Gx12;Hx20;Ix8;Lx9;M=inflorescence x0.5.

109. Lepidium campestre (L.) R.Br.

Pepperwort, Field Pepperwort

Annual (?biennial) (5-)10-40(-60) cm, with simple hairs below, often grey-green. Stems erect, branched mainly above. Basal rosettes rarely persisting, leaves to 7 cm, long-petiolate, elliptic to oblanceolate, obtuse; margins ± entire. Lower stem leaves shortly stalked, oblanceolate, obtuse at apex; margins entire. Upper stem leaves to 4(-5) cm, sessile, linear-lanceolate to triangular and clasping the stem with acute auricles, obtuse to acute at apex; margins toothed or entire. Inflorescence crowded. Sepals 1-1.8 mm, elliptic, green to yellow with purple tips, erect to ascending. Petals 1.5-2.6 x 0.2-0.4 mm; limb spathulate, white; claw distinct. Petals *c.* 1.5-2 times as long as sepals. Stamens 6; undehisced anthers yellow. Stigma simple. Pedicels 3-7 mm, short, hairy, patent. Fruits (4-)4.5-6.8 x (3.1-)3.7-5.5 mm, broadly ovate to broadly oblong, broadly winged above and notched at apex, flattened (angustiseptate, the septum 0.9-2.3 mm wide), patent to inclined, dehiscent. Valves (4-)4.5-6.6 mm, sparsely to densely covered in vesicles, especially below. Persistent style 0.1-0.7 mm, slender, shorter than or just exceeding notch. Seeds 1 per loculus, 2-2.8 mm, ovoid, unwinged (or rarely with very narrow wing), brown. 2n=16. Flowering April to August.

A casual or possibly native plant of sandy, stony, and gravelly ground, roadsides, banks, docks, arable fields, gardens, etc. Often either locally abundant or as isolated plants. Probably frequent in S. and E. England (map in Perring & Walters 1962), much rarer and probably often confused with *L. heterophyllum* in Ireland, Wales and Scotland. Frequent in Europe to the Caucasus, introduced to N. America, S. E. Asia and Australasia.

Overwintering plants are often larger and more robust with clasping lower leaves whilst spring-germinating plants are smaller. The fruit shape is slightly variable and the length of the style relative to the notch causes much confusion with *L. heterophyllum*. *L. heterophyllum* is usually perennial with many branches ascending from the base, and has purple-reddish anthers (at least on the margins), the style is usually longer and prominent, and the fruits are ovate.

Other than *L. heterophyllum*, only *L. perfoliatum*, *L. draba* and *L. chalepense* have clasping upper leaves. The former has yellow petals and finely divided lower leaves. The latter are perennials with creeping rhizomes and valves usually less than 4 mm long. *Thlaspi* species have more than 1 seed in each loculus and are usually glabrous.

Lepidium campestre 109

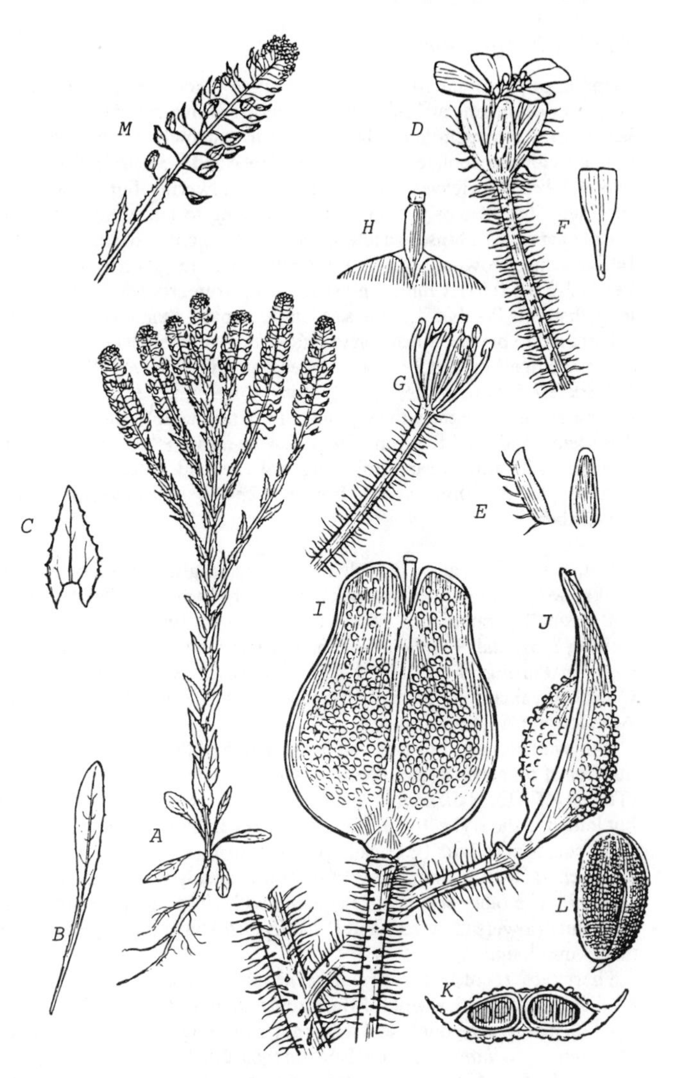

Ax0.2;B,Cx0.3;D-Gx12;Hx20;I,Kx8;Lx9;M=inflorescence x0.5.

110. **Lepidium heterophyllum** Bentham

Smith's Pepperwort, Smith's Cress

Biennial or perennial 10-50 cm, with simple hairs, grey-green. Stems ± erect to decumbent at base, often branched above and below. Rosette leaves to 8 cm, petiolate, oblanceolate to narrowly elliptic, simple or pinnate with a lanceolate, obtuse terminal lobe and 1-3 pairs of small, broad lateral lobes; the leaves do not persist to flowering but resprout after fruiting. Stem leaves to 5 cm, sessile, oblong to lanceolate with acute, clasping auricles, obtuse to acute at apex; margins entire or sharply toothed. Inflorescence crowded. Sepals 1.3-2.5 mm, oblong, green to purple, erect. Petals 2-3.6 x 0.5-1.1 mm; limb spathulate, white, rounded at apex; claw long, distinct. Petals *c.* 1.5 times as long as sepals. Stamens 6; undehisced anthers red to purple at least on margins (yellow when dehisced). Stigma entire. Pedicels in fruit 2-6 mm, slender, hairy, inclined to reflexed. Fruits 5-8.6 x 3.2-5 mm, ovate, broadly winged and notched or not at apex, flattened (angustiseptate, the septum *c.* 1-1.8 mm wide), patent to inclined, dehiscent. Valves 4.8-7.5 mm, with few or no vesicles. Persistent style (0.4-)0.5-1.2 mm, linear, projecting well beyond notch. Seeds 1 per loculus, 1.8-2.2 mm, ovoid, brown. 2n=16*. Flowering May to September.

Native on dry heaths, sand dunes, stony and gravelly places, shingle, banks, hedges, arable fields, rocks, ballast, railways, etc., often frequent but in small populations. Most common in W. England, E. and S. W. Ireland, S. W. and N. E. Scotland and Wales, scattered and rarer elsewhere (map in Perring & Walters 1962). Native in Europe from Spain to Czechoslovakia. Occasionally introduced elsewhere in Europe and in N. America and Australasia.

Fruits are variable in shape and in the length of the persistent style. The number of branches at the base is variable. One variety, var. *alatostylum* (Towns.) Thell., is distinct in that the wings of the valves are not notched but joined to the style. Plants in the Pyrenees are nearly glabrous and may merit subspecific rank (P.D. Sell, pers. comm.).

For characters distinguishing it from the closely related *L. campestre*, see 109. The other species of *Lepidium* with clasping upper stem leaves and white flowers (*L. draba* and *L. chalepense*) have unwinged fruits less than 4 mm long.

There are records for *Lepidium hirtum* (L.) Sm. from Britain, a Mediterranean species which differs from *L. heterophyllum* in having fruits hairy (at least when young) and wings *c.* 1/2 the length of the fruit. Material of *L. heterophyllum* with a few hairs on the fruits but wings only up to 1/3 the length of the fruit has been seen named *L. hirtum*, but the presence of the true species in Britain has not been confirmed.

Lepidium heterophyllum 110

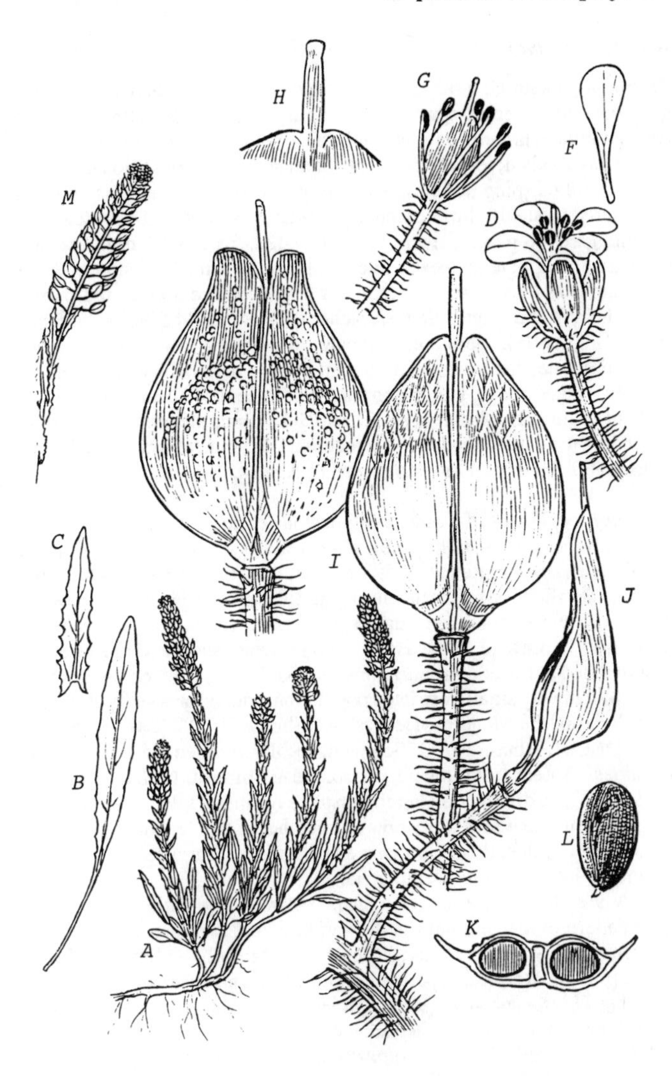

Ax0.2;B,Cx0.3;D-Gx12;Hx20;Ix8;Lx9;M=inflorescence x0.5.

111. **Thlaspi caerulescens** J. & C. Presl

Alpine Pennycress Map 111

Perennial (?biennial) 5-40(-55) cm, glabrous, pale green to purplish, often glaucous. Stems erect, sparingly branched or simple. Rosette leaves to 5 cm, spathulate, broadly petiolate, elliptic, rounded at apex; margins entire to sparsely toothed; leaves persistent. Stem leaves 3-8, ovate to lanceolate, sessile with clasping auricles, acute to obtuse at apex; margins entire or irregularly toothed. Inflorescence crowded. Sepals (1.1-)1.4-2.2 mm, elliptic, green to purple, erect. Petals (1.7-)2.2-5 x 0.5-1.3 mm; obovate, entire to emarginate at apex, white or lilac; claw short, indistinct. Petals 1.5-2 times as long as sepals. Stamens 6; undehisced anthers red-purple. Pedicels in fruit 3-7 mm, slender, patent. Fruits (3.5-)4.5-9.5 x 2-4.5 mm (including style), obovate, truncate to notched at apex, flattened (angustiseptate, the septum 1-1.9 mm wide), narrowly winged below, broadly winged above, inclined to patent, dehiscent. Valves (3-)4-8.5 mm, without veins. Persistent style (0.3-)0.5-1.5 mm, slender, equalling or exceeding the notch. Seeds 3-6 per loculus, 1-1.5 mm, elliptic, light brown, ± biseriate. 2n=14*. Flowering April to August.

A very local plant of alpine and limestone rocks, screes and grassland, river gravels and old mine spoil, almost always associated with heavy metal (lead/zinc) enriched soils (the metal status of its Scottish sites requires clarification). It is the most zinc-tolerant species of metalliferous mine soil and accumulates up to 2-3% zinc in its foliar dry matter (Shimwell & Lawrie 1972, Hajar 1987). Lead and cadmium are also accumulated at lower concentrations (A.J.M. Baker, pers. comm.).

Usually in very small, restricted populations, but occasionally abundant. Very local in the Mendips, Derbyshire (Shimwell 1968), W. Wales, the Lake District, Rhum and C. Scotland, locally frequent in the Northern Pennines. Absent from Ireland. Ingrouille & Smirnoff (1986) discuss the disjunct British distribution in relation to heavy metal tolerance. Native in S. and C. Europe where it is a notable member of the "galmei" (calamine) flora of zinc-rich soils in Germany and Belgium (Smith 1979), and in N. America. Introduced to Scandinavia.

Very variable in size and fruit shape between populations but there is little pattern or discontinuity in the variation and the plant is best treated as a single, polymorphic taxon (Riley 1955, Ingrouille & Smirnoff 1986).

The prominent style and perennial habit should easily distinguish it from the other *Thlaspi* species. *Draba* species have unwinged fruits. *Lepidium* species have only one seed in each loculus.

The more familiar name *T. alpestre* L. is predated by *T. alpestre* Jacq. (which applies to another European species) and is thus illegitimate.

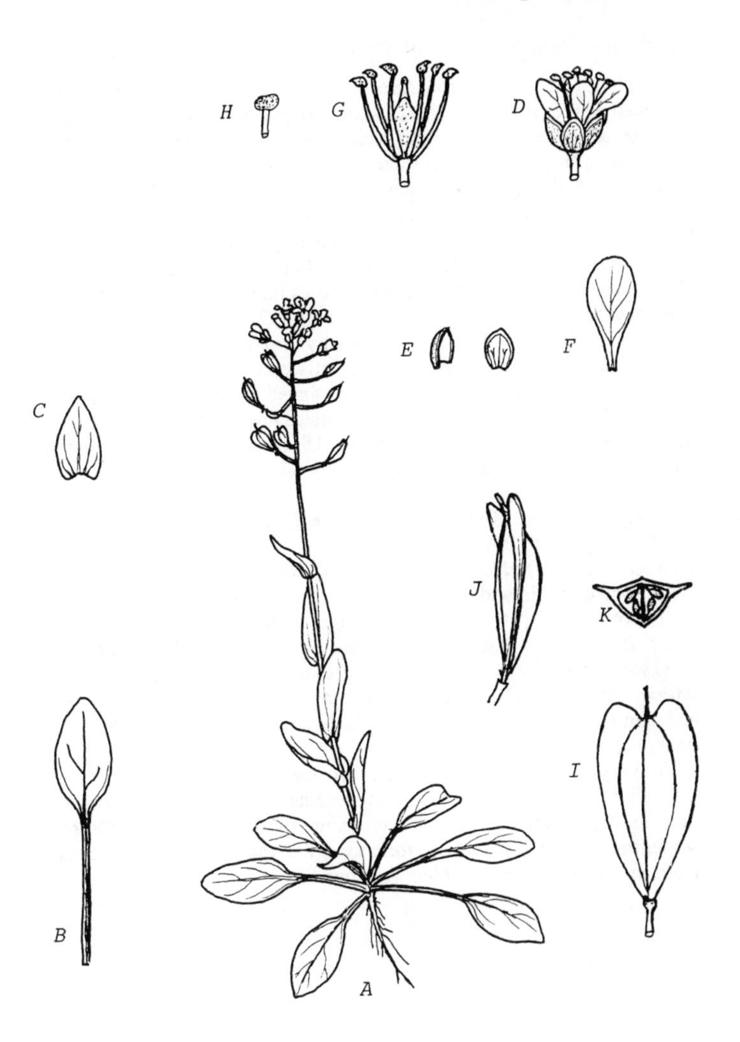

A-Cx1;D,Ex3;F,Gx4;Hx10;I,Jx5.

112. Thlaspi perfoliatum L.

Perfoliate Pennycress, Cotswold Pennycress Map 112

Annual 3-17(-20) cm, glabrous, green, often glaucous above. Stems erect, branched above. Rosette leaves to 4 cm, broadly petiolate, broadly elliptic to ovate, obtuse at apex; margins entire to sparsely toothed; leaves rarely persisting to flowering. Stem leaves (1-)2-4(-6), to 2.5 cm, sessile, ovate to lanceolate, the upper perfoliate with rounded auricles clasping stem, obtuse at apex; margins entire or sparsely toothed. Inflorescence crowded. Sepals 1-1.8 mm, oblong to ovate, green to purple, erect to ascending. Petals 1.4-3 x 0.7-1.3 mm; limb oblong to obovate, rounded at apex, white; claw short, indistinct. Petals about twice as long as sepals. Stamens 6; anthers yellow. Stigma entire. Pedicels in fruit 2.5-5 mm, slender, patent. Fruits 3-5.5(-7.5) x 3-4(-5.5) mm, broadly obovate (rarely ± orbicular), broadly notched at apex, flattened (angustiseptate, the septum *c.* 1.5 mm wide), narrowly winged below, broadly winged above, patent, dehiscent. Valves 3-5.2(-7.5) mm, without veins. Persistent style 0-0.3 mm, included in apical notch. Seeds 3-5 per loculus, 0.9-1.3 mm, ovoid, pale brown. 2n=42(?14,?70). Flowering March to May (-June).

A very rare native of open soils on oolite limestone in grassland, screes and banks in the Cotswolds. Often associated with other rare calcicoles such as *Pulsatilla vulgaris*, *Thesium humifusum* and *Astragalus danicus*. It also occurs more widely in England on railways, quarries, stone pits, walls, arable fields, etc., as a casual. A rare casual in Wales and Scotland, absent from Ireland. The distribution has been studied in detail by Rich *et al*. (1989). Native in most of S. Europe, extending to Afghanistan and N. Africa. The English localities are disjunct from its range on the continent, and similar disjunct sites occur in S. Scandinavia. Introduced to N. America.

Not very variable except in size. More variable in Europe.

Distinguished by the 2-4 ovate, perfoliate stem leaves, and small, broadly obovate fruits with a small style. Immature or depauperate *T. arvense* is sometimes confused with *T. perfoliatum* but in the former the ovaries and fruits are broadly winged all the way round.

Thlaspi perfoliatum 112

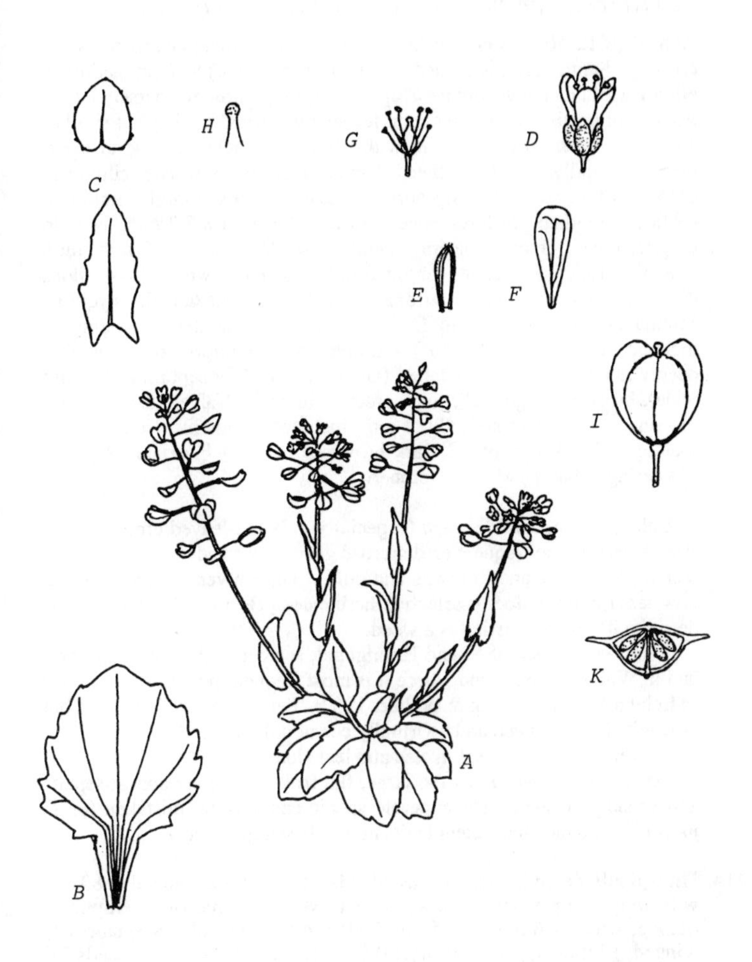

Ax1;B,Cx2;Dx3;E-Gx5;Hx10;I,Kx3.

113. **Thlaspi arvense** L.

Pennycress, Common Pennycress, Field Pennycress

Annual (8)12-50(-80) cm, glabrous, green, often glaucous, foetid when crushed. Stems erect, branched above. Rosette leaves to 7 cm, petiolate, elliptic to oblanceolate, obtuse at apex; margins ± entire to coarsely lobed; leaves not persisting. Lower stem leaves numerous, smaller but similar, elliptic to oblanceolate, sessile with clasping auricles, obtuse at apex; margins usually sinuate-toothed. Upper stem leaves sessile, elliptic to oblong, with acute, clasping auricles, acute at apex; margins entire or shallowly toothed. Inflorescence crowded. Sepals 1.8-3.3 mm, ovate to elliptic, green, erect to patent. Petals (2.4-)2.7-5 x 1.1-1.6 mm; limb obovate, truncate, rounded or emarginate at apex, white; claw short, distinct. Petals about twice as long as sepals. Stamens 6; anthers yellow. Stigma entire. Pedicels in fruit (5-)8-13 mm, slender, ascending to inclined. Fruits (6-)9-20 x (5-)7-20 mm, broadly elliptic to ± orbicular, deeply notched at apex, flattened (angustiseptate, the septum *c.* 1-2 mm wide), broadly winged all round, erect, dehiscent. Valves (6-)9-20 mm, unveined. Persistent style 0-0.3 mm, included in notch. Seeds 3-8 per loculus, 1.2-2.3 mm, ovoid, dark brown to black, ± biseriate. 2n=14*. Flowering (March-) May to October.

A characteristic arable weed (especially in broad-leaved crops such as strawberries), also frequent on disturbed ground, roadsides, waste ground, recent earthworks and in towns and cities. Once a very common weed, now less common due to selective herbicides. Best & McIntyre (1975) describe its characteristics as a weed.

Common in much of S. and E. England, scarcer in the west and north and in Wales and Scotland where it is most frequent near the coast, rarer in Ireland (map in Perring & Walters 1962). Perhaps native in C. Asia but now widely introduced and a virtually cosmopolitan weed.

Not very variable except in size and leaf shape.

Instantly recognisable by the large, flat fruits which are angustiseptate and broadly winged. Depauperate specimens may be mistaken for *T. perfoliatum* which has patent fruits narrowly winged below.

114. Thlaspi alliaceum L. is a rare casual (Map 114). It is an annual to 80 cm, with long, simple hairs below; stem leaves 4-11, narrowly elliptic to oblong; petals 2.6-5.3 mm; fruits 5-10 x 4.5-6 mm, obovate, narrowly winged, ± inflated; persistent style 0-0.3 mm, included in notch; seeds 2-4 mm. Distinguished from the other species by the faint garlic smell and inflated fruits (see also *Alliaria* and *Pachyphragma*).

Thlaspi arvense 113

Ax0.5;B,Cx1;D-Gx4;Hx10;I,Kx1.5.

115. Capsella bursa-pastoris (L.) Medicus

Shepherd's Purse

Annual (?biennial) 5-50(-70) cm, glabrous or with simple and/or branched hairs below. Stems erect, branched above. Leaves very variable. Rosette leaves to 15(-20) cm, with a broad petiole, oblanceolate in outline, simple to pinnatisect with an ovate, acute terminal lobe and 1-8 pairs of triangular lateral lobes; margins entire or toothed. Lower stem leaves sessile, oblanceolate, usually less divided than the rosette leaves. Upper stem leaves smaller, sessile, simple, ovate to lanceolate or narrowly oblong, with acute, clasping auricles, acute at apex; margins entire or irregularly toothed. Inflorescence crowded. Sepals 1.2-2.7 mm, oblong, green to purple or red, erect. Petals 1.5-3.5 x 0.4-1.4 mm; limb ovate to elliptic, rounded to emarginate at apex, white (rarely pinkish); claw short, indistinct. Petals *c.* 1-2 times as long as sepals. Stamens 6; anthers yellow. Stigma entire, ± capitate. Pedicels in fruit 3-15(-18) mm, slender, erect to patent. Fruits 4-10(-11) x 3.4-6.5(-7) mm, obtriangular, obcordate or obovate, flattened (angustiseptate, the septum *c.* 1 mm wide), emarginate, truncate or rounded at apex, dehiscent. Valves 4-10(-11) mm, veins weak or absent. Persistent style 0.2-0.8 mm, included in notch. Seeds numerous, 0.7-1.1 mm, ovoid, pale brown, ± biseriate. 2n=16,32. Flowering ± all year.

A very common weed of gardens, cultivated ground, paths and road sides, waste ground, sand dunes, docks, etc. One of the most ubiquitous of all crucifers throughout Britain and Ireland, uncommon only in the mountainous areas (map in Perring & Walters 1962). Introduced throughout the world.

Polymorphic and very variable in all characters. The plants are largely self-pollinating which results in many distinctive local populations which have often been given names (e.g. Almquist 1921).

Capsella rubella Reuter has occasionally been recorded as a casual but its current taxonomic status is unclear (Svensson 1983, Akeroyd 1986). For the time being, plants showing a combination of buds and/or sepals pink, red or reddish-purple, petals pink- or red-flushed and not longer than sepals, fruits with concave margins and rounded lobes can be recorded as *C. rubella* and those differing in one or two characters as "near *C. rubella*".

The ± triangular, angustiseptate fruits are distinctive and the species should not be confused with any other crucifers. Sterile plants may, however, give problems.

Capsella bursa-pastoris 115

Ax0.3;B,Cx0.4;D-Gx4;Hx20;I-Kx3;Lx10.

116. Isatis tinctoria L.

Woad

Biennial or perennial 40-150 cm, sparsely and softly hairy below with simple hairs, dark green, often glaucous. Stems erect, branched above. Rosette leaves to 30 cm, long petiolate, narrowly elliptic to lanceolate, obtuse at apex; margins entire to sinuate, sparsely toothed. Lower stem leaves sessile, oblanceolate with acute, clasping auricles, obtuse at apex. Upper stem leaves small, sessile, lanceolate to linear-oblong with large, acute to rounded, clasping auricles, acute at apex; margins entire. Inflorescence crowded. Sepals 2-2.8 mm, oblong, yellow to greenish, patent. Petals 2.5-4 x 0.9-1.5 mm; limb obovate, rounded at apex, yellow to pale yellow, indistinctly clawed. Petals about twice as long as sepals. Stamens 6; anthers yellow. Stigma entire. Pedicels in fruit 5-10 mm, very slender, reflexed to pendent. Fruits (9-)11-20(-21) x (3-)3.5-6 mm, oblong-lanceolate or elliptic-ovate, obtuse or emarginate at apex, flattened (angustiseptate, the septum *c.* 2 mm wide), with a thick, broad wing, indehiscent, turning brown when ripe, pendent. Persistent style sessile. Seeds 1(-2) per fruit, 3.2-4.8 mm, cylindrical, pale brown to yellow. 2n=14,28. Flowering May to August (-October).

Represented in Britain by subsp. *tinctoria*.

An introduced plant of chalk or marl cliffs, arable fields, docks, waysides and waste ground, persistent in a few localities (Lees 1859) and varying in abundance from year to year. Rarely found in England, very rarely in Wales, Scotland and Ireland and now only known on cliffs in Gloucestershire and Surrey with occasional casual plants elsewhere (map in Perring & Walters 1962). Possibly native from S. Europe and N. Africa to S. E. Asia. Introduced to Scandinavia, N. and S. America.

Variable in fruit shape, hairiness, etc., throughout its range, with a number of subspecies or close relatives in Europe. Records for *I. glauca* Aucher and *I. lusitanica* L. may refer to this species.

The pendent fruits are diagnostic. Occasionally confused with immature *Camelina* species which usually have at least some forked hairs.

The plant has been cultivated in Britain for the blue dye obtained by fermenting its leaves, a practice which apparently died out in the 1920s after the development of chemical dyes. Ancient Britains were reputed to paint their bodies with woad, possibly as warpaint or for religious rites. A fascinating, detailed account of the plant is given by Hurry (1930).

Isatis tinctoria 116

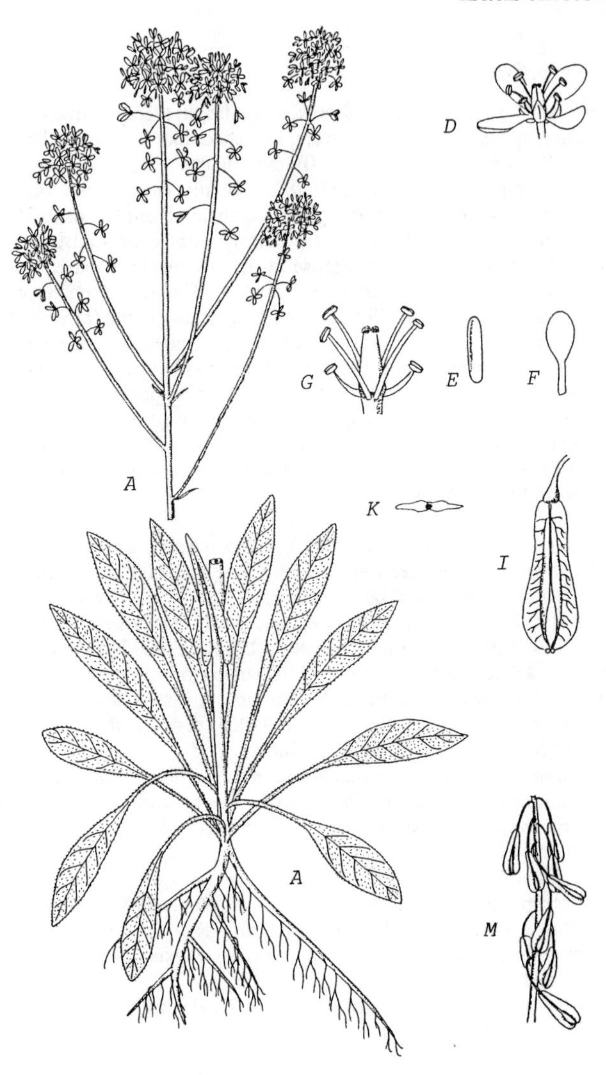

Ax0.2;Dx3;Ex5;Fx4;Gx10;I,Kx2;M=inflorescence x0.5.

117. **Armoracia rusticana** Gaertner, B. Meyer & Scherb.

Horseradish

Perennial 50-150(-200) cm, glabrous, waxy, bright green. Rhizomes often spreading aggressively. Stems erect, branched above. Rosette leaves to 100(-120) cm; petioles to 50(-60) cm; blades simple or irregularly pinnatifid, oblong, oblong-lanceolate or oblong-oblanceolate, obtuse at apex, cuneate at base; margins regularly crenate. Stem leaves to 15(-20) cm, simple (occasionally pinnatifid below), sessile or shortly stalked, oblong to oblong-lanceolate, obtuse at apex, cuneate at base; margins entire to crenulate. Inflorescence crowded. Flowers often imperfect. Sepals 2.6-4 mm, ovate to elliptic, green, erect to ascending (rarely patent). Petals 5.4-7.8 x 2.5-4.1 mm; limb obovate, rounded (to emarginate) at apex, white; claw indistinct. Petals *c.* 2-3 times as long as sepals. Stamens 6; anthers yellow. Stigma capitate, entire to emarginate. Pedicels 7-20 mm, slender, erect. Ovary rarely developing beyond *c.* 4 x 1 mm, ellipsoid, unveined. Ripe fruit not seen; apparently 1-2 seeds may develop per loculus even if the fruits do not develop (not seen). 2n=(28)32*. Flowering May to August.

A characteristic plant of roadsides, but also in fields, waste ground, river banks, railways, old gardens, etc., usually close to habitations. Naturalized and often abundant (especially in the east) throughout lowland England and Wales, less common in the west and north. Occasional in Scotland and Ireland (map in Perring & Walters 1962). Possibly native in S. E. Europe and S. Russia but its origin is unclear due to widespread cultivation. Introduced elsewhere in Europe, N. America, Australasia.

Variable in leaf shape. Rhodes *et al.* (1965) distinguish 3 cultivars but these probably scarcely merit recognition.

The large, erect leaves are distinctive and resemble Dock (*Rumex* spp.) leaves but lack the ochreae. The long racemes of white flowers are often puzzling when first seen, and the complete lack of fruit hinders identification.

The plant is highly sterile and seed set is very low. Plants are self-incompatible and pollen fertility is low (Weber 1949). About 1/3 of the ovules may be fertilized, but many then abort. Endosperm abnormalities cause further failure of seeds (Stokes 1955). The plant spreads and is propagated by root fragments.

Cultivated for many centuries (Courter & Rhodes 1969). The grated root is used in horseradish sauce.

A-Cx0.2;D-Fx3;Gx6;Hx10;I=immature fruit x 2.

118. Draba aizoides L.

Yellow Whitlowgrass

Perennial (2-)3-10 cm, densely tufted, with simple hairs on leaves, green. Stems erect, leafless, branched from below rosettes. Leaves in dense rosettes, to 13 mm, sessile, oblong, acute at apex; margins entire. Inflorescence crowded. Sepals 2.8-4 mm, oblong, green to yellowish, erect. Petals 3.5-6 x 2.5-4 mm; limb obovate, rounded to truncate at apex, yellow, indistinctly clawed. Petals about twice as long as sepals. Stamens 6; anthers yellow. Stigma entire. Pedicels in fruit 2-10 mm, erect to ascending. Fruits 5-9 x 3-4 mm, ellipsoid, flattened (latiseptate), erect to ascending, dehiscent. Valves 4-8 mm, unveined. Persistent style 1-2.5 mm, linear. Seeds numerous, 0.7-1.5 mm, obovate, flattened, pale brown, biseriate. 2n=16*. Flowering February to April (-June).

A very rare plant of south-facing crevices in carboniferous limestone rocks. The history, distribution and ecology is described in detail in the Biological Flora (Kay & Harrison 1970). Restricted to a short stretch of coastline on the Gower Peninsula (v.c. 41), S. Wales where it is locally abundant, and also introduced to the walls of Pennard Castle.

The Welsh locality is far removed from the main distribution range in S. and C. Europe where it is a rare plant of mountains from the Pyrenees to the Alps and Carpathians. There is another disjunct locality in Belgium. It is undoubtedly native in S. Wales and suggestions to the contrary are unfounded.

The continental material is variable in habit, size of flowers and fruits, hairiness, etc., but the infraspecific taxa described are of doubtful value as they grade into each other both morphologically and geographically. The Welsh plants are fairly uniform.

Unlikely to be confused with any other crucifer.

Draba aizoides is a very rare plant and must not be collected. In some of the more accessible sites, populations have been decimated by gardeners and botanists.

Draba aizoides 118

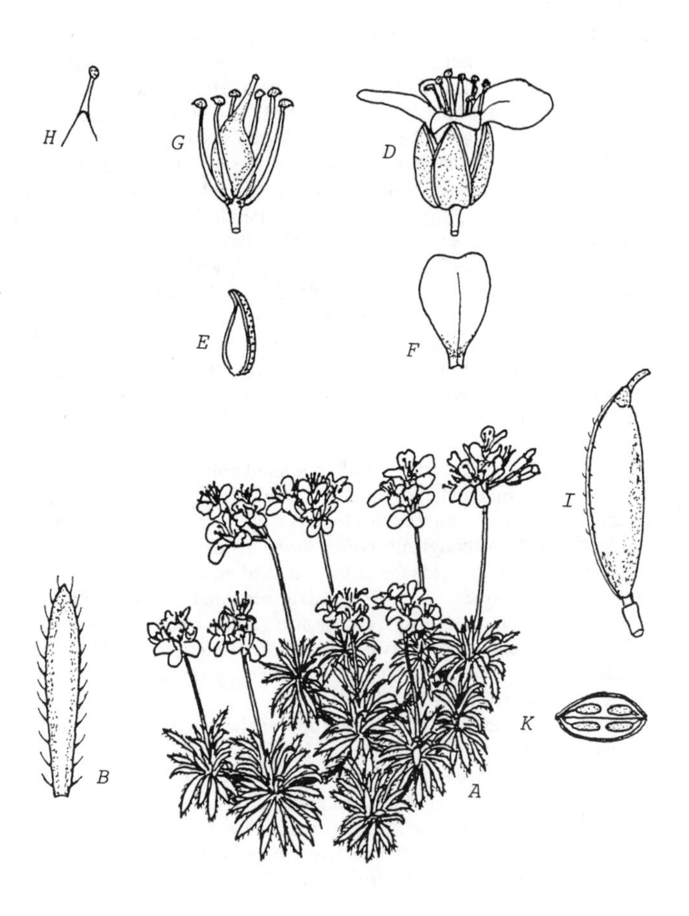

Ax1;Bx3;D-Gx4;Hx10;I,Kx4.

119. Draba muralis L.

Wall Whitlowgrass

Annual 5-50 cm, densely hairy with simple, forked and stellate hairs. Stems erect when young, often decumbent when older, branched above. Rosette leaves to 4 cm, ± sessile to broadly petiolate, oblanceolate to ovate, obtuse at apex; margins variably toothed; leaves rarely persistent to flowering. Stem leaves 4-many, to 2.5 cm, sessile, broadly ovate to suborbicular, base cordate to rounded and clasping stem, obtuse at apex; margins serrate. Inflorescence crowded. Sepals (0.8-)1-1.8(-2) mm, oblong, green, often purple-tipped, erect. Petals 1.4-3.2 x 0.6-1.1 mm; limb oblanceolate, rounded at apex, white, indistinctly clawed. Petals about twice as long as sepals. Stamens 4(-6); anthers yellow. Stigma entire. Pedicels in fruit 3-10 mm, slender, inclined to patent. Fruits (3-)3.5-5.5(-6) x (1.2-)1.5-2 mm (the upper often smaller), oblong to oblanceolate, flattened (latiseptate), inclined to patent, dehiscent. Valves (3-)3.5-5.5(-6) mm, without veins. Persistent style 0.1-0.2 mm. Seeds 4-8 per loculus, 0.6-0.8 mm, ovoid, pale brown, biseriate. 2n=32. Flowering April to August.

A rare plant as a native with small, scattered colonies principally on the limestone in W. England, the Peak District and the Pennines. It is a typical annual of therophyte communities found on open, unstable soils with a pH of more than 6.5. Although most often found in open sites with a southerly or westerly aspect, it avoids the driest soils and may occur occasionally in open, rocky ashwoods. It is also widely scattered through Britain and Ireland as a casual on railways, etc, or as a garden weed and on old walls accidentally spread with cultivated plants from nurseries. The ecology, similar in many respects to that of *Hornungia petraea*, is given in the Biological Flora (Ratcliffe, D. 1960).

Native in W., and S. and C. Europe eastwards to Turkey and the Caucasus, N. W. Africa and Madeira. Introduced to N. America.

Varying little except in size.

Distinguished from the other species of *Draba* by the broadly ovate, clasping stem leaves. Few other crucifers have 4 stamens in most flowers. *Cardamine hirsuta* and *Coronopus* have deeply divided leaves, *Lepidium* species have angustiseptate fruits with 1 seed in each loculus and *Arabidopsis* has linear fruits.

Draba muralis 119

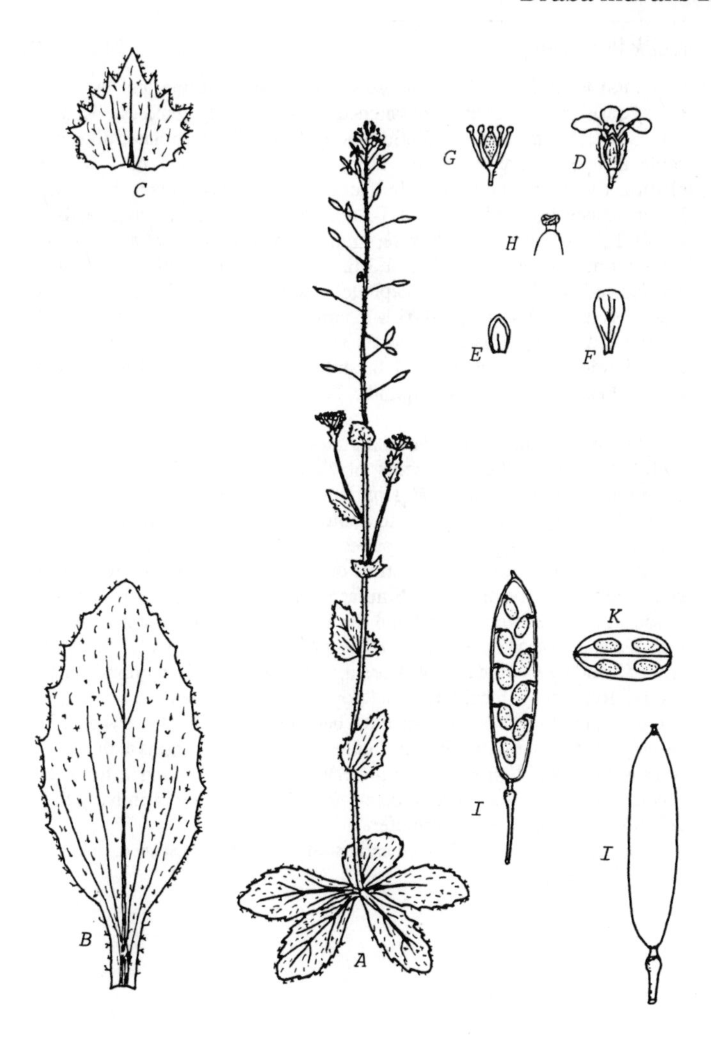

Ax1;B,Cx3;D-Gx3;Hx10;Ix6;Kx10.

120. **Draba norvegica** Gunnerus

Rock Whitlowgrass Map 120

Perennial to 5(-7)(-?12) cm, densely tufted, with simple and branched hairs, dark green. Stems unbranched, erect, usually leafless. Rosette leaves to 1.5 cm, sessile, elliptic-oblong to linear-lanceolate, obtuse to acute at apex; margins entire. Stem leaves absent or 1 or 2, sessile, elliptic-ovate, rounded and half-clasping at base; margins entire. Inflorescence crowded. Sepals 1.5-2 mm, ovate, green, erect to ascending. Petals 2-3(-?4) mm; limb obovate, rounded to emarginate at apex, white; claw short, indistinct. Petals *c.* 1.5 times as long as sepals. Stamens 6; anthers yellow. Stigma entire, capitate. Pedicels in fruit 2-4 mm, slender, erect. Fruits (3-)3.9-7(-8) x (1-)1.8-2.5 mm, elliptic, flattened (latiseptate), not twisted, erect, dehiscent. Valves 3.5-6.8 mm, with or without a weak central vein. Persistent style 0.1-0.5 mm. Seeds few, *c.* 0.8 mm, ovoid, brown, biseriate. 2n=48 (hexaploid). Flowering July to August?

A rare native alpine of some of the more base-rich rocks and ledges at high altitudes in the mountains of Scotland. Confirmed from the Grampians, Cairngorms, N. W. Highlands and Skye.

The description above applies to Scottish material only, where the plant varies little except perhaps in pubescence. Elsewhere there is much variation in height (to 12 cm), number of stem leaves (to 3 or more), flower size, and fruit shape and size. Similar plants to those in Scotland can be found in Scandinavia, Iceland and the Faeroes, and are best regarded as part of a widespread complex. *D. norvegica* s.l. is widespread in the Arctic from Canada eastwards to N. Russia. Further work is required on the group. In Norway, local forms and populations are noted.

Dwarf plants of *Draba incana* have been mistaken for *D. norvegica* but usually have 1-many stem leaves, toothed (not entire) margins to the leaves or twisted fruits. Some material in herb. Kew of *D. norvegica* from Ben Lawers may be *D. incana* x *D. norvegica*. *D. norvegica* is unlikely to be confused with other alpine crucifers.

D. norvegica is self-pollinated and usually sets abundant seed. It should not be collected.

Draba norvegica 120

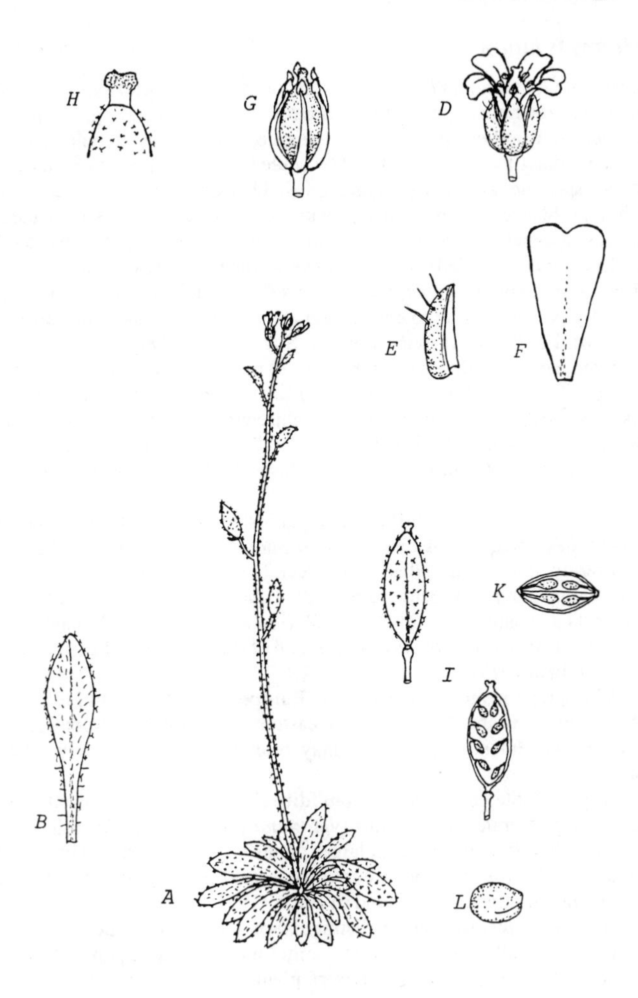

Ax1;Bx2;Dx5;E-Gx10;Hx20;Ix2.5;Kx5;Lx10.

121. Draba incana L.

Hoary Whitlowgrass Map 121

Biennial (?perennial) (2-)5-35(-50) cm, densely hairy with simple, forked and stellate hairs, grey-green. Stems erect to ascending, simple or branched. Rosette leaves to 3 cm, ± sessile, oblanceolate to oblong, acute at apex; margins entire or with a few acute teeth. Stem leaves 5-many (in dwarf specimens 0-4 and ± indistinguishable from rosette leaves), sessile, oblong-oblanceolate to narrowly ovate, acute at apex, rounded to cuneate at base (sometimes ± clasping stem); margins with a few acute teeth. Inflorescence crowded. Lowest flowers sometimes bracteolate. Sepals 1.5-2.7 mm, oblong, green, erect. Petals (2-)2.5-4.3(-5) x 1-2.1 mm; limb broadly obovate, entire to emarginate at apex, white; claw short, distinct. Petals *c.* 1.5-2 times as long as sepals. Stamens 6; anthers yellow. Stigma capitate, entire. Pedicels in fruit 1.5-5(-9) mm, slender, ascending. Fruits 5.5-10.5(-12) x 1.3-3 mm, elliptic to lanceolate, flattened (latiseptate), usually twisted, erect to ascending, dehiscent. Valves 5-10 mm, with a weak central vein. Persistent style 0.1-0.7 mm. Seeds 5-8 per loculus, 0.8-1.1 mm, elliptic, brown, biseriate. 2n=32*. Flowering May to August.

A calcicole of screes, cliffs, rocks, gullies, mica-schist, mine heaps and sometimes open turf. A rare alpine in Snowdonia and the Lake District, and rarely at lower altitudes on the Peak District limestone. Frequent in the Pennines and the mountains of C. Scotland, occasional at lower altitude on cliffs and sand dunes in N. and W. Scotland and N. W. Ireland. This disjunct distribution reflects a contraction of range since the last glaciation (see Godwin 1975).

Widespread in arctic and subarctic Europe and in the central European mountains from the Pyrenees to the eastern Alps. Records from C. Asia, Iceland, Greenland and Canada may refer to this or a closely related species.

Very variable, most populations differing slightly from each other. Small depauperate plants in turf (var. *nana* Lindbg.?) may be selectively genetically differentiated from larger plants on rocks nearby. The species as a whole is variable and is currently treated as an aggregate. Further details of variation can be found in Fearn (1971).

Typical *D. incana* can be distinguished from other crucifers by the characteristically twisted fruits (sometimes best seen on old fruits, sometimes on young ones). Dwarf plants have been mistaken for *D. norvegica*.

Draba incana 121

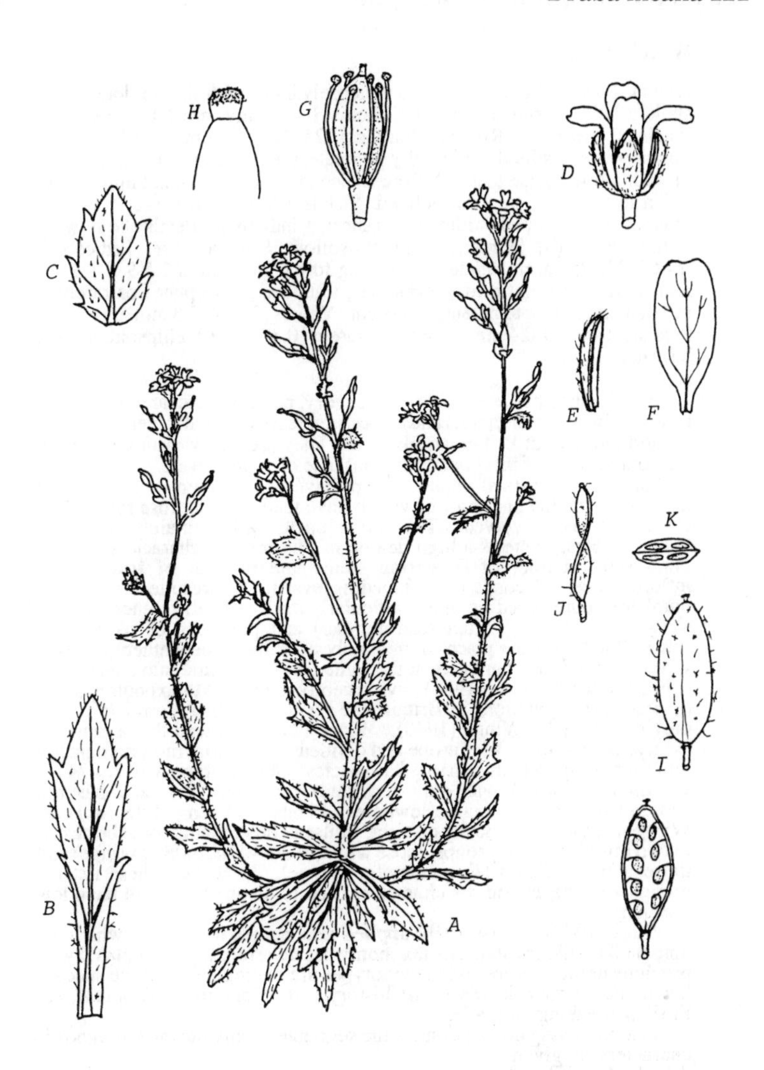

Ax1;B,Cx2;Dx5;E-Gx10;Hx20;I,Jx3.

122-124. **Erophila verna** group*

Whitlowgrass

Annual 1-15(-25) cm, ± glabrous to densely hairy with simple, forked and stellate hairs, green to greyish. Stems 1 to many, erect to ascending, leafless, simple. Rosette leaves 0.25-2(-3.5) cm, spathulate to oblanceolate, distinctly or broadly petiolate, obtuse at apex; margins entire or with a few coarse teeth. Inflorescence lax. Sepals 1.2-2.2 mm, ovate, green to purplish, erect to inclined. Petals 1.5-3.8 mm x 1-2(-3.3) mm; oblanceolate, 2-lobed, white; claws short, ± indistinct. Petals about twice as long as sepals. Stamens 6; anthers yellow. Stigma entire. Pedicels in fruit 3-20(-25) mm, slender, ascending to patent. Fruits 1.5-9 x 1.3-3.8 mm, narrowly elliptic to ± orbicular, flattened (latiseptate) or rarely ± inflated, ascending to patent, dehiscent. Valves (3.7-)4-7.3 mm, unveined. Persistent style 0-0.4 mm. Seeds numerous, 0.3-0.8 mm, ellipsoid, brown, biseriate.

Erophila species are annuals of open, dry, neutral to calcareous soils on rocks, railways, walls, paths and tracks, arable fields, sand dunes and sandy ground throughout Britain and Ireland. They are easily distinguished by the combination of bifid petals and absence of stem leaves.

Erophila is a critical genus. There have been a number of different taxonomic treatments, most of which have tended to confuse rather than clarify the taxa involved. The confusion has arisen mainly from two features. First, there is a high degree of plasticity in characters such as plant size, number of flowering stems and number of flowers per inflorescence. Second, the inbreeding system has resulted in a large number of pure-breeding lines particularly marked by differences in fruit shape and size (and often seed-number) and pubescence. Automatic self-pollination takes place immediately after the outer anthers dehisce, when they and the stigma are at the same level and come into contact.

Filfilan & Elkington (1988) have recently made a cytotaxonomic study of *Erophila* populations in Britain, the results of which correlate with a previous study by Winge (1940). Winge (1940) showed that a range of cytotypes exist in N. W. Europe and divided his material morphologically into four groups which he treated as species. One cytotype with 2n=24 (*E. semiduplex* Winge) was only recorded from Germany, and has not subsequently been found elsewhere. Plants with 2n=14 (*E. simplex* Winge) are morphologically distinct, whilst all other plants could be placed in two morphological groups: those with 2n=30-40 (*E. duplex* Winge) and those with 2n=52-64 (*E. quadriplex* Winge). In Britain, these latter 3 groups are present and the characters used to distinguish them have low plasticity.

Winge (1940) unfortunately applied new names to his taxa which are nomenclaturally invalid. His taxonomy has therefore been correlated with previous nomenclature and the cytotypes are distinguished at the specific level. For further details of the history and taxonomy of *Erophila*, see Filfilan & Elkington (1988).

In the following descriptions of the segregates, only the more diagnostic characters are given.

* by S. A. Filfilan & T. T. Elkington.

Erophila verna group **122-124**

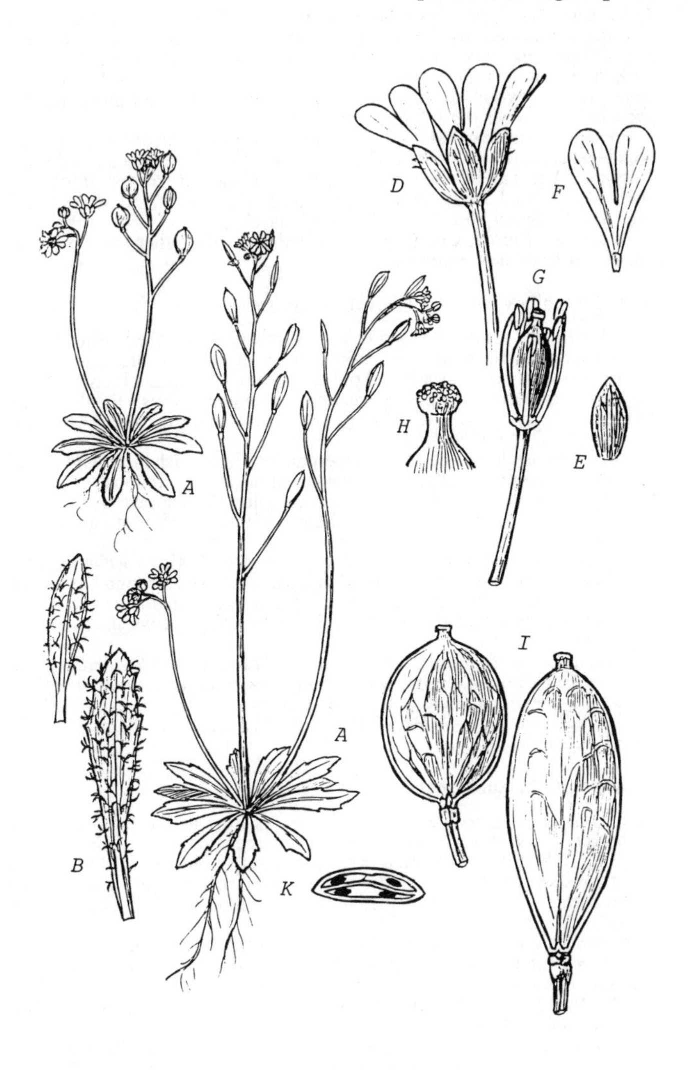

Ax1;Bx2;D-Gx8;Hx20;I,Kx10.

122. **Erophila majuscula** Jordan

Leaves to 1.6 cm; petioles relatively short, *c.* 0.2-0.5 times as long as the lamina; lamina usually densely pubescent with forked and stellate hairs, often appearing grey. Flowering stems densely hairy at the base with forked and stellate hairs, and usually with at least scattered hairs on the upper parts (beyond the first pedicel). Petals bifid to not more than 1/2 their length. Seeds 0.3-0.5 mm. 2n=14*.

Present in a range of habitats, though probably absent from coastal dunes. The least common of the British species described here, scattered through England, Wales and E. Scotland, local in Ireland. Recorded from v.c.7, 9,10,11,13-20,22-24,26,28,30,31,33-35,38,41,42,44-47,50,56-58,61,62, 64,78,82,83,86,89,95,H5,9,17,19,22,37 and S.

A distinct diploid, easily distinguished from the other species by the small seeds and dense pubescence.

123. **Erophila verna** (L.) Chevall. *sensu stricto*

Leaves to 2(-3.5) cm; petioles *c.* 0.5-1.7 times as long as the lamina; lamina with scattered to moderate pubescence of mainly branched hairs. Flowering stems with a scattered pubescence of simple and branched hairs on the lower parts, the upper parts and pedicels glabrous. Petals usually bifid for 1/2-3/4 of their length. Seeds 0.5-0.8 mm. 2n=30*,32*,34*,36*,40*,42*,44*.

Two varieties based on fruit shape can be distinguished, though Smith (1968) showed that fruit shape varies more or less continuously.

a. var. **verna.** Fruits oblong to oblanceolate. Widespread in a range of habitats including sand dunes. Plants with inflated fruits are present sporadically in normal populations.

b. var. **praecox** (Steven) Diklic. Fruits broadly elliptic to orbicular, most typically twice as long as broad or less. Scattered through Britain and Ireland, and in the north and west particularly characteristic of sand dunes; this is the taxon mapped as *E. verna* subsp. *spathulata* by Perring (1968).

E. verna is the commonest and most widely distributed *Erophila* species in Britain and Ireland, probably occurring in all vice-counties in which the genus has been recorded. Recorded from v.c. 1-3,5-24,26-35,37-47, 49-70,72,78-81,83,85-90,92,94,103,106-111,H1,2,4,6,8-10,12,15-28,30, 31,33-40 and S.

124. **Erophila glabrescens** Jordan

Leaves to 1.5 cm; petioles 0.5-2.5 times as long as the lamina; lamina glabrescent or with scattered branched and occasional simple hairs, often ± ciliate only, the leaves often appearing shiny. Flowering stems with a few forked or simple hairs on the lower parts or often nearly glabrous, the upper parts glabrous. Petals bifid to not more than 1/2 their length. Seeds 0.5-0.8 mm. 2n=48*,52*,54*,56*.

Present in a wide range of habitats, including coastal dunes. Generally distributed throughout Britain, but less common than *E. verna*. Local in Ireland. Recorded from v.c. 3,5,6,8-11,13-17,19,20,23,26,29,30,34-36, 38-42,46,47,49,51,54-60,62-66,69,73,78,82,85,86,88,90,96,100, 101,103,106-110,H9,11,15,17,20,39 and S.

E. glabrescens usually has distinctly long petioles, and the ± glabrous stems and leaves are best seen in young material as the hairs in *Erophila* may be lost with age.

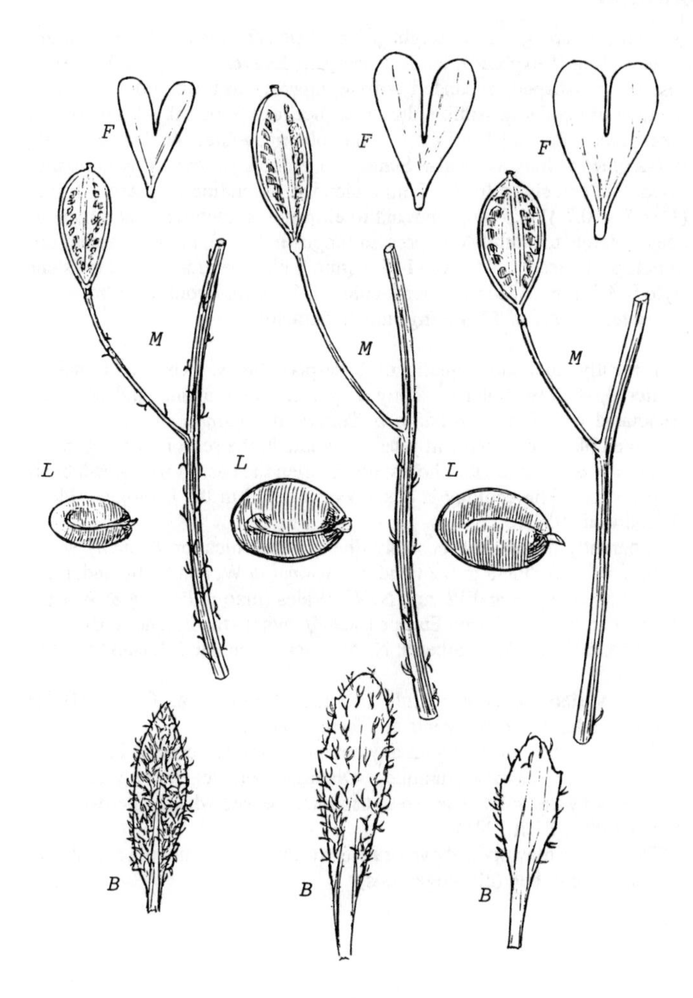

E. majuscula **E. verna s.s.** **E. glabrescens**

Bx2;Fx8;Lx20;M=hairs at first pedicel.

125. Subularia aquatica L. subsp. aquatica

Awlwort

Annual 2-12 cm, glabrous, bright green. Stems erect (sometimes prostrate to ascending if exposed), leafless, simple. Leaves to 6 cm, all in basal rosette, awl-shaped, expanded at base, tapering to a fine point, solid, ± terete to triangular in section. Inflorescence lax. Sepals 0.7-1 mm, ovate, green, erect. Petals 1.1-1.3 x *c.* 0.5 mm, oblong, white, not clawed. Petals *c.* 1-2 times as long as sepals. Stamens 6; anthers yellow. Stigma sessile, discoid. Pedicels in fruit 1-5 mm, slender, ascending to patent. Fruits (1-)2-5 x (0.7-)1-2.5 mm, obovoid to ellipsoid, sometimes emarginate at apex, ± terete to slightly compressed (angustiseptate), inflated, ascending to patent, dehiscent. Valves (1-)2-5 mm, with 1 central vein. Persistent style 0-0.3 mm. Seeds 2-7 per loculus, 1-1.4 mm, ovoid, pale brown, ± biseriate. 2n=*c.* 36. Flowering June to September.

A locally abundant aquatic of base-poor lakes. Usually found in sheltered, shallow water on stony or gravelly lake shores and bottoms, associated with *Littorella uniflora*, *Lobelia dortmanna* and *Isoetes* spp. Uprooted plants can frequently be found late in the season floating in the water or washed up on the shores; this is often the easiest way to establish its presence. The ecology is described in detail in the Biological Flora (Woodhead 1951).

Apparently decreasing, possibly due to eutrophication of many water bodies. Most frequent in Scotland, occasional in W. and N. Ireland, rarer in the Lake District and W. and N. W. Wales (map in Perring & Walters 1962). Circumboreal from Europe (Scandinavia to the Pyrenees, Bulgaria and the Urals), N. Asia, Siberia, N. America, Greenland, Iceland and the Faeroes.

Little variable except slightly in size. Mulligan & Calder (1964) distinguish N. American plants as subsp. *americana*.

Unlikely to be confused with any other crucifer, the aquatic habitat and awl-shaped leaves being unique. Distinguished vegetatively from its associates by the solid, ± terete to trigonous leaves which taper to a fine point (Rich & Rich 1988).

The flowers may open above or below water, or may not open at all and be pollinated in bud (cleistogamous).

Subularia aquatica 125

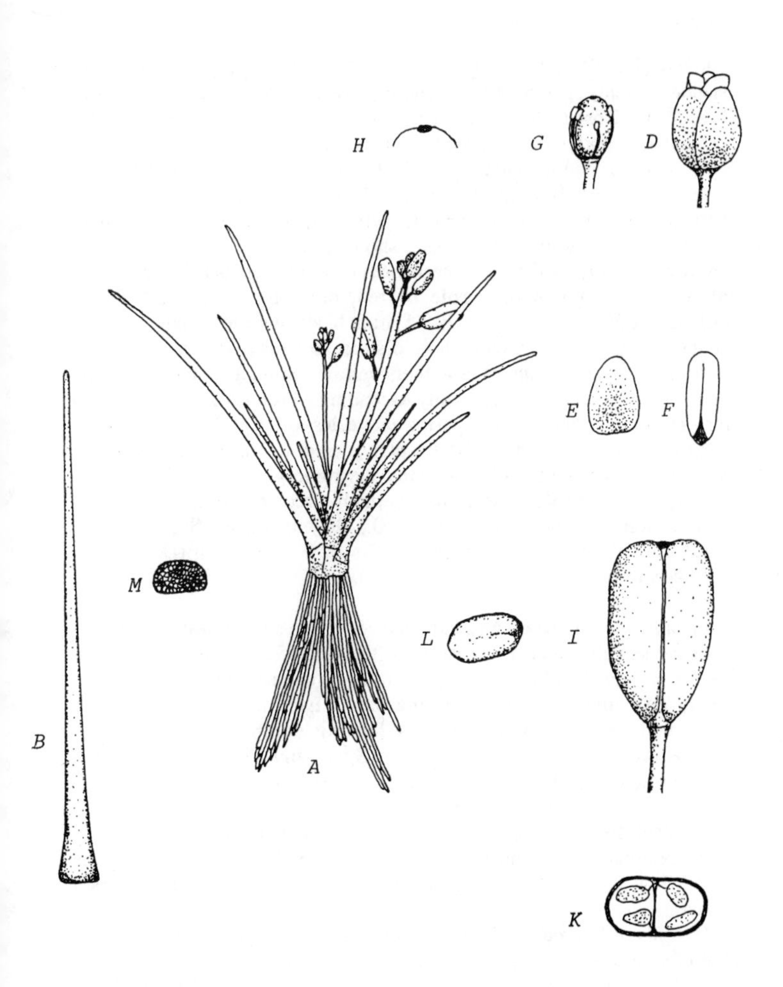

Ax3;Bx5;D-Gx12;Hx20;I,Kx8;Lx10;M=leaf T.S.x8.

126. Cochlearia danica L. *

Danish Scurvygrass

Annual (2-)5-25(-50) cm, glabrous, dark green or purplish. Tap-root slender. Stems spreading to ascending (erect in small plants), simple or branched. Rosette leaves to 5(-9) cm, petiolate, the blades to 1 cm, orbicular, suborbicular or rounded-triangular, cordate at base, obtuse at apex; margins entire, distinctly toothed or shallowly lobed; rosette leaves rarely persistent beyond flowering. Lower and middle stem leaves petiolate, palmately 3- to 7-lobed. Uppermost stem leaves sessile, without auricles (except sometimes on side shoots), oblong-lanceolate; margins entire or sparsely toothed. Inflorescence crowded. Sepals 1.7-3(-3.5) mm, oblong-ovate, green to purple or red, erect to ascending. Petals (1.8-)2.5-4.5(-5.5) x (0.7-)1.1-2(-3.5) mm; blade ovate to elliptic, rounded at apex, purple to white; claw short, distinct, greenish. Petals *c.* 1.5 times as long as sepals. Stamens 6; anthers yellow. Stigma capitate, ± entire. Pedicels in fruit 3-10 mm, usually ascending. Fruits 3-5(-7) x 2.5-4(-5) mm, ovoid and often clearly tapering above, less often globose or ellipsoid, ± terete to slightly flattened (latiseptate), erect to ascending, dehiscent. Valves 3-5(-6.5) mm, with fine veins, drying to give an areolate pattern closely associated with the veins. Septum circular to elliptic, less than twice as long as wide. Persistent style 0.1-0.7 mm, linear. Seeds 2-many, 1-1.3 mm, elliptic, brown, biseriate. 2n=42*. Flowering (December-) February to June.

Locally very abundant on sandy and rocky sea shores, walls and banks near the sea all round the coast of Britain and Ireland. In England and Wales widely distributed inland on railway ballast and roadsides, also rarely on buildings in cracks in masonry. Especially characteristic of disturbed and temporary habitats lacking competition from other vegetation. N. and W. Europe from Spain and Portugal to southern Norway and in the Baltic to the Åland archipelago.

Varying little except in size.

A rather distinctive species, characterised by its usually slender habit, by its petiolate and palmately lobed lower and middle stem leaves, by the small flowers very often being pale lilac, and by the ovoid fruits tapering above.

Hybrids of *C. danica* with the *C. officinalis* aggregate are occasionally found where the two species meet (see Stace 1975). The hybrids are partially fertile and introgression may occur (Fearn 1977). F1 hybrids are best recognised using petal and sepal size.

* by D. H. Dalby.

Cochlearia danica 126

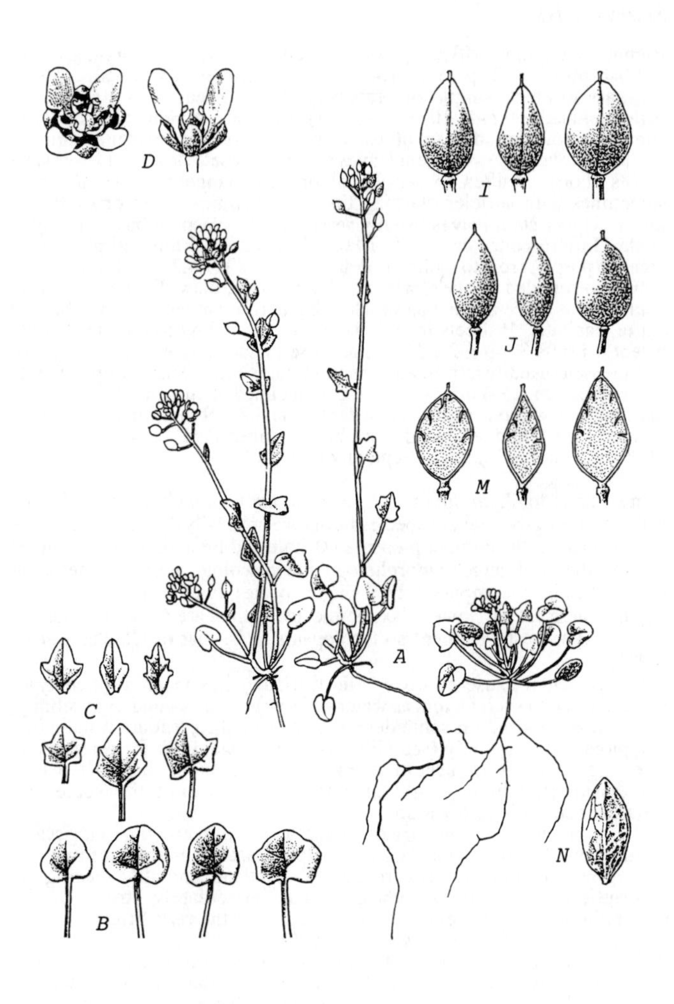

Ax0.75;B,Cx1;Dx3;I,Jx2;M=septum x2;N=ripe valves x2.

127-131. Cochlearia officinalis group *

Scurvygrass

Biennial-perennial 2-40 cm, glabrous, matt or shiny, dark to light green or reddish-purple. Tap-root prominent, sometimes thickened. Stems prostrate to erect, simple or branched. Rosette leaves to 10(-20) cm, petiolate; blades 0.5-4(-10) x 0.6-5(-13) cm, ovate, orbicular or reniform, cordate to rounded at base, obtuse or rounded at apex; margins entire to shallowly lobed. Lowest stem leaves similar to rosette leaves. Lower stem leaves shortly stalked to sessile, oblong to triangular-ovate, the base sometimes with auricles clasping the stem; margins entire or shallowly lobed. Upper stem leaves usually sessile with a clasping base, oblong to ovate. Inflorescence crowded or lax. Sepals 2.2-4.5 mm, elliptic-ovate, green to purple, erect to inclined. Petals 3.5-9.5 x 1.5-4.2 mm; limb elliptic, obtuse to rounded at apex, white or lilac-purple; claw distinct, greenish. Petals about twice as long as sepals. Stamens 6; anthers yellow. Stigma entire, capitate. Pedicels in fruit 4-14 mm, ascending to inclined (rarely patent). Fruits 3.5-9 x 2.7-5 mm, globose to narrowly ellipsoid, terete or compressed (usually latiseptate, or sometimes slightly angustiseptate, with a broad septum 2.5-4 mm wide), erect to inclined, dehiscent. Valves 3-8.5 mm, with or without veins. Persistent style 0.3-0.9 mm, linear. Seeds 2 to many, 1-1.8 mm, ellipsoid, dark brown, tuberculed. 2n=12*,24*,26*. Flowering mainly March to September.

In contrast to *C. anglica* and *C. danica*, the taxa included here form a very complex group where species limits are not easily drawn, and authors do not agree on the number present in Britain and Ireland. I am convinced that the observed range in morphology and in ecological preferences is far too great to be encompassed by a single species, but the species grouped together here are of "lower" taxonomic status than are *C. anglica* and *C. danica*. The best treatment so far appears to be that of Clapham *et al.* (1981).

The principal cause of taxonomic difficulty lies in great phenotypic plasticity due to such factors as water and nutrient stress, and high salinity, being superimposed on some degree of genetic differentiation including polyploidy and euploidy (see Gill 1971, 1973 and 1975 for references). The aggregate is in urgent need of a full, experimentally based reassessment, which would also take into account the scale and consequences of hybridization.

I have accepted *C. atlantica* as a distinct species after seeing living plants beside Loch Linnhe, which I have matched against herbarium material named by E. G. Pobedimova from N. W. Scotland, and after reading her description (Pobedimova 1968). I was immediately struck by the morphological distinctness of these plants, and by the very different habitat from that usually favoured by *C. officinalis* s.s.

C. scotica offers greater problems, with herbarium material named as this taxon being so diverse as to be more confusing than of assistance. The species is based on plants from Lochinver, Sutherland and Tain, Easter

* by D.H. Dalby.

Ross (Marshall 1892), originally identified as *C. groenlandica* L., but later renamed *C. scotica* by Druce (1929). Marshall attributes the first discovery of this taxon in Britain to Beeby (1887), who collected plants from near Ollaberry, Shetland. Scott & Palmer (1987), who know the Shetland flora well, consider that *C. scotica* is a dwarf coastal form of *C. officinalis* s.s. (note that they also include *C. alpina* under *C. officinalis* in Shetland). Plants from Gluss, 2.5 km from Ollaberry, approximate to Marshall's description, but merge imperceptibly into the *C. pyrenaica* group discussed below. They appear to lose their succulence in cultivation in non-saline conditions, further weakening a possible independent taxonomic status. As in other members of the genus, petal shape is unreliable as a taxonomic guide (unless related to a specified stage of flower development) since the limb elongates markedly with age. The taxonomic significance of the lilac coloration (sometimes strong) is not known - it may be related to nutrient stress.

C. pyrenaica and *C. alpina* are regarded here as conspecific (in spite of differences in chromosome number), partly for convenience (both are inland species for most of their ranges), and partly because they are morphologically indistinguishable or almost so (Clapham *et al.* 1987). Robust plants and those growing nearer the sea appear to merge into *C. officinalis* s.s. Some plants on northern mountains seem to merge with *C. micacea* in fruit form (these need checking cytologically).

Hybrids may be common in the field - the majority of species can be crossed with great facility in cultivation (Gill 1975). They are not considered further here because the putative parent species are not well enough understood at present.

127. **Cochlearia officinalis** L. *sensu stricto*

Common Scurvygrass

Perennial to 30 cm, usually robust. Stems branched, erect or decumbent at base. Rosette and lowest stem leaves with petioles to 8 cm; blades to 3 cm, mostly orbicular or slightly angular, cordate at base and sometimes with overlapping lobes or sinuate, less often (especially on small leaves) truncate; margins often somewhat recurved, entire or with a few, indistinct (rarely prominent), blunt teeth, fleshy, yellow or mid-green; under-surface often purple-coloured. Lower stem leaves shortly petiolate, oblong to triangular-ovate, usually with a few coarse teeth. Upper stem leaves sessile, generally clasping stem, with a sinuously toothed margin which may be very slightly recurved. Flowers 10-15 mm in diameter, sweetly scented. Petals white. Fruits 3-6 mm, globose to ovoid, rounded or less often slightly narrowed below, somewhat narrowed above. Valves of mature fruit with prominent veins and the central vein almost always well-defined and reaching the valve apex, drying to give a strong areolate pattern determined by the veins; septum orbicular to ovate. Seeds 1.1-1.5 mm. $2n=24 \pm 0\text{-}5$ accessory chromosomes *.

Upper, drier parts of salt marshes, crevices and ledges on rocky sea cliffs (especially those associated with sea bird colonies), and very frequently in grass on cliff top turf and roadside banks near the sea in the west and south-west where it competes well with taller vegetation. Distributed all round the British Isles, but less common in the south-east. Coasts of N. W. Europe, eastwards to Poland.

This species is interpreted here as consisting of *C. officinalis* subsp. *officinalis* of Clapham *et al.* (1987). It varies considerably in size, and in basal leaf and fruit shape. Smaller plants from inland mountain areas are generally best placed under *C. pyrenaica*. Plants from Little Pentland Skerry, Orkney, referred to *C. islandica* Pobed. by E. Pobedimova, with large, cordate basal rosette leaves, middle and lower stem leaves elongated, petiolate and cuneate at the base, and with fruits narrowed at both ends, match closely the type gathering from Iceland but are thought of here as variants of *C. officinalis.*

The species is characterised by the large basal leaves which are cordate at the base, by the large flowers and by the globose to ovoid fruits with valves strongly veined and reticulately patterned when mature.

The juice and fresh foliage of this species (and perhaps also *C. anglica*) were used by sailors in past times as a prevention against scurvy, the active agent being ascorbic acid (vitamin C).

128. Cochlearia atlantica Pobed.

Atlantic Scurvygrass

Perennial to 20 cm in fruit. Stems erect or decumbent at base. Rosette leaves with petioles to *c.* 4 cm; blades to 1.5 cm, flat or slightly concave, ovate to cordate, normally truncate at base but sometimes slightly cordate and less often slightly cuneate, dark green, frequently with purplish colouring beneath, matt or slightly shiny above. Lower stem leaves shortly petiolate or sessile, oblong to triangular-ovate, usually with a few coarse teeth. Upper stem leaves sessile, generally entire or only very slightly toothed. Flowers *c.* 10 mm in diameter, sweetly scented, though not as strongly as in *C. officinalis.* Petals white (rarely purple). Fruits 2.5-4 mm, globose to ovoid, rounded or less often slightly narrowed below, somewhat narrowed above. Valves of mature fruit with prominent veins, drying to give a clearly areolate pattern determined by the veins; septum orbicular to ovate. Seeds 1.4-1.8 mm. Chromosome number not known.

Stony, sandy or silty sea shores, often in association with *Puccinellia maritima*. Recorded from Mull, Arran, Loch Linnhe and Outer Hebrides, but the total range is probably more extensive along the west coast sea lochs. Probably endemic to the British Isles.

Although this species is accepted here, alternative opinions are (a) that it is a hybrid between *C. officinalis* and *C. scotica*, and (b) that gatherings named as *C. atlantica* are actually mixtures of those same two species. The Loch Linnhe population seems to be homogeneous and without either putative parent; the plants also set good seed.

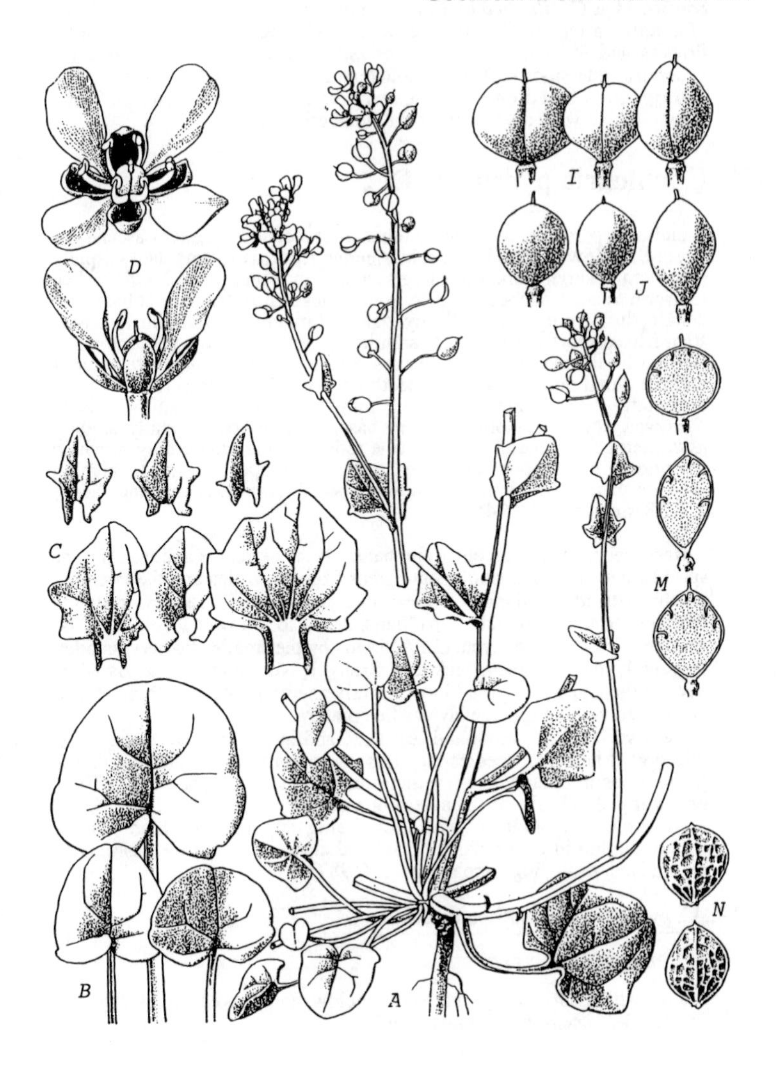

Ax0.75;B,Cx1;Dx3;I,Jx2;M=septum x2;N=ripe valves x2.

Living plants (corresponding closely with herbarium specimens confirmed as *C. atlantica* by E. Pobedimova), differ significantly from *C. officinalis* in the smaller, tougher and darker green leaf blades, smaller flowers and in their habitat. *C. atlantica* differs from *C. scotica* (as generally understood) in leaf size and texture, and in its larger flowers. The description is based on authenticated herbarium material and on living plants (which were also used for the illustration).

129. Cochlearia pyrenaica DC.

Biennial or perennial to 30 cm. Stems prostrate, decumbent or ascending. Rosette and lowest stem leaves with slender petioles to 5 cm; blades usually to 1.5 cm (though sometimes larger, and then resembling *C. officinalis*), reniform to heart-shaped, cordate, sometimes almost truncate at base, and deltate, thin or only slightly fleshy, pale to dark green, often shiny. Lower stem leaves shortly petiolate, triangular to ovate; margins with indistinct teeth. Middle and upper stem leaves sessile, ovate to lanceolate, with or without small auricles with 1-4 teeth on each side. Flowers 5-8 mm in diameter. Petals white. Fruits 3-5 mm, subglobose to ovoid-ellipsoid, narrowed above and often also at base. Valves occasionally slightly asymmetrical, at maturity with a fine reticulate vein patterning in which the mid-vein does not always reach the valve apex, drying to give an areolate pattern which may sometimes be rather weakly evident. Seeds 1.3-1.8 mm. 2n=12*, 24*.

Basic seepage sites on cliffs and banks, beside streams and on old mine spoil heaps. Somerset and upland areas from N. Wales and the Pennines northwards to Shetland. Widespread in mountain ranges from the Pyrenees to the Alps and Carpathians, and also at lower levels in N. W. Europe. Distinguished from *C. officinalis* by the smaller flowers, smaller, frequently darker green, and less fleshy leaves, and by the ovoid to ovoid-ellipsoid fruits (sometimes slightly asymmetrical), with much less developed veining on the valves when mature.

Very variable in stature and leaf size, and in leaf and fruit shape. As interpreted here, the species includes the majority of inland *Cochlearia* populations in upland areas (sometimes put under *C. officinalis*). These are very difficult to treat taxonomically, perhaps because of environmentally-induced variation being superimposed on genetic differentiation in geographically isolated sites. Two subspecies may be recognised, following Clapham *et al.* (1987).

129a. C. pyrenaica DC. subsp. pyrenaica

Pyrenean Scurvygrass

Plants with fruits always narrowed below and leaves not fleshy, and 2n=12+0-2 accessory chromosomes*.

Cochlearia atlantica 128

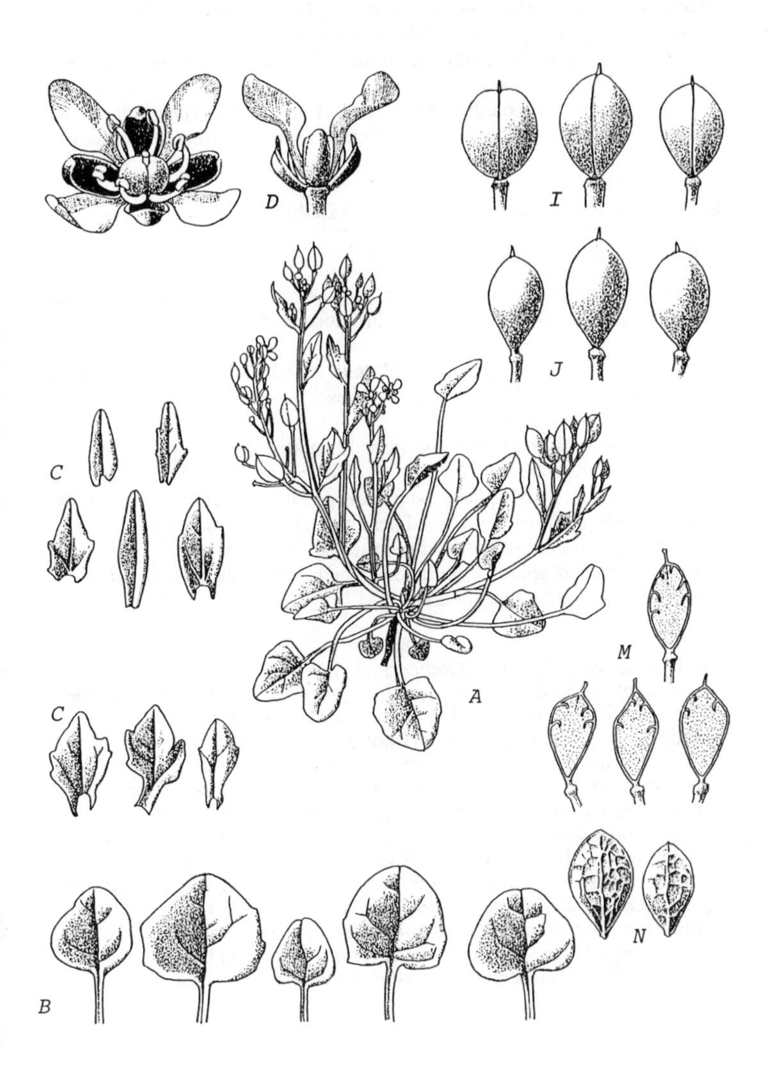

Ax0.75;B,Cx1;Dx3;I,Jx2;M=septum x2;N=ripe valves x2.

Old lead- or zinc-rich spoil heaps, streamside banks and cliffs. S. Lancashire and Derbyshire northwards to Cumberland and Durham, in Scotland only in Skye. Widespread in mountain ranges in the Pyrenees, Alps and Carpathians, also in scattered localities in Belgium, the Netherlands and Germany.

The illustration includes leaves and fruits of this subspecies from Derbyshire; less well-developed plants may not be so different from subsp. *alpina*.

129b. C. pyrenaica DC. subsp. alpina (Bab.) Dalby

Alpine Scurvygrass

Plants with the fruits more rounded at the base and leaves often thicker and fleshier, and 2n=24+0-5 accessory chromosomes*.

Somerset (Cheddar), N. Wales, W. Ireland and Scotland. More western and northern in the British Isles than subsp. *pyrenaica*. Probably also in Faeroes and Scandinavia.

The majority of the leaves and fruits illustrated are from plants derived from material grown in Surrey for eight years, but originating from Shetland. Most small plants with dark green, very shiny foliage from Shetland are probably best placed here, though some from coastal sites in Shetland and northern Scotland seem to be transitional to *C. officinalis*. For northern and upland populations which intergrade with *C. micacea*, see discussion under that species.

130. C. scotica Druce (*C. groenlandica* auct., non L.), *Scottish Scurvygrass*, is closely related to *C. pyrenaica*, but its precise status remains uncertain. The name is usually applied to small plants from coastal sites, often on sandy ground, from Shetland and northern Scotland, with fleshy leaves which are slightly cordate or truncate at the base, and short, square-cut, often lilac petals. These distinctions are not absolute however, since some specimens of *C. pyrenaica* subsp. *alpina* can have truncate leaf bases, and petal shape in this genus is rather unreliable with the limb elongating as it ages. Compact forms with very fleshy leaves certainly look distinct in the field, but these frequently intergrade through to plants very similar to *C. pyrenaica* subsp. *alpina*. The view adopted here, based on limited cultivation studies, is that *C. scotica* is an environmentally-induced form from saline coastal sites in the north. Some specimens from W. Scotland named as *C. scotica* seem to be better placed in *C. atlantica*.

Cochlearia pyrenaica subsp.**alpina 129b**
***subsp.pyrenaica 129a**

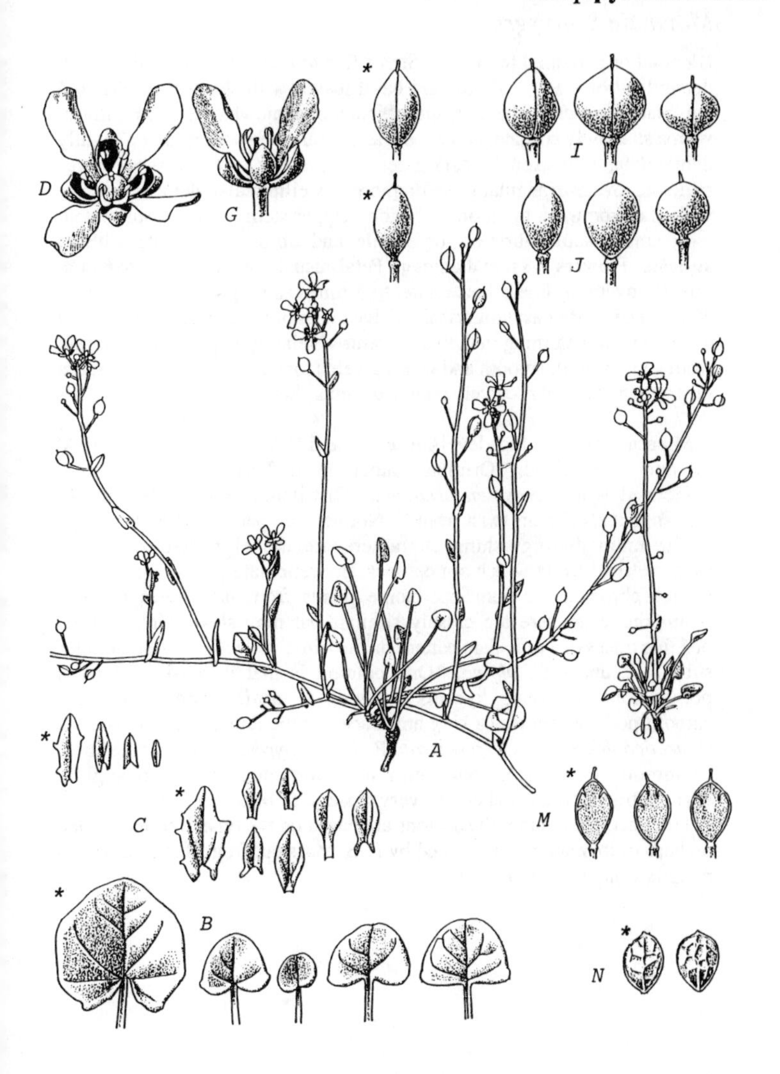

Ax0.75;B,Cx1;Dx3;I,Jx2;M=septum x2;N=ripe valves x2; (*same scales).

131. Cochlearia micacea E. S. Marshall

Mountain Scurvygrass

Biennial or perennial to 10 cm. Stems few to many, simple or branched, decumbent or erect. Rosette and basal leaves with slender petioles to 4 cm; blades to 1 cm, orbicular, suborbicular or somewhat deltate-reniform, with a shallowly cordate, sinuate or rarely truncate base, entire or slightly denticulate, rather tough, dark green, very shiny. Lower stem leaves petiolate, ovate-triangular to oblong, broadly elliptical or deltate, entire or with one prominent tooth on each side. Upper stem leaves clasping stem with small, acute auricles, or sessile and oblong-lanceolate without auricles. Flowers 5-8 mm diameter. Petals usually white. Fruits to 6 mm, broadly ovate- to linear-lanceolate, to 3 times as long as wide, narrowed at both ends, often asymmetrical. Valves with only very slight traces of a delicate vein patterning in the valve centre (veining may be totally absent from some fruits), smooth and lacking veins towards margins, the pattern on drying not or only scarcely related to veins. Seeds 1.6-1.8 mm. 2n=26*.

Micaceous soils at high levels in central and N. W. Scotland (above 1000 m) and in Shetland. Original gatherings in Perth and Argyll were associated with *Cerastium arcticum.* Distribution outside Britain not known, but specimens exist from N. Norway and perhaps elsewhere.

The main distinguishing characters are the elongated and often asymmetrical fruits which almost lack any reticulate patterning, and the unique chromosome number. Some plants from basic sites on high mountains elsewhere are closely similar, but may show slightly more positive traces of a reticulate pattern on the fruits. *C. micacea* is sometimes subsumed under *C. alpina.* One opinion is that it consists of relict populations involving hybrids between *C. alpina* and a northern smooth-podded species lacking any trace of reticulate patterning such as *C. arctica* Schlecht. or *C. fenestrata* R.Br. *C. pyrenaica,* considered here to include *C. alpina,* sometimes has elongated (and also slightly asymmetrical) fruits, and comes very close to *C. micacea.*

This account and the illustrations are based on the type description, and on herbarium material confirmed by E. S. Marshall from Ben Lawers and neighbouring mountains.

Cochlearia micacea 131

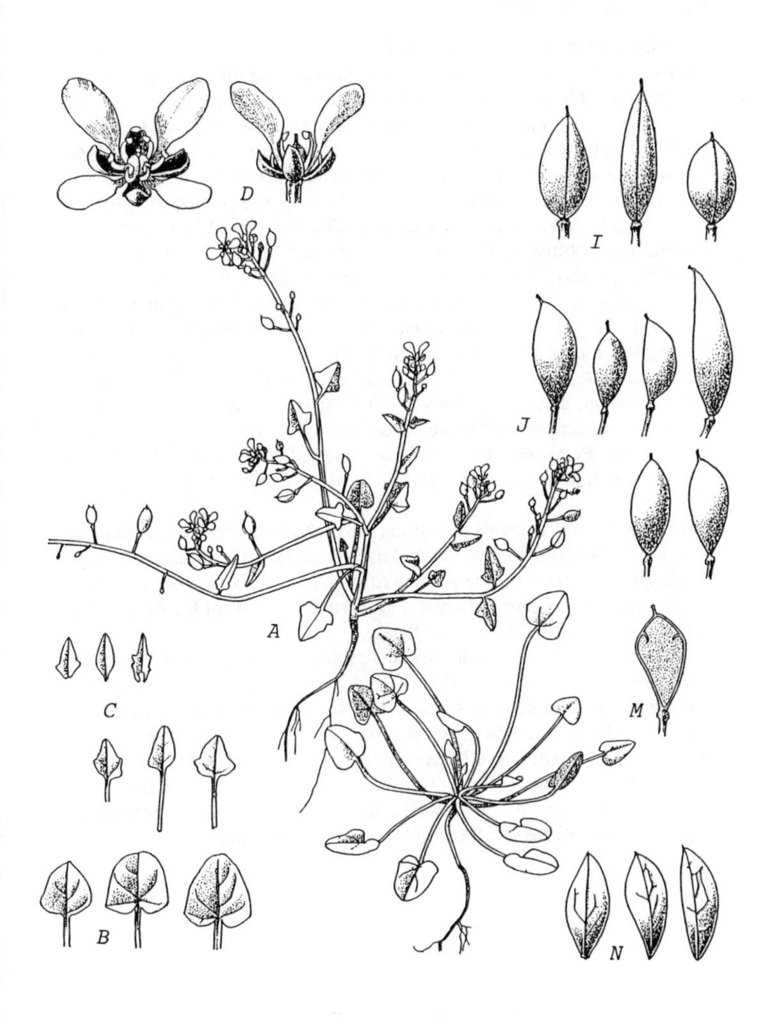

Ax0.75;B,Cx1;Dx3;I,Jx2;M=septum x2;N=ripe valves x 2.

132. Cochlearia anglica L. *

English Scurvygrass

Biennial (-perennial) 7-40 cm, glabrous, light green. Tap-root strong. Stems erect to decumbent, branched or simple. Rosette leaves to 10(-20) cm, petiolate; blade to 3 cm, ovate to obovate, cuneate or less often rounded at base, obtuse at apex; margins entire or with a few blunt teeth. Lower stem leaves petiolate, ovate-elliptical; margins entire or with blunt teeth. Upper stem leaves sessile, with or without auricles clasping stem, oblong. Inflorescence lax. Sepals (2.5-)3-4.9(-5.3) mm, elliptic, green to purple-tipped or red, erect to ascending. Petals 5-10(-12.1) x 2.4-5.7 mm; blade elliptic, obtuse at apex, white, rarely pale mauve; claw short, distinct, greenish. Petals *c.* 2-3 times as long as sepals. Stamens 6; anthers yellow. Stigma ± entire. Pedicels in fruit 5-15 mm, ascending to patent. Fruits 7-14 x 4-11(-12) mm, ± orbicular and rounded at the ends to narrowly elliptic and tapering at the ends, flattened and constricted at the septum (angustiseptate, the septum 1.3-2.5 mm wide, 3-5 times as long as wide), erect to patent, dehiscent. Valves 6-12.5 mm, with many fine veins, but drying to give an often elongated areolate pattern only weakly determined by the veins. Persistent style 0.7-2 mm, linear. Seeds 2-many, 1.5-2.3 mm, ovoid, brown, biseriate. 2n=48*(54,60). Flowering March to July.

Soft mud at lower levels of saltmarshes and along creek edges at higher levels in S. England, Wales and Ireland. Records for Scotland (cf. map in Perring & Walters 1962) require verification, most being referable to the *C. officinalis* aggregate. Atlantic and North Sea coasts of Europe, north to Finnmark and, in the Baltic, to Öland.

Often the most robust of the British members of the genus. It is probably less variable than *C. officinalis* s.l., and is generally characterised by the flattened fruits, large seeds and basal leaves with cuneate bases. *C. officinalis*, however, sometimes has stem leaves which are cuneate at the base, and such plants have been named as *C. islandica* Pobed.

Hybrids of *C. anglica* and *C. officinalis* s.l. (*C.* x *hollandica* Henrard) may be over-recorded due to morphological variation of the parents, and many reports require verification by chromosome counts.

* by D. H. Dalby.

Cochlearia anglica 132

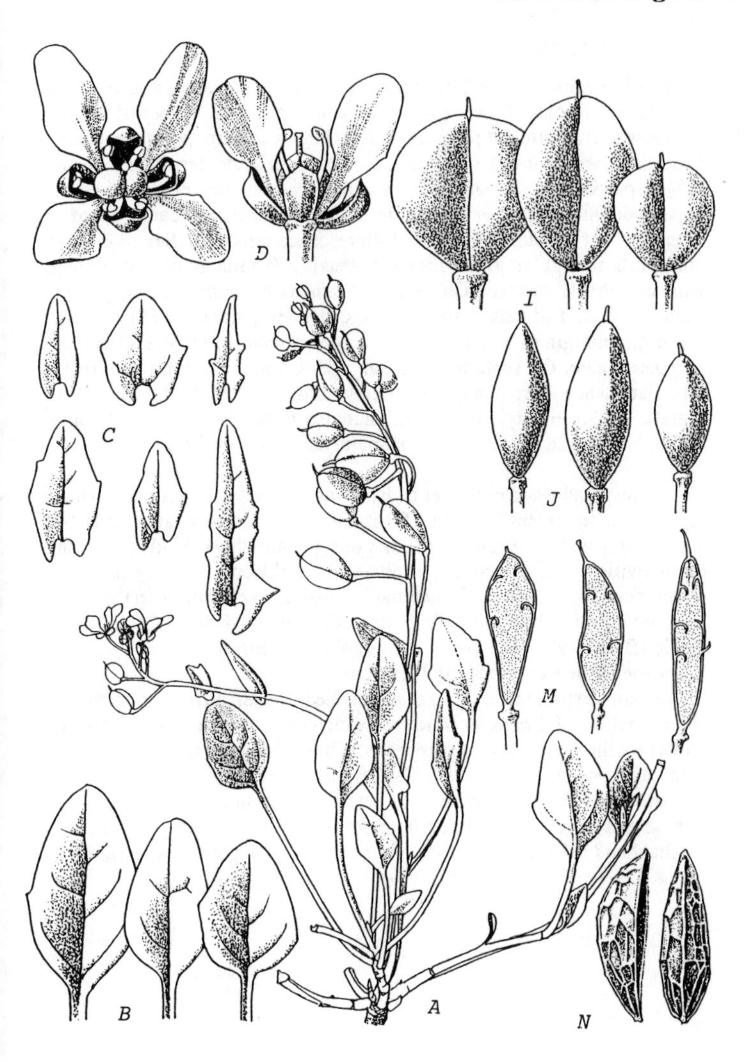

Ax0.75;B,Cx1;Dx3;I,Jx2;M=septum x2;N=ripe valves x2.

133. Coronopus didymus (L.) Sm.

Lesser Swinecress

Annual (?biennial) 5-35(-50) cm, glabrous or with sparse, simple hairs, light green, foetid. Stems procumbent or ascending, branched mainly above. Basal rosette not persisting. Stem leaves to 5 cm, petiolate, deeply pinnatisect with a linear-oblanceolate to oblong, acute, terminal lobe and 2-5(-6) pairs of similar lateral lobes which are often acutely lobed on the upper side; margins otherwise entire. Racemes terminal, leaf-opposed or in the axils of the stem branches. Inflorescence crowded. Sepals *c.* 0.5-1 mm, ovate-triangular, green, inclined. Petals *c.* 0.5 mm or absent, oblong, whitish. Petals shorter than sepals. Stamens 6; anthers 2(-4), yellow. Stigma entire. Pedicels in fruit 1-4 mm, slender, patent. Fruits 1.3-1.7 x 2-3 mm, emarginate at apex and base (constricted in middle), flattened (angustiseptate, the septum *c.* 0.2 mm wide), smooth when fresh, finely reticulate when dry, patent, separating into 2 indehiscent, subglobose nutlets. Persistent style 0-0.2 mm, included in notch. Seeds *c.* 0.7 mm ovoid, brown. 2n=32*. Flowering mainly June to October.

A widespread weed of flower beds, lawns, roadsides, tips, docks, ballast fields, waste ground, paths, farmyards, etc., rarely abundant but surprisingly widespread in the countryside, presumably spread by man and his activities. First recorded in the early eighteenth century and now widespread in S. England, Wales and S. Ireland, less common in the north. Commoner now than shown in Perring & Walters (1962). Of unknown origin (Baker 1972) and now a cosmopolitan weed.

It varies little except in length of stem.

Coronopus species are easily distinguished from all other crucifers (with the exception of *Carrichtera annua* which has large petals and distinct fruits) by the lateral inflorescences which are opposite the leaves on the stem and not in their axils. *C. didymus* is easily separated from *C. squamatus* by the usual absence of petals and the small fruits emarginate at base and apex (like a miniature dumb-bell).

Chauhan (1979) has suggested the plants may be pollinated by ants.

Ax0.3;B,Cx1;D-Gx10;Hx30;I-Kx10;Lx15.

134. **Coronopus squamatus** (Forskål) Ascherson

Swinecress, Wart Cress

Annual (?biennial) (3-)5-50 cm, glabrous, dark green. Stems procumbent to ascending, branched above and below. Leaves in basal rosette and on stem to 10(-20) cm, petiolate, oblong in outline, pinnatisect with a linear-oblong, acute terminal lobe and 2-7(-9) pairs of linear-oblong lateral lobes which are usually acutely to obtusely lobed on the upperside (the leaves thus ± bipinnatisect); margins otherwise entire. One raceme in the centre of the basal rosette terminating the main stem, lateral stems with leaf-opposed racemes. Inflorescence crowded. Sepals 1-1.5 mm, oblong, green, inclined to patent. Petals 1-2 x 0.5-1.1 mm, obovate to oblong, rounded at apex, white, unclawed. Petals 1-2 times as long as sepals. Stamens 6; anthers purple (?yellow). Stigma capitate to entire. Pedicels in fruit 0.5-3 mm, stout, inclined to patent. Fruits 2.3-3.5 x 3-4.7 mm, reniform, flattened and somewhat constricted at the septum (angustiseptate), apiculate above with a persistent, ± conical-triangular style *c.* 0.3-1 mm, emarginate below, with usually persistent sepals, coarsely reticulate with raised ridges, patent, indehiscent. Seeds 1 per loculus, *c.* 1.5 mm, ovoid, brown. 2n=32*. Flowering May to October.

A plant of gateways, paths, gardens, waste ground, shingle, farmyards, damp patches and pastures. Tolerant of trampling and preferring nutrient-rich ground. Common in the south and east of England, frequent in Wales, N. England, Scotland and Ireland usually near the coast (map in Perring & Walters 1962). Probably native near the coast in S. E. England. Native in W. and C. Europe and around the coasts of the Mediterranean. Introduced to N. America, southern Africa and Australasia.

Variable in leaf dissection and in overall size. Dwarf plants may lack an aerial stem and the inflorescence develops in the axil of the basal rosette.

Easily distinguished by the leaf-opposed lateral racemes and apiculate fruits. The large white petals will distinguish it from *C. didymus* when fruits are not available.

Ax0.3;B,Cx1;D-Gx10;Hx30;I-Kx10;Lx15.

135. **Carrichtera annua** (L.) DC.

Annual 15-40(-60) cm, with simple hairs. Stem leaves to 10 cm, petiolate, pinnatisect to bipinnatisect with linear, obtuse lobes. Lateral inflorescences arising opposite leaves. Petals 6-8 mm, pale yellow (fading white), *c.* 1.5-2 times as long as sepals. Fruits 6-8 x 1.5-3.5 mm, dehiscent; valves 2-4 mm, with rough hairs on the 3 veins; terminal segment 3-4 mm, ovate, flattened, cupped, sterile. Seeds 3-4 per loculus.

A rare casual of ports, fields, waste ground, etc. (Map 135). The lateral inflorescence arising opposite a leaf and not in its axils is a diagnostic character otherwise only found in *Coronopus* (which has small flowers and quite different fruits), and the wonderful little fruits (rather like a miniature *Sinapis alba* fruit) are distinctive. Native in the Mediterranean region.

136. **Pachyphragma macrophyllum** (Hoffm.) N. Busch

Perennial 15-40 cm, glabrous, rhizomatous and forming patches, smelling of garlic when crushed. Basal leaves petiolate, ovate to reniform; margins sinuate to crenate. Stem leaves smaller, petiolate. Petals *c.* 8-9 mm, white, *c.* 3-4 times as long as sepals. Fruits 8-12 x 10-16 mm, ± transversely elliptic, flattened (angustiseptate), broadly winged, notched, dehiscent. Loculi 1-2 seeded.

Rarely naturalized as a garden escape (there are 2 post-1950 records). Easily distinguished by the garlic smell which otherwise occurs only in *Alliaria* (which has linear, long fruits) and *Thlaspi alliaceum* (which has upper stem leaves with clasping auricles). For full details see Davie & Akeroyd (1983).

137. **Myagrum perfoliatum** L.

Annual 20-60(-100) cm, glabrous, glaucous. Stem leaves sessile, lanceolate, with clasping auricles; margins entire or shallowly toothed. Petals 3-5 mm, yellow, about twice as long as sepals. Fruits 5-8 x *c.* 3-5 mm, broadly club-shaped to obovate, compressed, erect, indehiscent with 3 loculi: the upper 2 loculi side by side, sterile, the lower loculus with 1 seed. Persistent style *c.* 1 mm, conical.

A rare casual imported with grain (Map 137). The fruits are distinct and quite unlike those of any other crucifer. There is an excellent illustration by G.M.S. Easy in *BSBI News* **47**:34-35 (1987).

138. **Cardamine heptaphylla** (Vill.) O. E. Schulz

Perennial 30-60 cm, with sparse, simple hairs. Rhizome 4-10 mm in diameter, creeping. Leaves 3-5, all on upper part of stem, petiolate, pinnate with a sessile, lanceolate to oblanceolate, acute, terminal lobe and 2-5 pairs of similar lateral lobes; margins acutely toothed. Petals 11-20 x 5-9 mm, white, pink or purple, *c.* 2-3 times as long as sepals. Fruits 35-85 x 3-5 mm, flattened, linear, dehiscent. Persistent style 3-11 mm.

A garden plant rarely established (there are 3 post-1950 records). Somewhat similar to *C. bulbifera* but without the bulbils in the leaf axils.

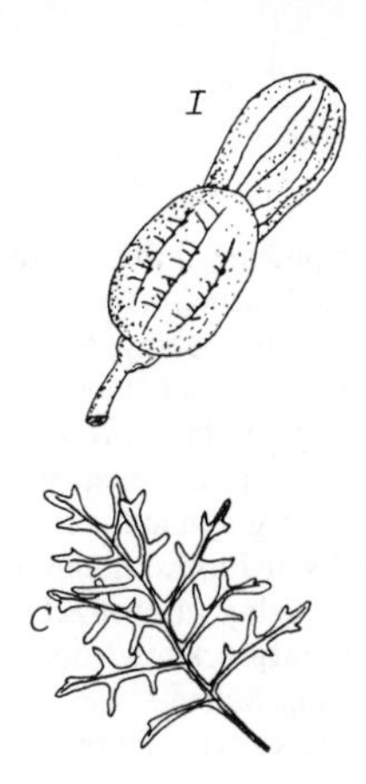

Carrichtera annua
Cx0.5;Ix4.

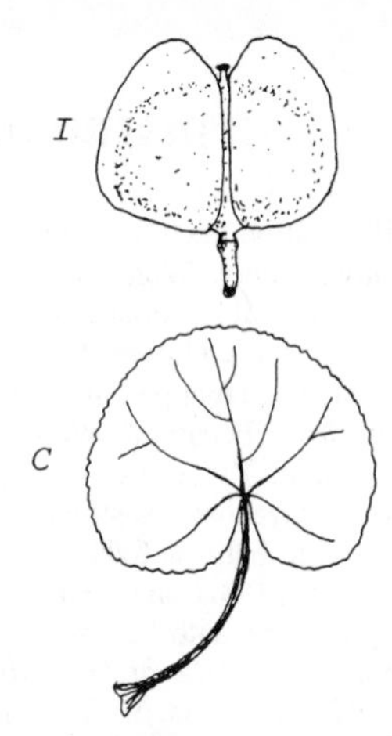

Pachyphragma macrophyllum
Cx0.3;Ix2.

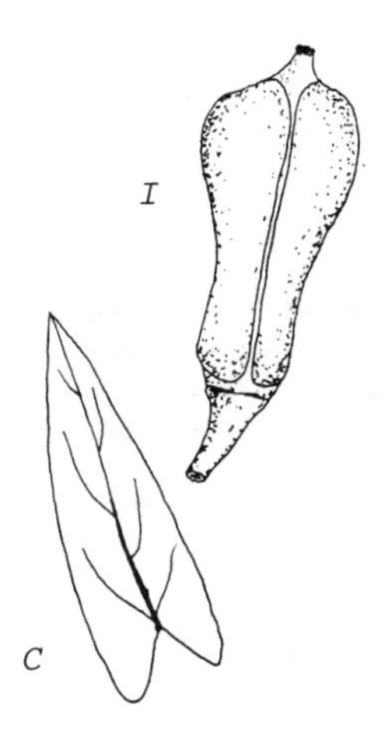

Myagrum perfoliatum
Cx1;Ix7.

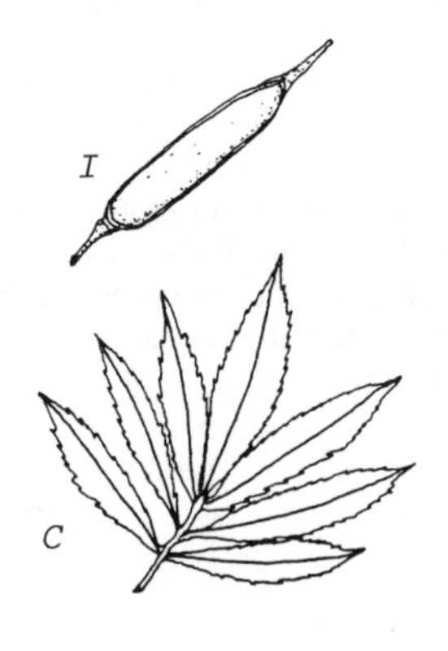

Cardamine heptaphylla
Cx0.3;Ix0.5.

DISTRIBUTION MAPS

The distribution maps were prepared with the help of Jane Croft assisted by Claire Appleby, Wendy Forest, Chris Preston and Rosemary Woodruff at the Biological Records Centre, Institute of Terrestrial Ecology, Monks Wood. Species chosen for mapping include those which have not previously been mapped, have poor or out-dated maps in the *Atlas of the British Flora* (Perring & Walters 1962) or where a map may help with identification. References to maps of other species are given in the text.

Records have been abstracted from about 100 local Floras, about 10 herbaria (the help of Sylvia Reynolds and Stan Beesley in abstracting records from **DBN** and **BEL** respectively is gratefully acknowledged), correspondence with botanists, the BSBI Monitoring Scheme, the data collected for the *Atlas of the British Flora* and other miscellaneous records held at BRC, and from the field. The maps have not been checked by BSBI Recorders or other botanists.

The symbols used on the maps are as follows:-

□ 1660 - 1949; record confirmed by expert

■ 1950 - 1989; record confirmed by expert

○ 1660 - 1949; other acceptable records

● 1950 - 1989; other acceptable records

The maps on average probably include about 70% of the total number of available records. For instance, there are few or no historical records for Norfolk, Suffolk or Northamptonshire, or recent records for counties with modern tetrad Floras, and many records in herbaria have not been abstracted.

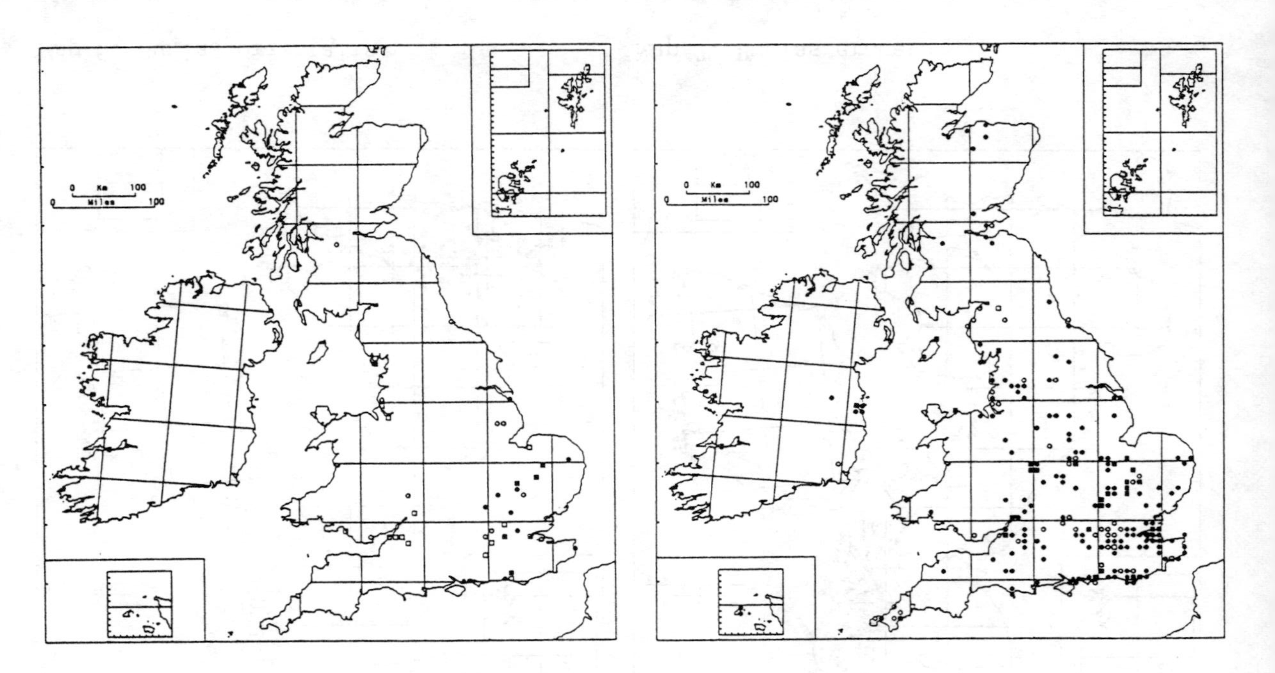

Map 4. Rapistrum rugosum

Map 5. Rapistrum perenne

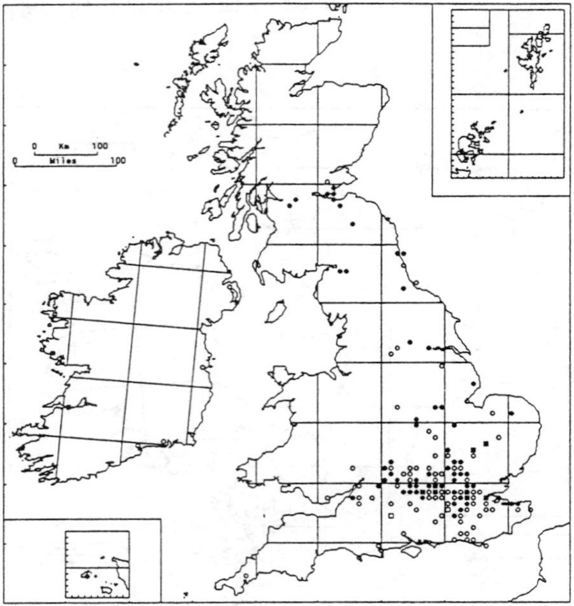

Map 6. Bunias orientalis

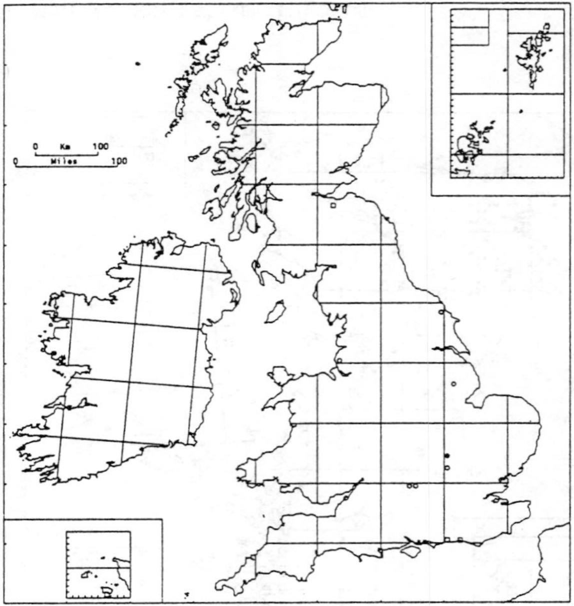

Map 7. Bunias erucago

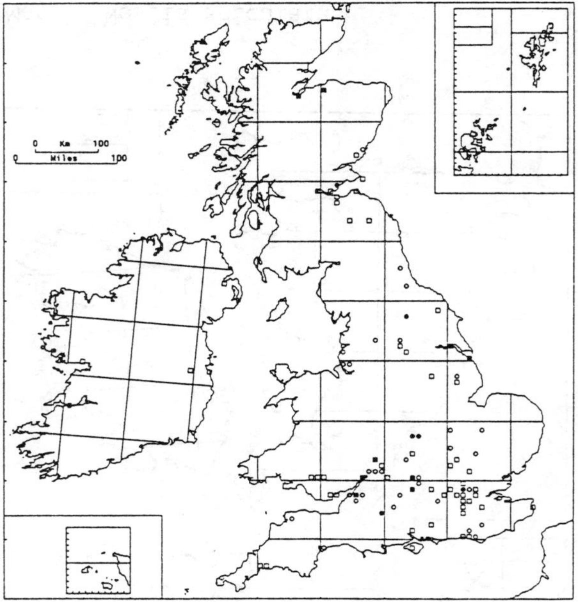

Neslia (all records)

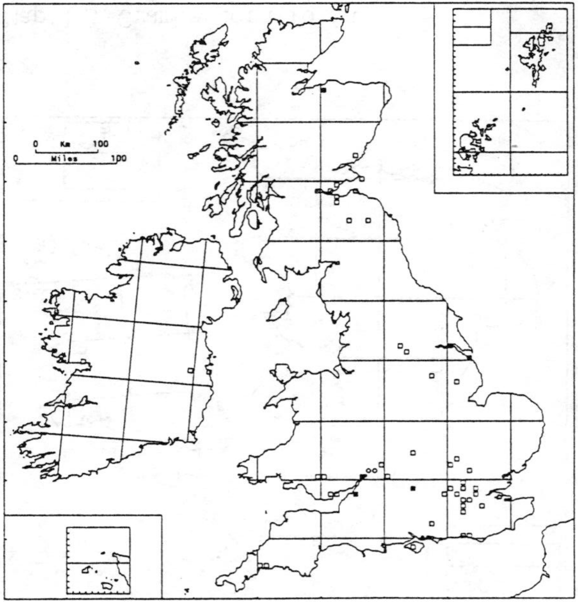

Map 8. Neslia paniculata s.s.

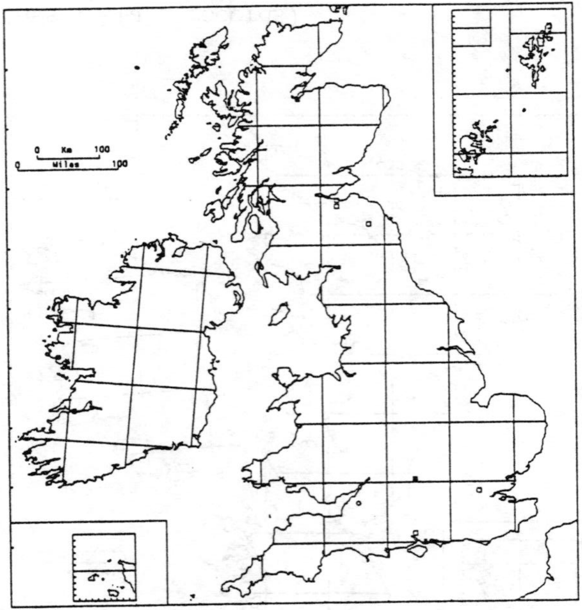

0 Km 100
0 Miles 100

Map 9. Neslia apiculata

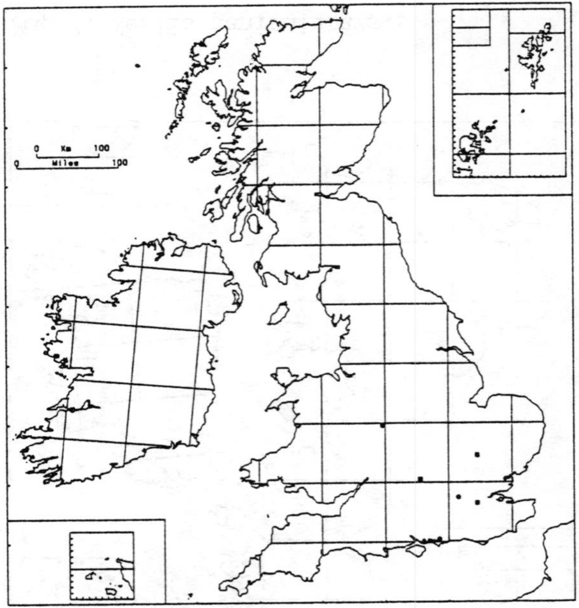

0 Km 100
0 Miles 100

Map 11. Crambe cordifolia

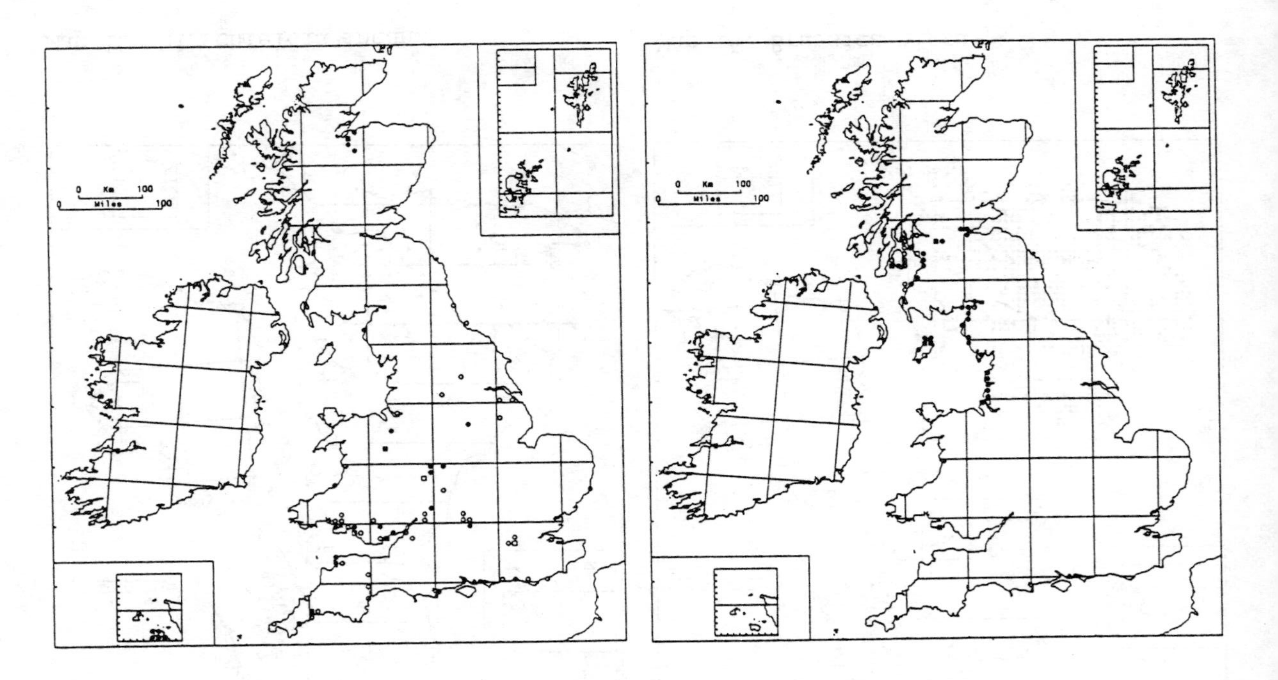

Map 14b. Coincya monensis subsp. recurvata

Map 14a. Coincya monensis subsp. monensis

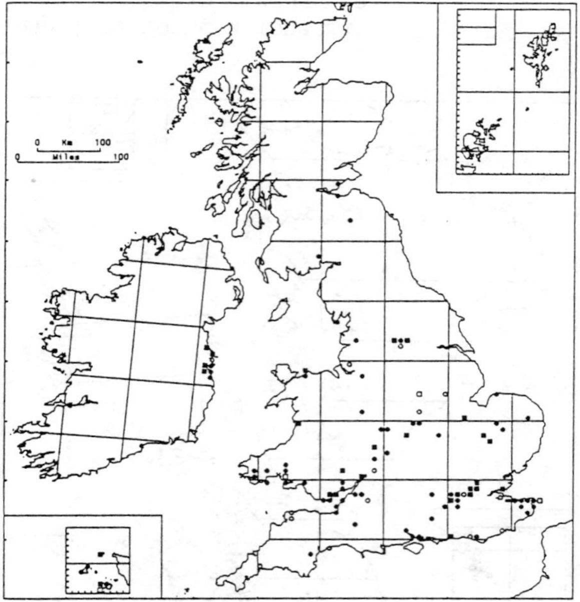

Map 15. Hirschfeldia incana

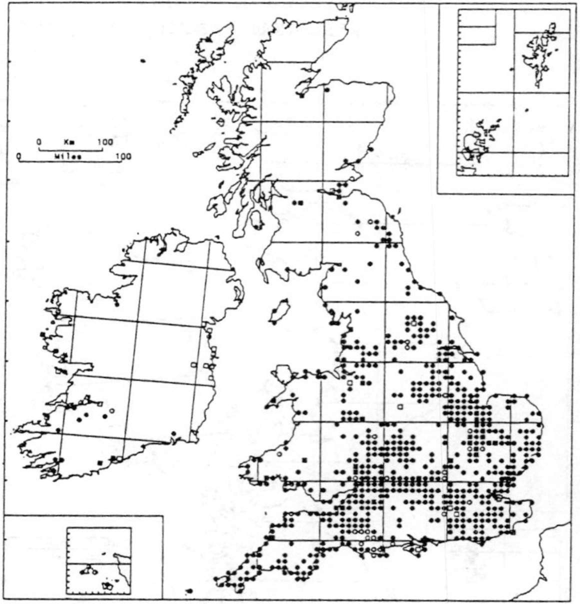

Map 16. Brassica nigra

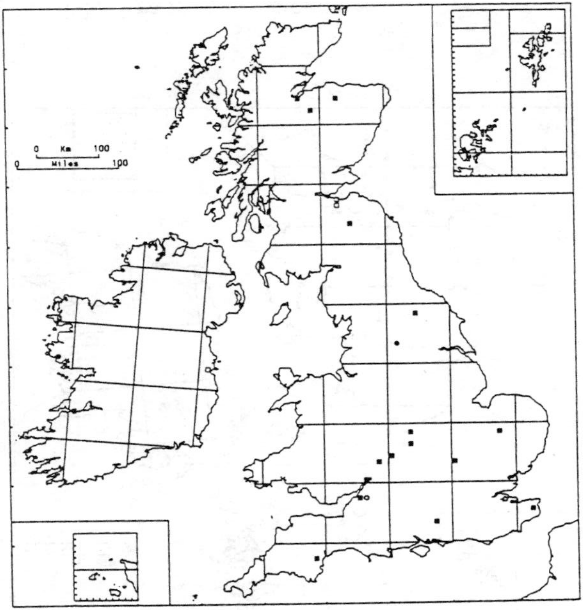

0 Km 100
0 Miles 100

Map 21. Brassica tournefortii

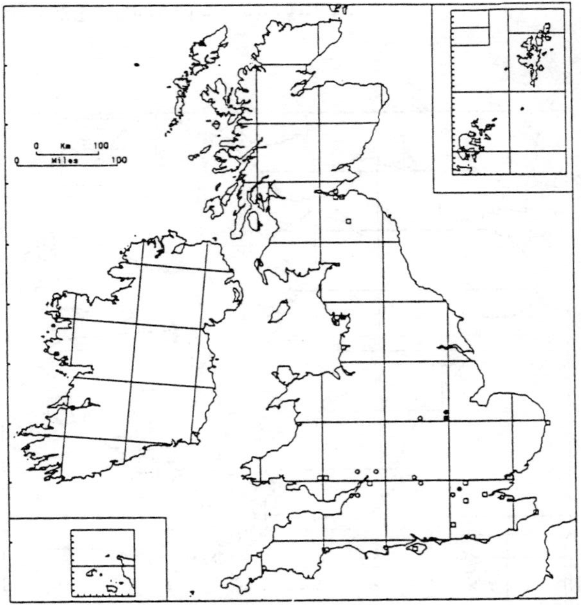

0 Km 100
0 Miles 100

Map 20. Brassica elongata

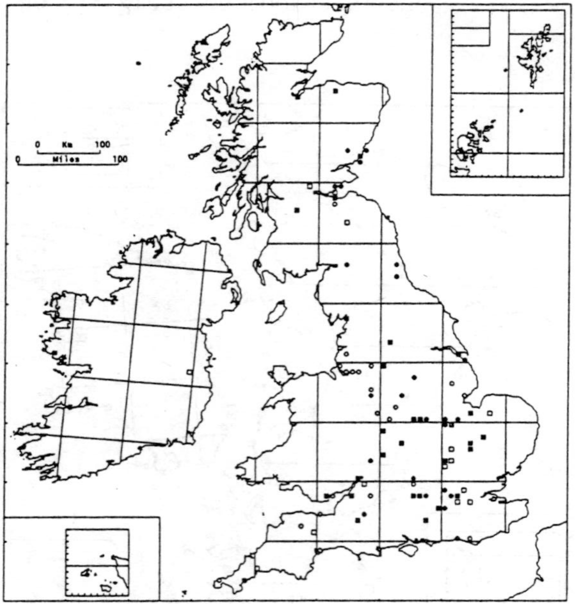

0 Km 100
0 Miles 100

Map 22. Brassica juncea

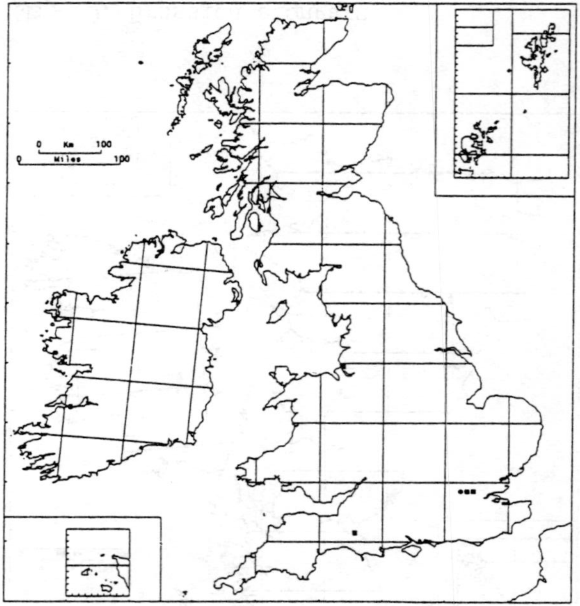

0 Km 100
0 Miles 100

Map 23. Brassica carinata

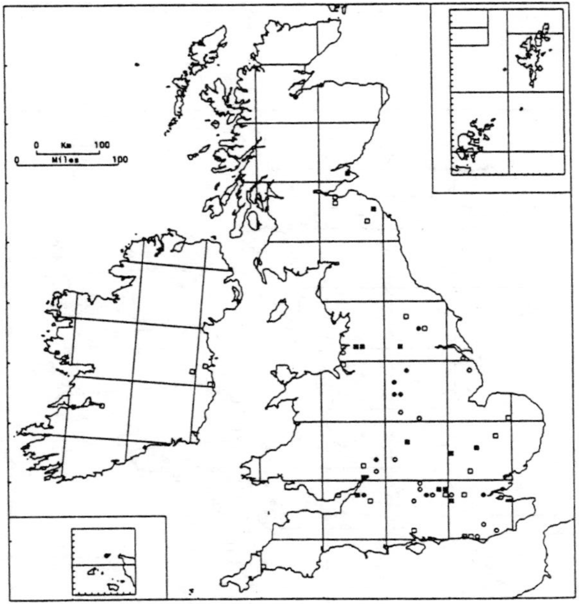

Map 26. *Eruca vesicaria*

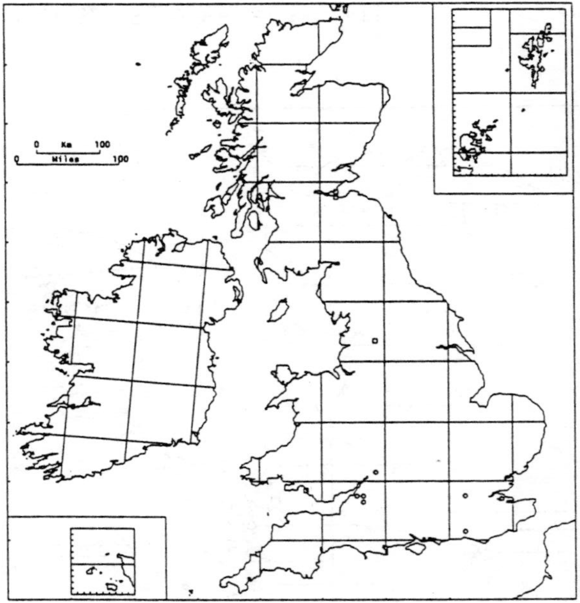

Map 25b. *Sinapis alba* subsp. *dissecta*

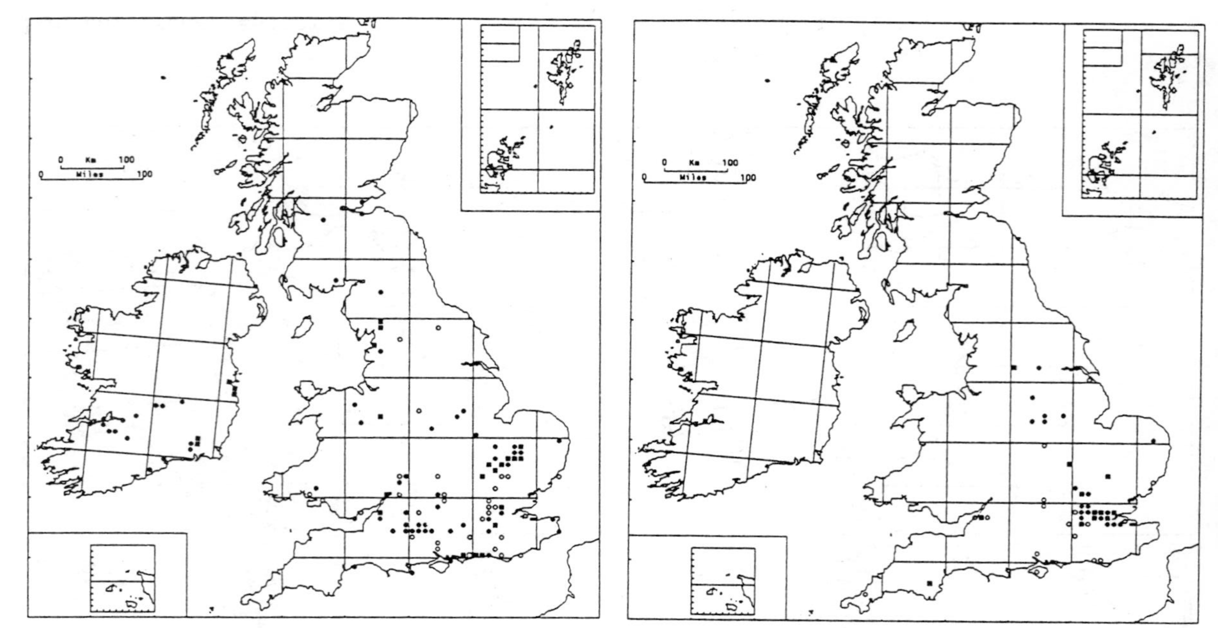

Map 29. Erucastrum gallicum

Map 32. Sisymbrium loeselii

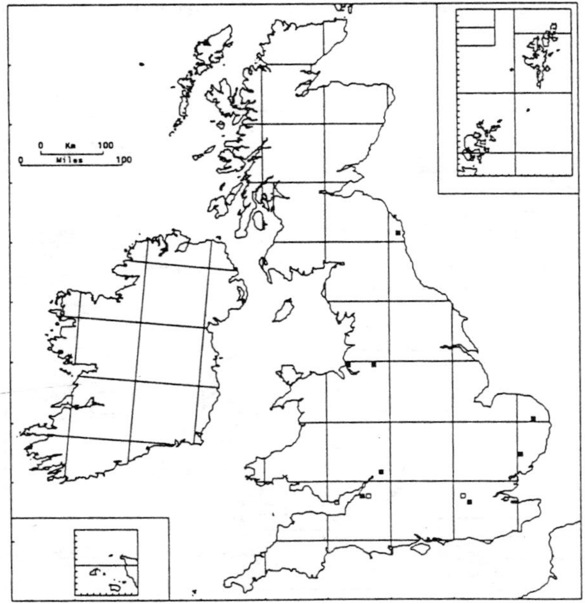

0 Km 100
0 Miles 100

Map 37. Sisymbrium volgense

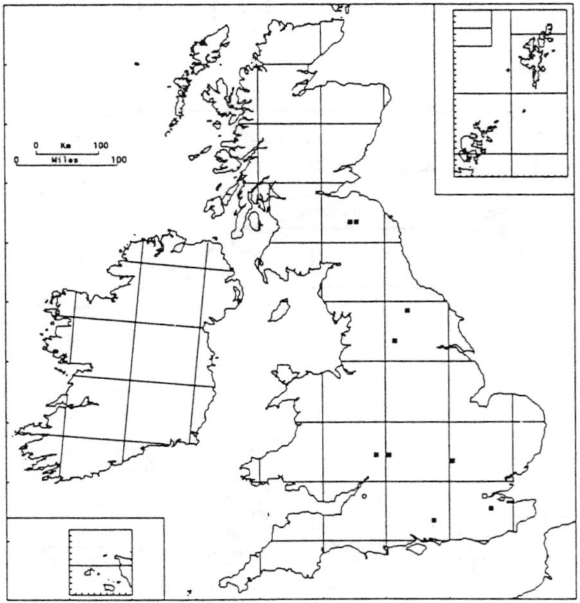

0 Km 100
0 Miles 100

Map 33. Sisymbrium erysimoides

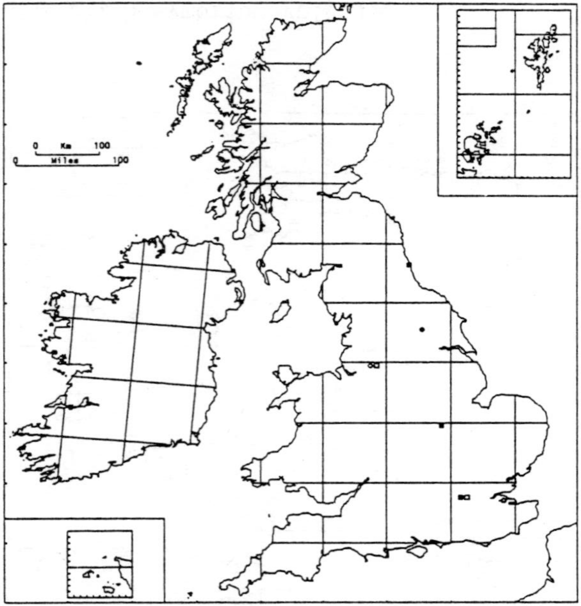

Map 38. Sisymbrium strictissimum

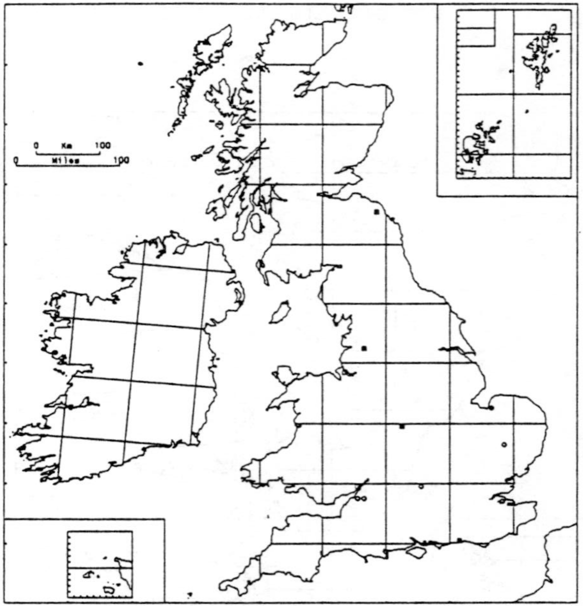

Map 39. Sisymbrium polyceratium

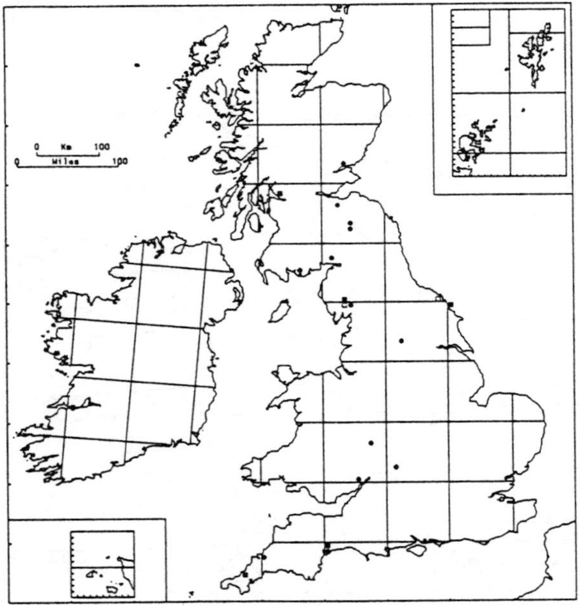

0 Km 100
0 Miles 100

Map 56. Cardamine raphanifolia

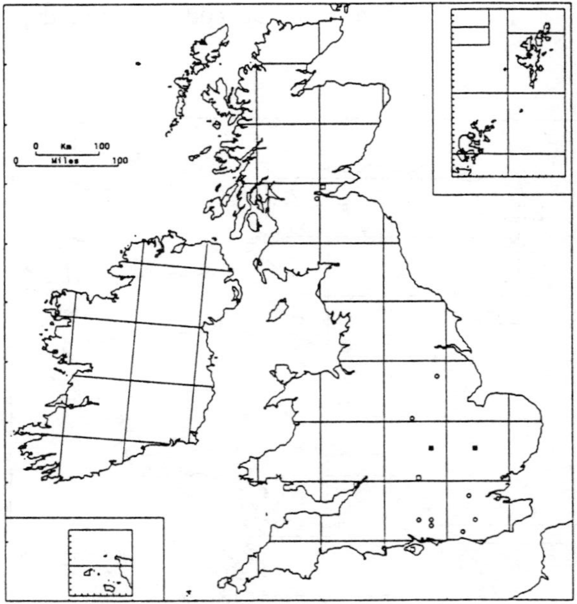

0 Km 100
0 Miles 100

Map 46. Arabis turrita

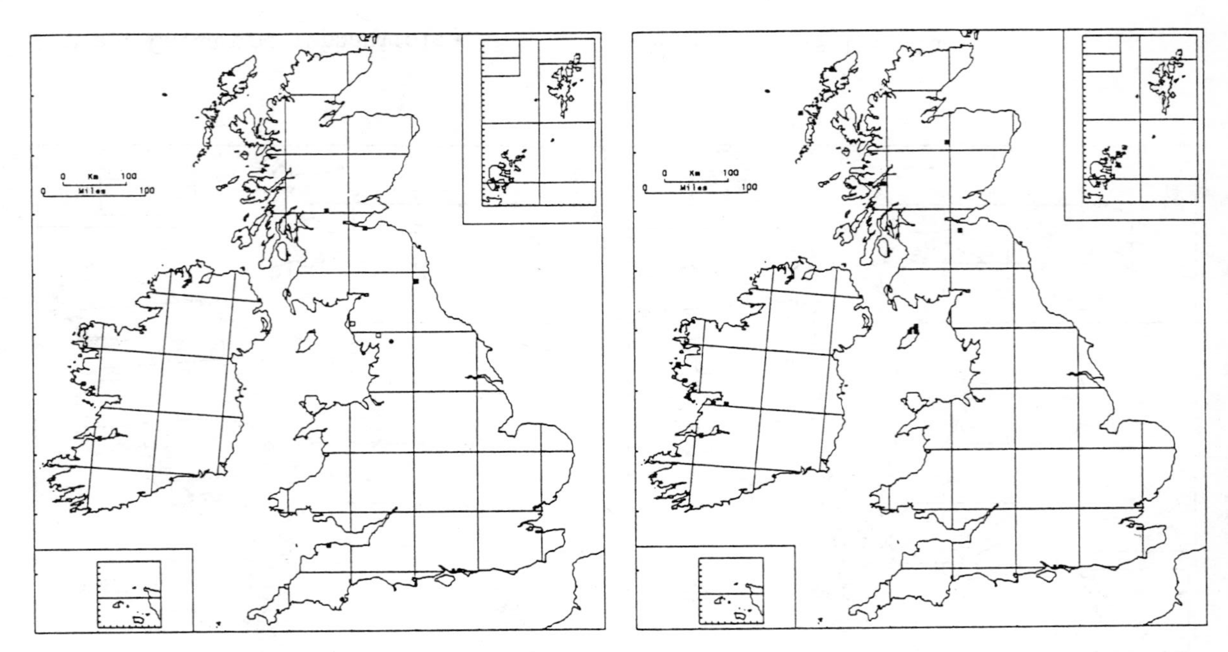

Map 57. Cardamine trifolia

Map 62. Rorippa islandica s.s.

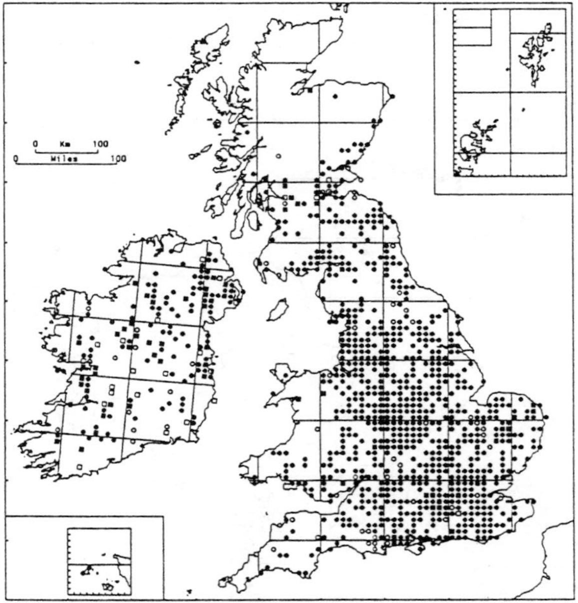

0 Km 100
0 Miles 100

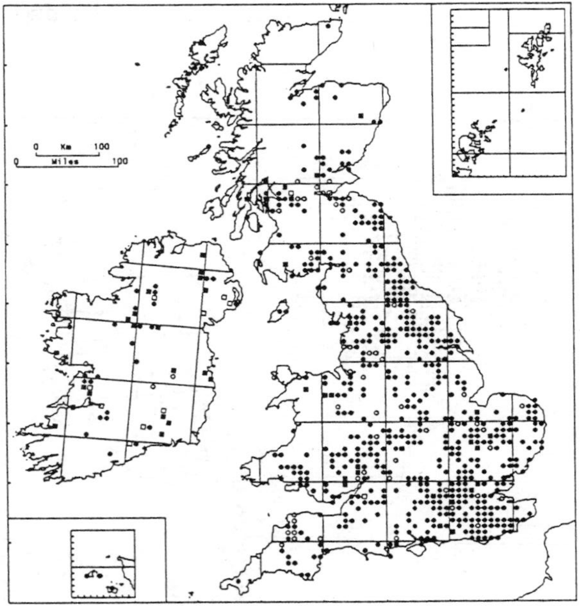

Map 64. Rorippa sylvestris

0 Km 100
0 Miles 100

Map 63. Rorippa palustris

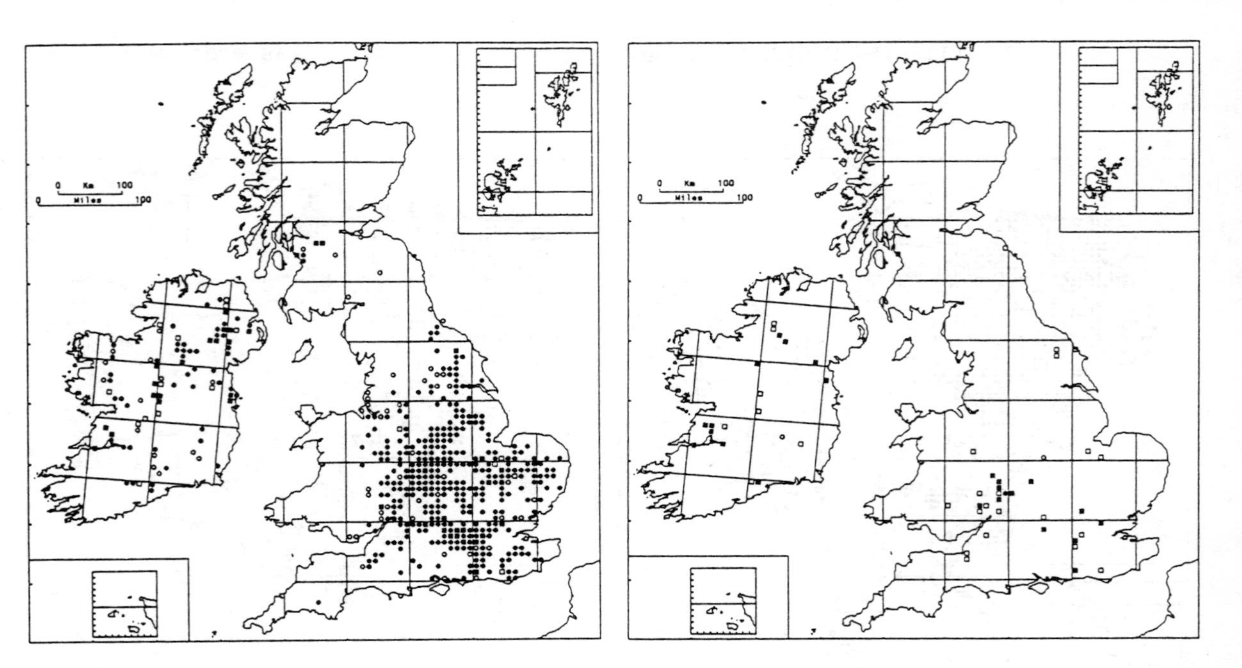

Map 65. Rorippa amphibia

Map 66. Rorippa x anceps
(R. amphibia x sylvestris)

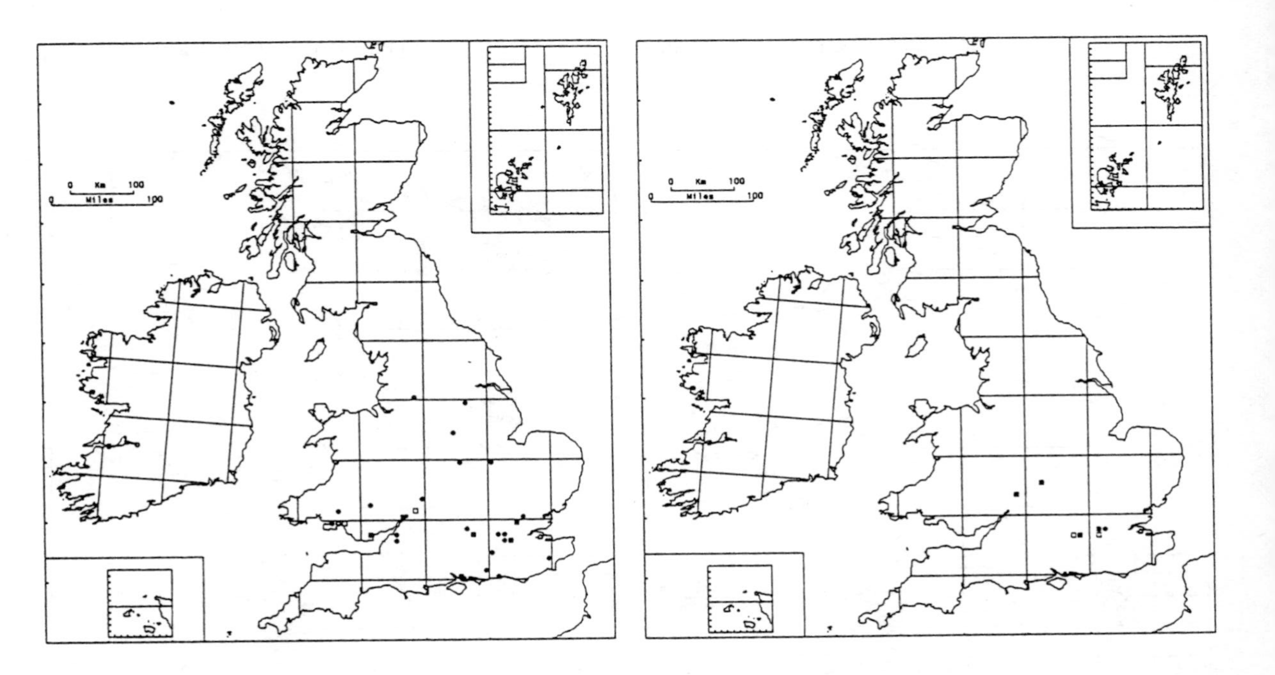

Map 68. Rorippa austriaca

Map 67. Rorippa x erythrocaulis
(R. amphibia x palustris)

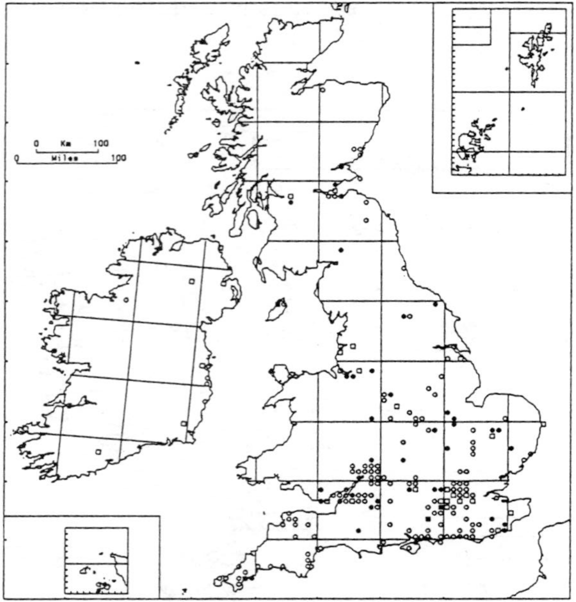

Map 73. Conringia orientalis

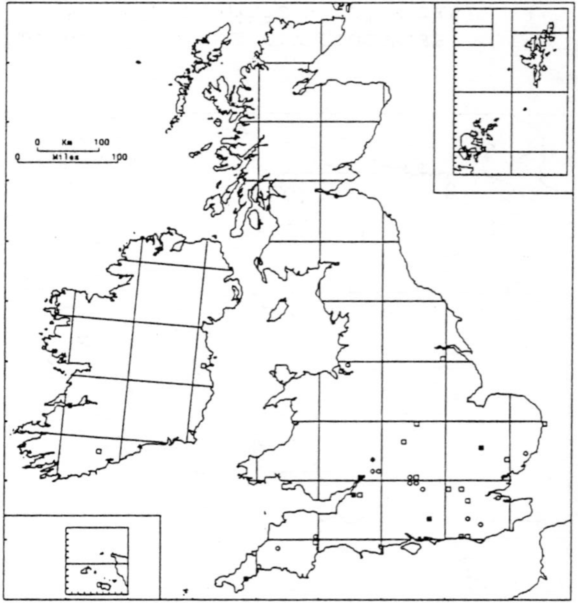

Map 75. Erysimum repandum

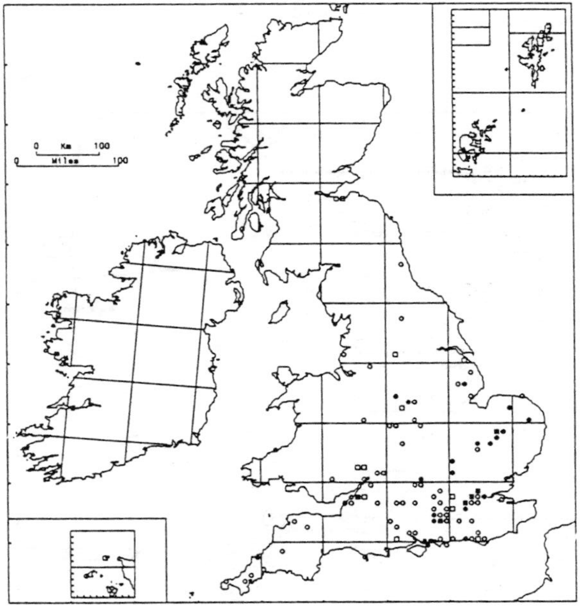

0 Km 100
0 Miles 100

Map 88. Berteroa incana

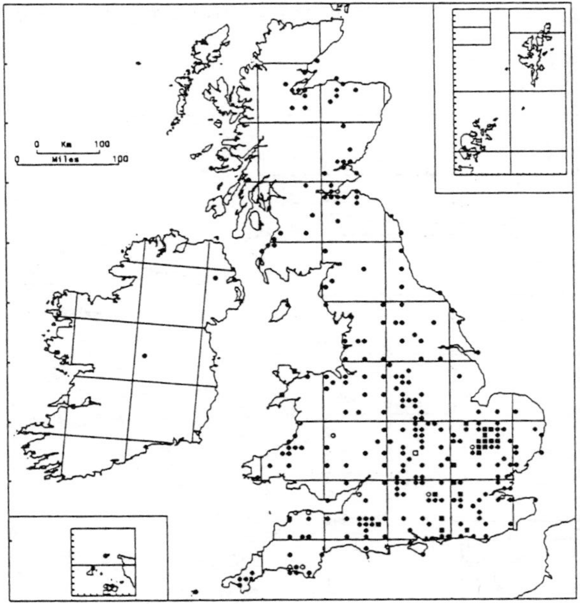

0 Km 100
0 Miles 100

Map 84. Lunaria annua

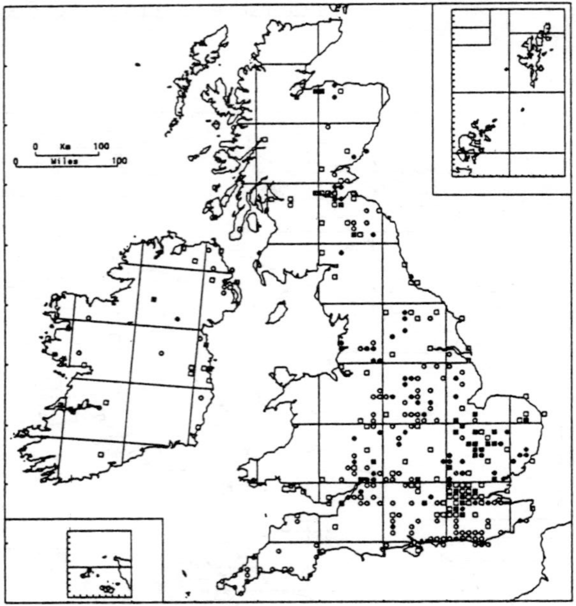

Camelina (all records)

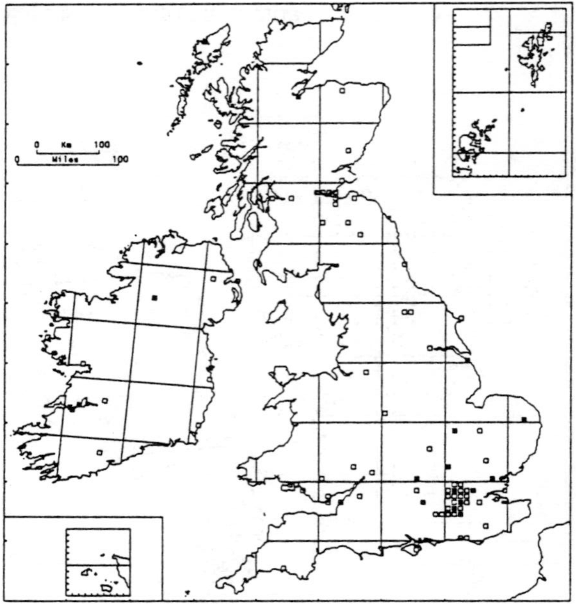

Map 91. Camelina sativa s.s.

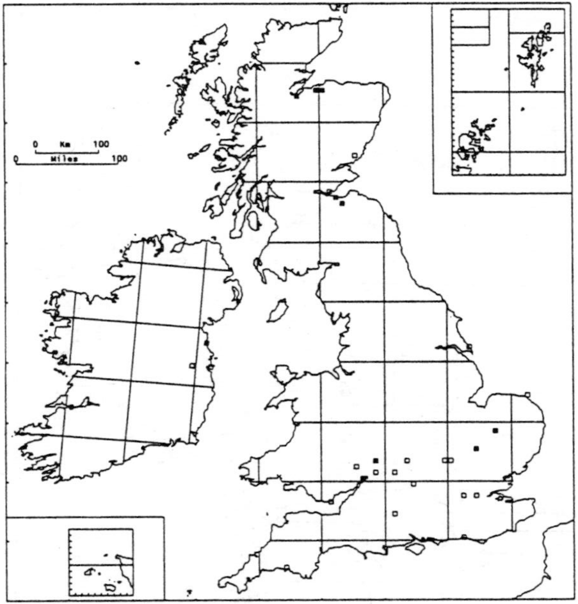

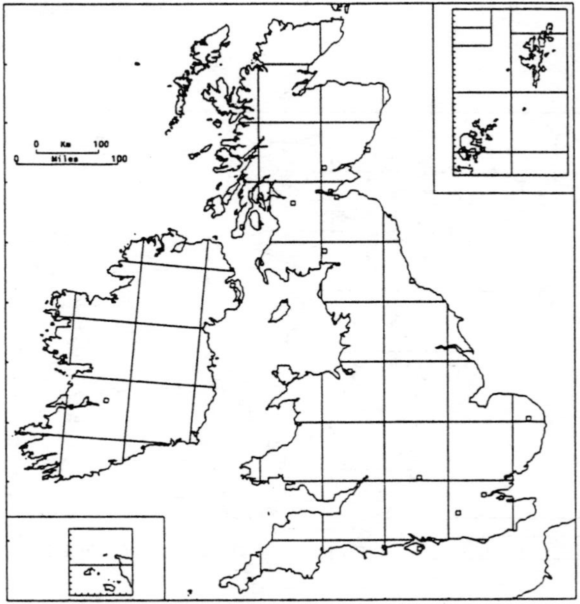

Map 92. Camelina alyssum

Map 93. Camelina microcarpa

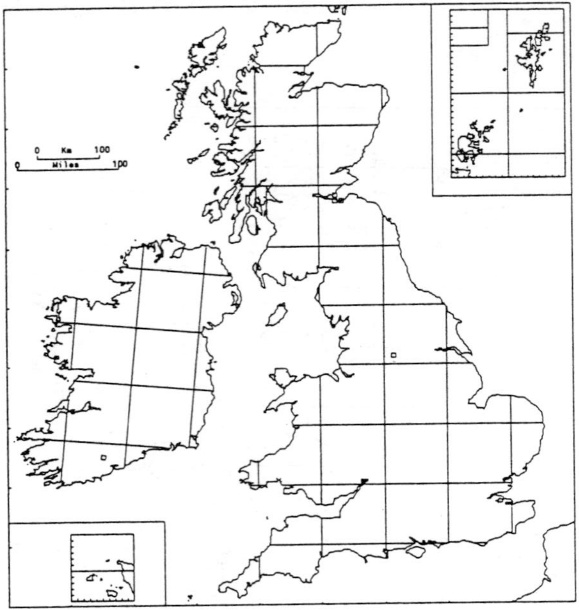

Map 94. *Camelina rumelica*

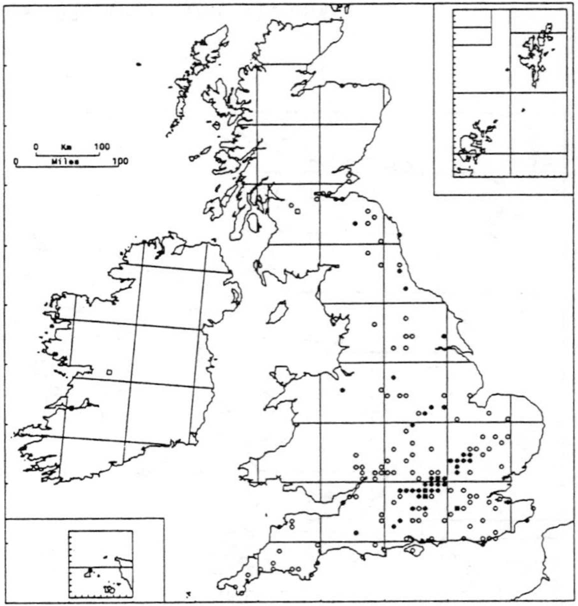

Map 95. *Iberis amara*

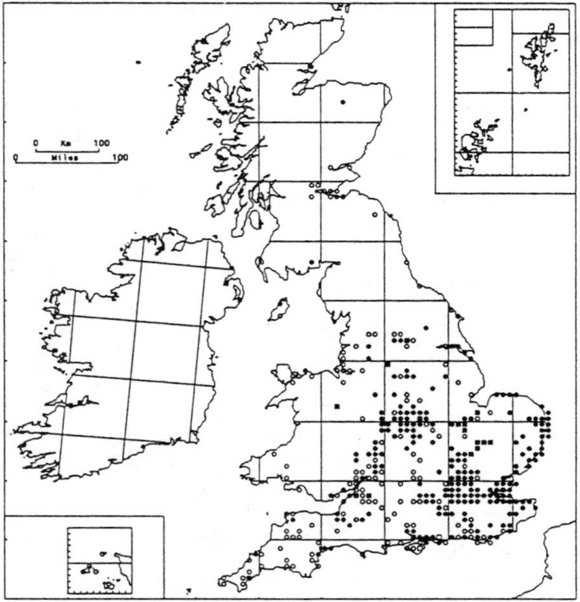

Map 100. Lepidium ruderale

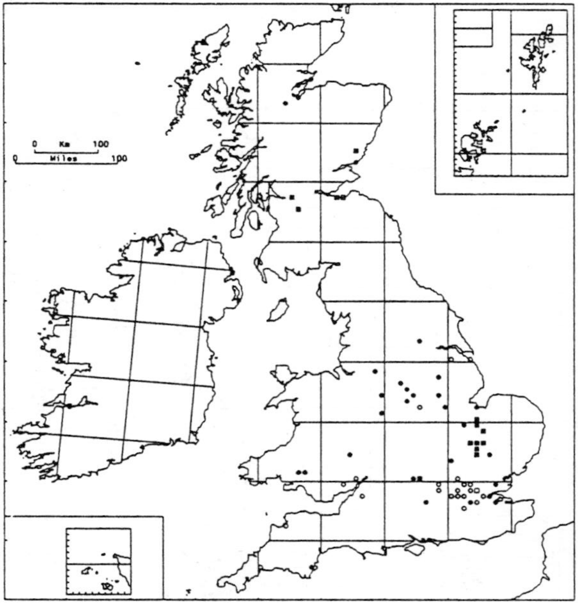

Map 96. Iberis umbellata

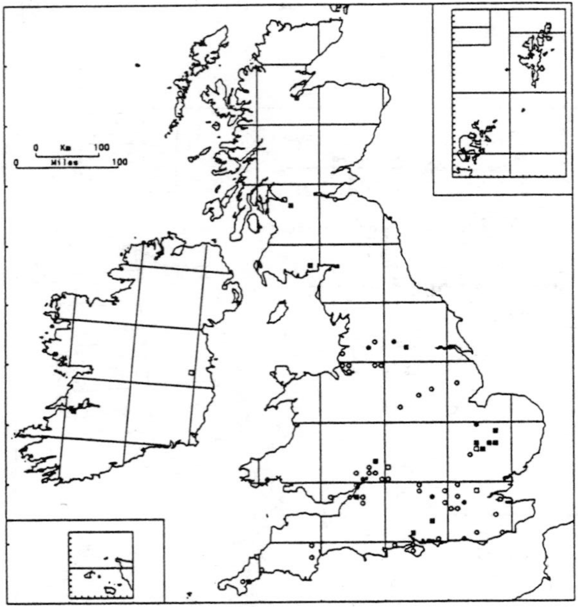

Map 101. Lepidium densiflorum

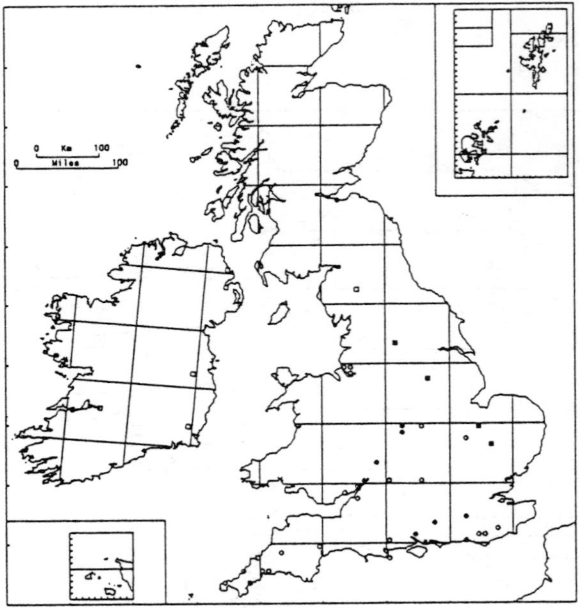

Map 102. Lepidium virginicum

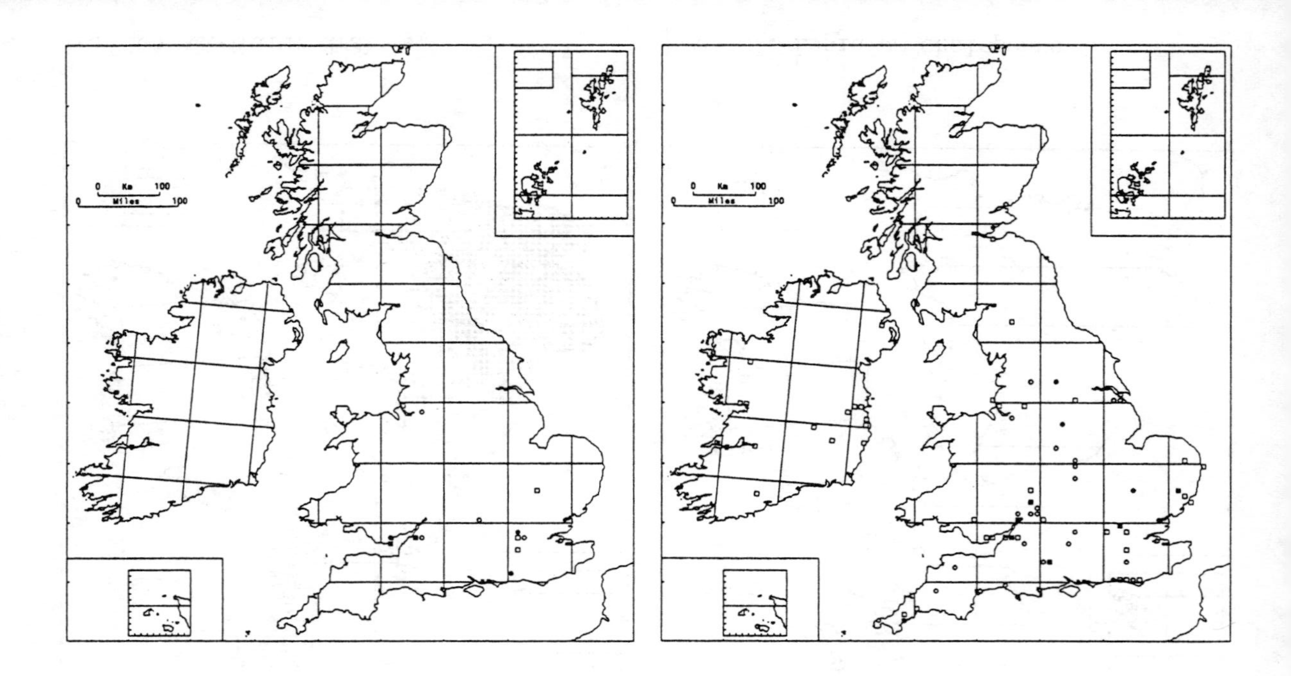

Map 104. Lepidium graminifolium

Map 103. Lepidium perfoliatum

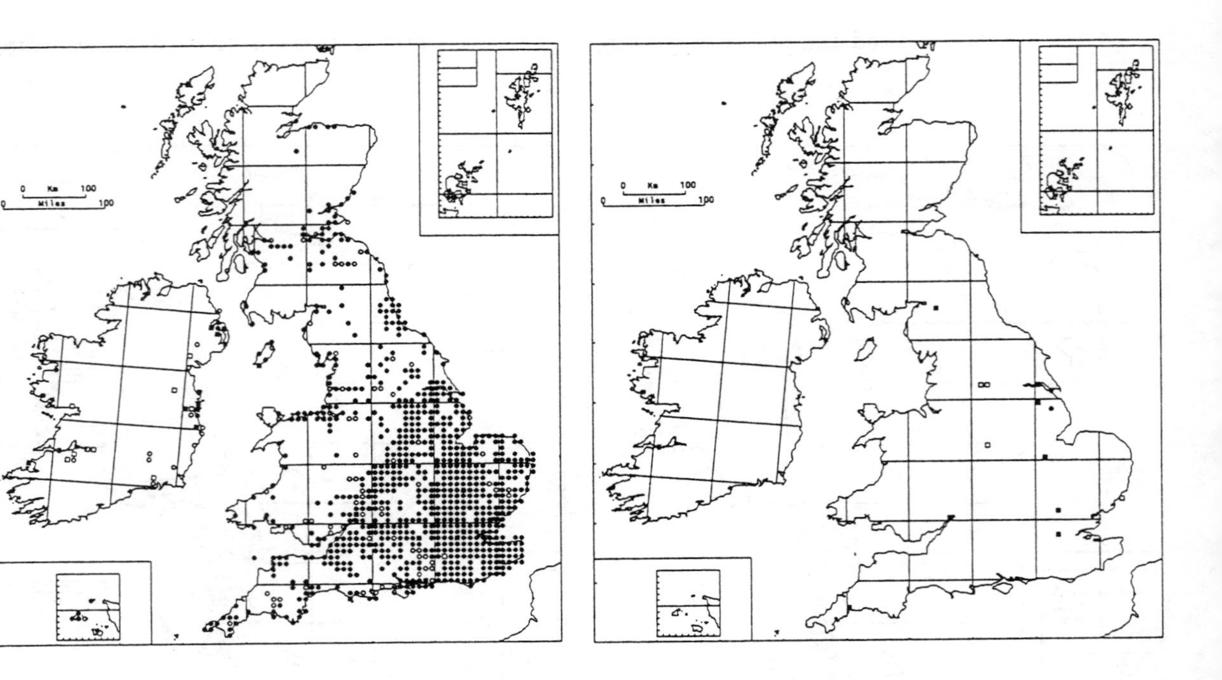

Map 106. Lepidium draba

Map 107. Lepidium chalepense

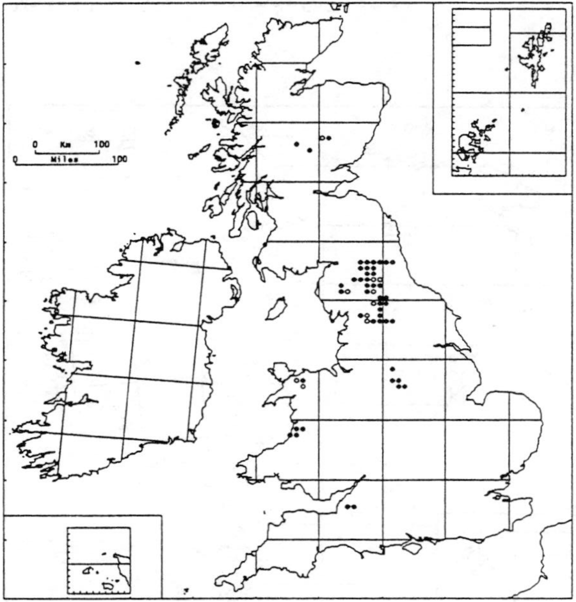

Map 111. Thlaspi caerulescens

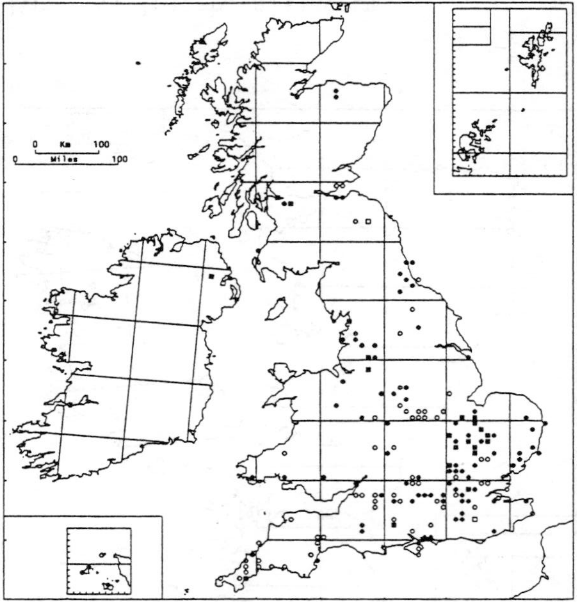

Map 108. Lepidium sativum

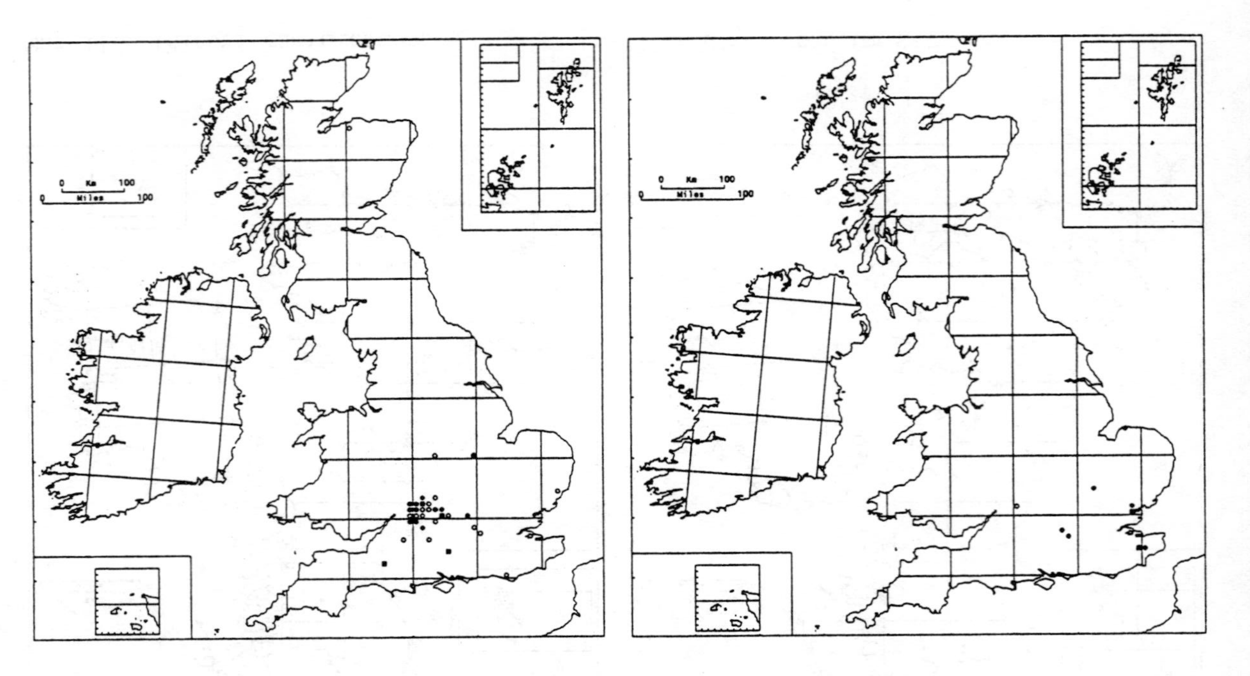

Map 112. Thlaspi perfoliatum

Map 114. Thlaspi alliaceum

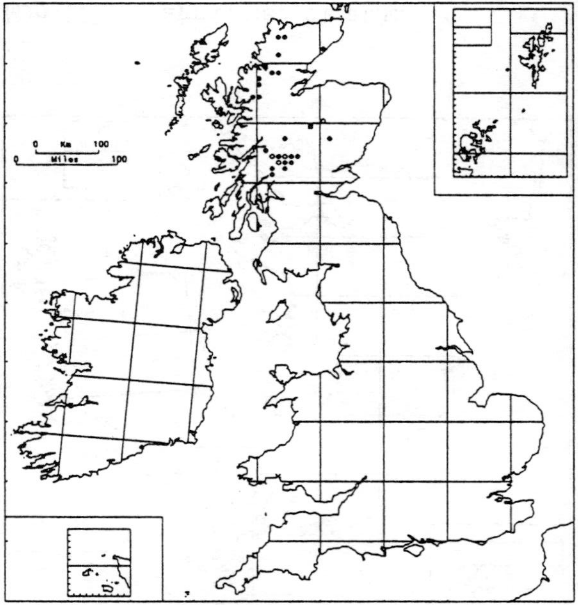

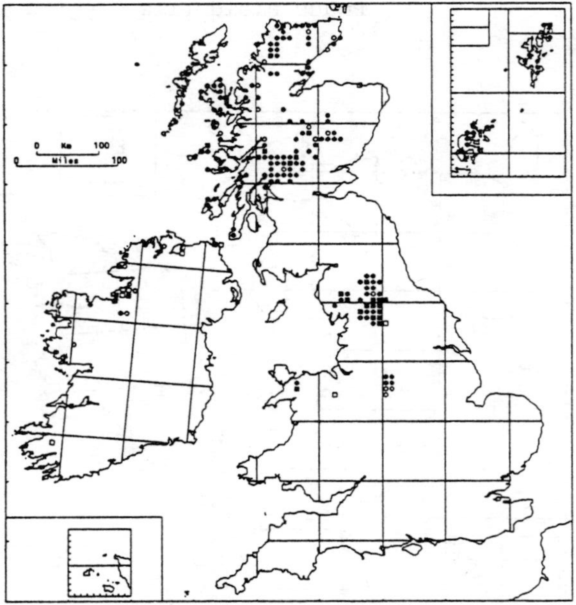

Map 121. Draba incana

Map 120. Draba norvegica

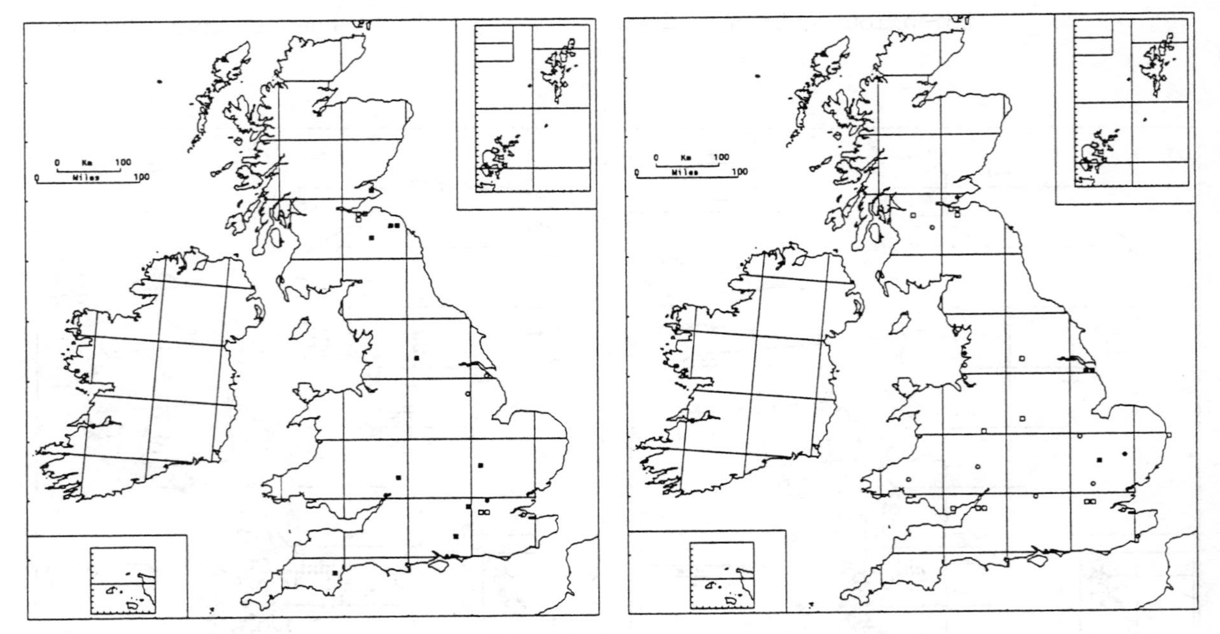

Map 135. Carrichtera annua

Map 137. Myagrum perfoliatum

REFERENCES

AHTI, T. 1961. On the taxonomy of *Erysimum cheiranthoides* L. (Cruciferae). *Archivum Societatis Zoologicae Botanicae Fennicae "Vanamo"*, **16**:22-35.

AKEROYD, J.R. 1986. Med-checklist Notulae 12. *Capsella rubella* Reuter. *Willdenowia*, **15**:417.

AL-SHEHBAZ, I.A. 1985. The genera of Brassiceae (Cruciferae; Brassicaceae) in the southeastern United States. *Journal of the Arnold Arboretum*, **66**:279-351.

ALLEN, D.E. 1981. *Cardamine pratensis* agg., in Reports. Vice-county Recorders' Conference, Rogate Field Centre, West Sussex, 5th-8th October, 1979. *Watsonia*, **13**:247-248.

ALLEN, D.E. 1984. *Flora of the Isle of Man*. Douglas.

ALMQUIST, E. 1921. *Bursa pastoris* Weber. *Report of the Botanical Society and Exchange Club of the British Isles*, **6**:179-207.

BAKER, H.G. 1972. Migration of weeds, in Valentine, D.H., ed., *Taxonomy, phytogeography and evolution*. London.

BALL, P.W. 1961. The taxonomic status of *Neslia paniculata* (L.) Desv. and *N. apiculata* Fisch., Mey. & Avé.-Lall. *Feddes Repertorium*, **64**:11-13.

BALL, P.W. 1963. A review of *Malcolmia maritima* and allied species. *Feddes Repertorium*, **68**:179-186.

BALL, P.W. 1964. A revision of *Cakile* in Europe. *Feddes Repertorium*, **69**:35-40.

BANGERTER, E.B. & WELCH, B. 1952. The London Rocket and its allies of the London area. *London Naturalist*, **31**:13-17.

BARBOUR, M.G. & RODMAN, J.E. 1970. Saga of the West Coast sea-rockets: *Cakile edentula* ssp. *californica* and *C. maritima*. *Rhodora*, **72**:370-386.

BATEMAN, A.J. 1954. Self-incompatibility systems in Angiosperms II. *Iberis amara*. *Heredity*, **8**:305-332.

BEEBY, W.H. 1887. On the flora of Shetland. *Scottish Naturalist (Perth)*, **9**:20-32.

BEST, K.F. 1977. The biology of Canadian Weeds. 22. *Descurainia sophia* (L.) Webb. *Canadian Journal of Plant Science*, **57**:499-507.

BEST, K.F. & McINTYRE, G.I. 1975. The biology of Canadian Weeds. 9. *Thlaspi arvense* L. *Canadian Journal of Plant Science*, **55**:279-292.

BREEN, C., CURTIS, T.G.F. & SCANNELL, M.J.P. 1984. *Cardamine impatiens* L. in Co. Westmeath (H23) - an addition to the Irish flora. *Irish Naturalists' Journal*, **21**:344-345.

BRITTEN, J. 1900. The genus *Mathiola* in Britain. *Journal of Botany*, **38**:168-169.

BRITTON, C.E. 1909. A *Radicula*-hybrid. *Journal of Botany*, **47**:430.

BRUNKER, J.P. 1952. *Sisymbrium irio* L. in the Dublin District. *Irish Naturalists' Journal*, **10**:319-320.

BUNTING, E. S. 1988. Exploited plants. Oilseed rape. *Biologist*, **35**:95-100.

BURTON, R.M. 1983. *Flora of the London area.* London.

CASSIDI, M.D. 1981. Status of the Lundy Cabbage, *Rhynchosinapis wrightii. Annual Report of the Lundy Field Society,* **31**:64-67.

CAVERS, P.B., HEAGY, M.I. & KOKRON, R.F. 1979. The biology of Canadian weeds. 35. *Alliaria petiolata* (M. Bieb.) Cavara and Grande. *Canadian Journal of Plant Science*, **59**:217-229.

CHAUHAN, E. 1979. Pollination by ants in *Coronopus didymus* (L.) Sm. *New Botanist*, **6**:39-40.

CLAPHAM, A.R., TUTIN, T.G. & WARBURG, E.F. 1952. *Flora of the British Isles*, 1st ed. Cambridge.

CLAPHAM, A.R., TUTIN, T.G. & WARBURG, E.F. 1962. *Flora of the British Isles*, 2nd ed. Cambridge.

CLAPHAM, A.R., TUTIN, T.G. & WARBURG, E.F. 1981. *Excursion Flora of the British Isles*, 3rd ed. Cambridge.

CLAPHAM, A.R., TUTIN, T.G. & MOORE, D.M. 1987. *Flora of the British Isles*, 3rd ed. Cambridge.

CLEMENT, E.J. 1979. *Sisymbrium volgense* Bieb. ex E. Fourn. in Britain. *Watsonia*, **12**:311-314.

CLEMENT, E.J. 1982. Adventive News 22. *Sisymbrium* spp. in Britain. *BSBI News*, **30**:10-12.

CLEMENT, E.J. & FOSTER, M.C. 1983. *Alphabetical tally-list of alien and adventive plants*, 3rd ed. London.

CORNER, R.W.M. 1988. *Cardamine impatiens* L.: a native Scottish locality. *Watsonia*, **17**:91-92.

COURTER, J.W. & RHODES, A.M. 1969. Historical notes on Horseradish. *Economic Botany*, **23**:156-164.

DALE, A. & ELKINGTON, T.T. 1974. Variation within *Cardamine pratensis* L. in England. *Watsonia*, **10**:1-17.

DANDY, J.E. 1969. Nomenclatural changes in the *List of British vascular plants. Watsonia*, **7**:157-178.

DAVIE, J.H. & AKEROYD, J.R. 1983. *Pachyphragma macrophyllum* (Hoffm.) Busch (Cruciferae), a Caucasian species naturalized in Co. Avon, England. *Botanical Journal of the Linnean Society,* **87**:77-82.

DRUCE, G.C. 1929. Notes on the second edition of the British Plant List. *Report of the Botanical Society and Exchange Club of the British Isles*, **8**:867.

DUDLEY, T.R. 1964. Synopsis of the genus *Aurinia* in Turkey. *Journal of the Arnold Arboretum,* **45**:390-400.

DUDLEY, T.R. 1965. Studies in *Alyssum*: Near Eastern representatives and their allies, II. Section *Meniocus* and Section *Psilonema. Journal of the Arnold Arboretum,* **46**:181-213.

DUNN, S.T. 1905. *Alien Flora of Britain.* London.

DUVIGNEAUD, J. & LAMBINON, J. 1975. Le groupe de *Lepidium ruderale* en Belgique et dans quelques regions voisines. *Dumortiera*, **2**:27-32.

DVORAK, F. 1967. A note on the genus *Turritis* L. *Österreichische Botanische Zeitschrift*, **114**:84-87.

ELLIS, R.P. & JONES, B.M.G. 1969. The origin of *Cardamine flexuosa* With.: evidence from morphology and geographical distribution (corrected title). *Watsonia*, **7**:92-103.

ESTELLE, M.A. & SOMERVILLE, C.R. 1986. The mutants of *Arabidopsis*. *Trends in Genetics*, **2**:89-93.

FEARN, G.M. 1971. *Biosystematic studies of selected species on the Teesdale flora*. Ph.D. Thesis (unpublished). University of Sheffield.

FEARN, G.M. 1977. A morphological and cytological investigation of *Cochlearia* populations on the Gower Peninsula, Glamorgan. *New Phytologist*, **79**:455-458.

FERROUSSAT, K. 1982. *Cardamine bulbifera* at Old Park Wood. *Herts and Middlesex Trust Nature Conservation Newsletter*, **56**:4.

FILFILAN, S. & ELKINGTON, T.T. 1988. *Erophila* DC. in Rich, T.C.G. & Rich, M.D.B. (1988). *Plant Crib*. London.

FOGG. G.E. 1950. Biological Flora of the British Isles. *Sinapis arvensis* L. *Journal of Ecology*, **38**:415-429.

FORD, M.A. & KAY, Q.O.N. 1985. The genetics of incompatibility in *Sinapis arvensis* L. *Heredity*, **54**:99-102.

GATES, R.R. 1953. Wild cabbages and the effects of cultivation. *Journal of Genetics*, **51**:363-372.

GERDEMANN, J. W. 1968. Vesicular-arbuscular mycorrhiza and plant growth. *Annual Review of Phytopathology*, **6**:397-418.

GILL, J.J.B. 1971. Cytogenetic studies in *Cochlearia* L. The chromosomal homogeneity within both the 2n=12 diploids and the 2n=14 diploids and the cytogenetic relationship between the two chromosome levels. *Annals of Botany, new series*, **35**:947-956.

GILL, J.J.B. 1973. Cytogenetic studies in *Cochlearia* L. (Cruciferae). The origins of *C. officinalis* L. and *C. micacea* Marshall. *Genetica*, **44**:217-234.

GILL, J.J.B. 1975. *Cochlearia* L., in Stace, C.A. (ed.) *Hybridization and the flora of the British Isles*. London.

GODWIN, H. 1975. *History of the British flora. A factual basis for phytogeography*, 2nd ed. Cambridge.

GREEN, P.S. 1955. Pollen grain size in *Nasturtium* and *Cakile*. *Transactions and Proceedings of the Botanical Society of Edinburgh*, 36:289-304.

GRIGSON, G. 1958. *The Englishman's Flora*. London.

HAJAR, A.R.M. 1987. Comparative ecological studies on *Minuartia verna* (L.) Hiern. and *Thlaspi alpestre* L. in the Southern Pennines, with particular reference to heavy metal tolerance. Ph.D. Thesis (unpublished). University of Sheffield.

HANSON, C.G. & MASON, J.L. 1985. Bird seed aliens in Britain. *Watsonia*, **15**:237-252.

HARBERD, D.J. 1975. *Brassica* L., in Stace, C.A. (ed.) *Hybridization and the flora of the British Isles*. London.

HARBERD, D.J. & McARTHUR, E.D. 1972. The chromosome constitution of *Diplotaxis muralis* (L.) DC. *Watsonia*, **9**:131-135.

HARBERD, D.J. & KAY, Q.O.N. 1975. *Raphanus* L., in Stace, C.A. (ed.) *Hybridization and the flora of the British Isles*. London.

HART, H.C. 1887. *Arabis alpina* in Skye. *Journal of Botany*, **25** :247.

HEDGE, I.C. 1976. A systematic and geographical survey of the old world Cruciferae. In, J.G. Vaughan *et al.* (eds). *The Biology and Chemistry of the Cruciferae*. London. 1976.

HEWSON, H.J. 1981. The genus *Lepidium* L. (Brassicaceae) in Australia. *Brunonia*, **4**:217-308.

HITCHCOCK, L.C. 1936. The genus *Lepidium* in the United States. *Madroño*, **3**:265-320.

HITCHCOCK, L.C. 1946.The South American species of *Lepidium*. *Lilloa*, **11**:75-134.

HJELMQVIST, H. 1950. The flax weeds and the origin of cultivated flax. *Botaniska Notiser*, **1950** :257-298.

HOCKING, P.J. 1982. Salt and mineral nutrient levels in fruits of two strand species, *Cakile maritima* and *Arctotheca populifolia*, with special reference to the effect of salt on the germination of *Cakile*. *Annals of Botany, new series*, **50**:335-343.

HODGKIN, E. 1971. The double rockets. *Journal of the Royal Horticultural Society*, **96**:188-189.

HOLLAND, S. C., CADDICK, H.M., DUDLEY-SMITH, D.S. & LUDBROOK, K.E. 1986. *Supplement to the Flora of Gloucestershire*. Bristol.

HOROVITZ, A. & GALIL, J. 1972. Gynodioecism in East Mediterranean *Hirschfeldia incana*, Cruciferae. *Botanical Gazette (Crawfordsville)*, **133**:127-131.

HOWARD, H.W. & LYON, A.G. 1952. Biological Flora of the British Isles. *Nasturtium* R.Br. *Journal of Ecology*, **40**:228-245.

HOWARD, H.W. & MANTON, I. 1946. Autopolyploid and allopolyploid watercress with the description of a new species. *Annals of Botany, new series*, **10**:1-13.

HURRY, J.B. 1930. *The Woad plant and its dye*. Oxford.

INGROUILLE, M.J. & SMIRNOFF, N. 1986. *Thlaspi caerulescens* J. & C. Presl. (*T. alpestre* L.) in Britain. *New Phytologist*, **102**:219-233.

JACKSON, A.B. 1908. in, Sprague, T.A. & Hutchinson, J. 1908. Note on *Barbarea stricta* Andrz. *Journal of Botany*, **46**:106-109.

JACKSON, P.W. 1981. *Rapistrum rugosum* (L.) All. in Ireland. *Bulletin of the Irish biogeographical Society*, **5**:15-18.

JANCHEN, E. 1942. Das System der Cruciferen. *Österreichische Botanische Zeitschrift*, **91**:1-28.

JEHLíK, V. 1981. Chorology and ecology of *Sisymbrium volgense* in Czechoslovakia. *Folia Geobotanica Phytotaxonomica, Praha*, **16**:407-421.

JONES, B.M.G. 1959. Distribution of *Bunias orientalis* in Britain. *Proceedings of the Botanical Society of the British Isles*, **3**:330.

JONES, B.M.G. 1963. *Arabis hirsuta* and *A. brownii*. Ph.D. Thesis (unpublished). University of Leicester.

JONSELL, B. 1968. Studies in the north-west European species of *Rorippa* s.str. *Symbolae botanical upsalienses*, **19**:1-222.

JONSELL, B. 1975a. *Lepidium* L. (Cruciferae) in Tropical Africa. A morphological, taxonomical and phytogeographical study. *Botaniska Notiser*, **128**:20-46.

JONSELL, B. 1975b. Hybridization in yellow-flowered European *Rorippa* species, in Walters, S.M. & King, C.J., eds., *European floristic and taxonomic studies*. BSBI conference report 15. Faringdon.

KAY, Q.O.N. 1976. Preferential pollination of yellow-flowered morphs of *Raphanus raphanistrum* by *Pieris* and *Eristalis* spp. *Nature*, **261**:230-232.

KAY, Q.O.N. & HARRISON, J. 1970. Biological Flora of the British Isles. *Draba aizoides* L. *Journal of Ecology*, **58**:877-888.

KIERNAN, J.A. 1971. *Rhynchosinapis* - the Worcestershire records. *Watsonia* **8**:293.

KINGTON, M. 1983. *Nature made ridiculously simple*. Harmondsworth.

KLEMOW, K.M. & RAYNALL, D.J. 1983. Population biology of an annual plant in a temporally variable habitat. *Journal of Ecology*, **71**:691-703.

KHOSHOO, T.N. 1958. Biosystematics of *Sisymbrium irio* complex I: Modifications in phenotype. *Caryologia*, **11**:109-132.

KHOSHOO, T.N. 1966a. Biosystematics of *Sisymbrium irio* complex XII: Distributional pattern. *Caryologia*, **19**:143-150.

KHOSHOO, T.N. 1966b. Variation and evolution in some species of the genus *Sisymbrium*. *Indian Journal of Genetics and Plant Breeding*, **26A**:247-257.

LAMBERT, D.S. 1971. *Teesdalia nudicaulis* R.Br. in Co. Londonderry. *Irish Naturalists' Journal*, **17**:95-96.

LEADLAY, E.A. & HEYWOOD, V.H. 1990. The biology and systematics of the genus *Coincya* Porta et Rigo ex Rouy (Cruciferae). *Botanical Journal of the Linnean Society*, **102**:313-398.

LEES, E. 1859. On certain localities for Woad. *Phytologist, new series*, **3**:230-232.

LEWIS-JONES, L.J., THORPE, J.P. & WALLIS, G.P. 1982. Genetic divergence in four species of the genus *Raphanus*: implications for the ancestry of the domestic radish *R. sativus*. *Biological Journal of the Linnean Society*, **18**:35-48.

LHOTSKA, M. 1975. Notes on the ecology of germination of *Alliaria petiolata*. *Folia Geobotanica Phytotaxonomica (Praha)*, **10**:179-183.

LOUSLEY, J.E. 1953. The recent influx of aliens into the British flora, in Lousley, J.E. (ed.) *The changing flora of Britain*. Oxford.

LOUSLEY, J.E. 1961. A census list of wool aliens found in Britain 1946-1960. *Proceedings of the Botanical Society of the British Isles*, **4**:221-247.

LÖVKVIST, B. 1956. *The Cardamine pratensis* complex. Outlines of its cytogenetics and taxonomy. *Symbolae Botanicae Upsalienses*, **14**:1-131.

MacGOWRAN, B. 1979. *Rorippa islandica* (Oeder ex Murray) Borbás in turloughs of South East Galway (H15). *Irish Naturalists' Journal*, **19**:326-327.

MANTON, I. 1932. Introduction to the general cytology of the Cruciferae. *Annals of Botany*, **46**:509-556.

MARREN, P.R. 1972. The Lundy Cabbage. *Annual Report of the Lundy Field Society*, **22**:27-31.

MARREN, P.R. 1973. Addenda to the Lundy Cabbage. *Annual Report of the Lundy Field Society*, **23**:51-52.

MARSHALL, E.S. 1892. On *Cochlearia groenlandica* L. *Journal of Botany*, **30**:225-226.

MATTFELD, J. 1939. The species of the genus *Aubrieta* Adanson. *Bulletin of the Alpine Garden Society Great Britain*, **7**:157-181.

McCLINTOCK, D. 1955. Plants notes. *Matthiola sinuata* (L.) R.Br. *Proceedings of the Botanical Society of the British Isles*, **1**:320.

MEDVE, R.J. 1983. The mycorrhizal status of the Cruciferae. *American Midland Naturalist*, **109**:406-408.

MENNEMA, J. 1973. Zeekool (*Crambe maritima* L.) in Nederland. *Natura (Amsterdam)*, **70**:1-4.

MESSENGER, K.G. 1968. A Railway Flora of Rutland. *Proceedings of the Botanical Society of the British Isles*, **7**:325-344.

MEYER, F.K. 1973. Conspectus der "*Thlaspi*"-Arten Europas, Afrikas und Vorderasiens. *Feddes Repertorium*, **84**:449-469.

MIREK, Z. 1981. Genus *Camelina* in Poland. *Fragmenta Floristica et Geobotanica*, **27**:445-507.

MITCHELL, N.D. 1976. The status of *Brassica oleracea* L. subsp. *oleracea* (Wild Cabbage) in the British Isles. *Watsonia*, **11**:97-103.

MITCHELL, N.D. & RICHARDS, A.J. 1979. Biological Flora of the British Isles. *Brassica oleracea* L. ssp. *oleracea* (*B. sylvestris* (L.) Miller). *Journal of Ecology*, **67**:1087-1096.

MULLIGAN, G.A. 1961. The genus *Lepidium* in Canada. *Madroño*, **16**:77-90.

MULLIGAN, G.A. & BAILEY, L.G. 1975. The biology of Canadian Weeds. 8. *Sinapis arvensis* L. *Canadian Journal Plant Science* **55**:171-183.

MULLIGAN, G.A. & CALDER, J.A. 1964. The genus *Subularia* (Cruciferae). *Rhodora*, **66**:127-135.

MULLIGAN, G.A. & FINDLAY, J.N. 1974. The biology of Canadian Weeds. 3. *Cardaria draba*, *C. chalepensis*, and *C. pubescens*. *Canadian Journal Plant Science*, **54**:149-160.

MULLIGAN, G.A. & FRANKTON, C. 1962. Taxonomy of the genus *Cardaria* with particular reference to the species introduced to North America. *Canadian Journal Botany*, **40**:1411-1425.

MULLIN, J.M. 1988. A colony of *Crambe cordifolia* Steph. in West London. *BSBI News*, **49**:30.

MULLIN, J.M. 1989 *Crambe cordifolia* Steven: Additions and corrections. *BSBI News*, **52**:30-31.

NEWMAN, E.I. 1964. Factors affecting the seed production of *Teesdalia nudicaulis* I. Germination date. *Journal of Ecology*, **52**:391-404.

NEWMAN, E.I. 1965. Factors affecting the seed production of *Teesdalia nudicaulis* II. Soil moisture in spring. *Journal of Ecology*, **53**:211-232.

NOVO, F.G. 1976. Ecophysiological aspects of the distribution of *Elymus arenarius* and *Cakile maritima* on the dunes of Tents-Muir Point (Scotland). *Oecologia Plantarum*, **11**:13-24.

PERRING, F.H. (ed.) 1968. *Critical supplement to the Atlas of the British flora*. London.

PERRING, F.H. & WALTERS, S.M. (eds.) 1962. *Atlas of the British flora*. London.

PHILP, E.G. 1982. *Atlas of the Kent flora*. Kent.

PIGOTT, C.D. & WALTERS, S.M. 1954. On the interpretation of the discontinuous distributions shown by certain British species of open habitats. *Journal of Ecology*, **42**:95-116.

POBEDIMOVA, E.G. 1968. Species novae generis *Cochlearia* L. *Novitates Systematicae Plantarum Vascularium* (Leningrad), **5**:130-135.

PRING, M.E. 1961. Biological Flora of the British Isles. *Arabis stricta* Huds. *Journal of Ecology*, **49**:431-437.

PROCTOR, M.C.F. & YEO, P.F. 1973. *The pollination of flowers*. London.

PUGSLEY, H.W. 1936. The *Brassica* of Lundy Island. *Journal of Botany*, **74**:323-326.

RADFORD, A.E., *et al*. 1974. *Vascular plant systematics*. New York.

RANDALL, R.E. 1974. *Rorippa islandica* (Oeder) Borbás *sensu stricto* in the British Isles. *Watsonia*, **10**:80-82.

RATCLIFFE, D. 1959. Biological Flora of the British Isles. *Hornungia petraea* (L.) Rchb. *Journal of Ecology*, **47**:241-247.

RATCLIFFE, D. 1960. Biological Flora of the British Isles. *Draba muralis* L. *Journal of Ecology*, **48**:737-744.

RATCLIFFE, D.A. 1959. The mountain flora of Lakeland. *Proceedings of the Botanical Society of the British Isles*, **4**:1-25.

RHODES, A.M., COURTER, J.W. & SHURTLEFF, M.C. 1965. Identification of Horseradish types. *Transactions of the Illinois State Academy of Science*, **58**:115-122.

RICH, T.C.G. 1987a. Cabbage Patch I. *Brassica rapa* L. and *B. napus* L. *BSBI News*, **45**:6-8.

RICH, T.C.G. 1987b. Cabbage Patch II. What-a-cress?, or, How to do *Nasturtium* R.Br. *BSBI News*, **46**:18-19.

RICH, T.C.G. 1987c. Cabbage Patch III. Parallels between a pair of casual cabbages or Dual Cabbage-ways. *BSBI News* **47**:25-27.

RICH, T.C.G. 1987d. The genus *Barbarea* R.Br. (Cruciferae) in Britain and Ireland. *Watsonia*, **16**:389-396.

RICH, T.C.G. 1987e. Identifying yellow Crucifers II. *Rorippa* Scop. *Wild Flower Magazine*, **410**:19-21.

RICH, T.C.G. 1988a. Cabbage Patch IV. *Cardaria chalepensis* (L.) Handel-Mazzetti in the British Isles. *BSBI News*, **48**:12-14.

RICH, T.C.G. 1988b. A little Cabbage Patch V. Food for thought or the Rape of Mustard and Cress. *BSBI News*, **49**:12-13.

RICH, T.C.G. 1988c. *Hirschfeldia incana* (L.) Lagrèze-Fossat present in Ireland. *Irish Naturalists' Journal*, **22**:531-532.

RICH, T.C.G. 1988d. Identifying yellow Crucifers III. *Brassica* and a couple of relatives. *Wild Flower Magazine*, **411**:29-32.

RICH, T.C.G. 1989. Cabbage Patch VI. Recognizing Radishes. *BSBI News*, **51**:13-15.

RICH, T.C.G., KITCHEN, M.A.R. & KITCHEN, C. 1989. *Thlaspi perfoliatum* L. (Cruciferae) in the British Isles: distribution. *Watsonia*, **17**:401-407.

RICH, T.C.G. & RICH, M.D.B. 1988. *Plant Crib.* London.

RICH, T.C.G. & WURZELL, B. 1988. *Rorippa* x *hungarica* Borbás (*R. amphibia* x *R. austriaca*) (Cruciferae) new to the British Isles. *Watsonia*, **17**:174-176.

RILEY, R. 1955. Genecological studies in *Thlaspi alpestre* L. Ph.D. Thesis (unpublished). University of Sheffield.

ROBERTS, H.A. & BODDRELL, J.E. 1983. Seed survival and periodicity of seedling emergence in eight species of Cruciferae. *Annals of Applied Biology*, **103**:301-309.

RODMAN, J.E. 1974. Systematics and evolution of the genus *Cakile* (Cruciferae). *Contributions from the Gray Herbarium*, **205**:3-146.

RODMAN, J.E. 1986. Introduction, establishment and replacement of sea-rockets (*Cakile*, Cruciferae) in Australia. *Journal of Biogeography*, **13**:159-171.

ROLLINS, R.C. 1980. Another cruciferous weed establishes itself in North America. *Contributions from the Gray Herbarium*, **210**:1-3.

ROSE, F. 1966. Distribution maps of Kent plants. *Proceedings of the Botanical Society of the British Isles*, **6**:279-281.

RYVES, T.B. 1977. Notes on wool-alien species of *Lepidium* in the British Isles. *Watsonia*, **11**:367-372.

SALISBURY, E.J. 1961. *Weeds and aliens*. London.

SALISBURY, E.J. 1965. The reproduction of *Cardamine pratensis* L. and *Cardamine palustris* Peterman particularly in relation to their specialized foliar vivipary, and its deflexion of the constraints of natural selection. *Proceedings of the Royal Society of London*, **163**:321-342.

SAMPSON, D.R. 1957. The genetics of self-incompatibility in the radish. *Journal of Heredity*, **48**:26-29.

SANDWITH, C.I. 1933. The adventive flora of the Port of Bristol. *Report of the Botanical Society and Exchange Club of the British Isles*, **10**:314-363.

SANDWITH, C.I. & SANDWITH, N.Y. 1935. *Vogelia apiculata. Report of the Botanical Society and Exchange Club of the British Isles*, **11**:102.

SAUNDERS, E.R. 1928. *Matthiola. Bibographica Genetica*, **4**:141-170.

SCANNELL, M.J.P. 1973. *Rorippa islandica* (Oeder ex Murray) Borbás in Ireland. *Irish Naturalists' Journal*, **17**:348-349.

SCHULZ, O.E. 1936. Cruciferae. In, Engler, A. & Prantl, K. *Die naturlichen Pflazenfamilien, ed. 2*, **17b**:227-658.

SCOTT, G.A.M. & RANDALL, R.E. 1976. Biological Flora of the British Isles. *Crambe maritima* L. *Journal of Ecology*, **64**:1077-1091.

SCOTT, W. & PALMER, R. 1987. *The flowering plants and ferns of the Shetland Islands*. Lerwick.

SCURFIELD, G. 1962. Biological Flora of the British Isles. *Cardaria draba* (L.) Desv. *Journal of Ecology*, **50**:489-499.

SHIMWELL, D.W. 1968. Notes on the distribution of *Thlaspi alpestre* L. in Derbyshire. *Proceedings of the Botanical Society of the British Isles*, **7**:373-376.

SHIMWELL, D.W. & LAURIE, A.E. 1972. Lead and zinc contamination of vegetation in the Southern Pennines. *Environmental Pollution*, **3**:291-301.

SHOWLER, A.J. 1988. Fruiting in Coralroot, *Cardamine bulbifera* (L.) Crantz. *BSBI News*, **48**:26-28.

SIMMONDS, N.W. (ed.) 1976. *Evolution of crop plants*. London.

SMITH, G. (1968). *Variation in* Erophila verna *(L.) Chevall*. Ph.D. Thesis (unpublished). University of Aberdeen.

SMITH, R.F. 1979. The occurrence and need for conservation of metallophytes on mine wastes in Europe. *Minerals and the Environment*, **1**:131-147.

SNOGERUP, S. 1967a. Studies in the Aegean Flora VIII. *Erysimum* sect. *Cheiranthus*. A. Taxonomy. *Opera Botanica*, **13**:1-70.

SNOGERUP, S. 1967b. Studies in the Aegean Flora IX. *Erysimum* sect. *Cheiranthus*. B. Variation and evolution in the small-population system. *Opera Botanica*, **14**:1-86.

SPRAGUE, T.A. & HUTCHINSON, J. 1908. Note on *Barbarea stricta* Andrz. *Journal of Botany*, **46**:106-109.

STACE, C.A. 1975. *Hybridization and the flora of the British Isles*. London.

STELFOX, A.W. 1970. The forms of *Cardaminopsis petraea* (L.) in Ireland. *Irish Naturalists' Journal*, **16**:308-309.

STIRLING, A. McG. 1984. *Rhynchosinapis monensis* in the Glasgow area. *Glasgow Naturalist*, **20**:374-375.

STOKES, G.W. 1955. Seed development and failure in Horseradish. *Journal of Heredity*, **46**:15-21.

SVENSSON, S. 1983. Chromosome numbers and morphology in the *Capsella bursa - pastoris* complex (Brassicaceae) in Greece. *Willdenowia*, **13**:267-276.

TITZ, W. 1978. Experimentelle systematik und genetik der kahlen sippen in der *Arabis hirsuta*-gruppe (Brassicaceae). *Botanische Jahrbucher*, **100**:110-139.

TSUNODA, S., HINATA, K. & GOMEZ-CAMPO, C. (eds.) 1980. *Brassica crops and wild allies*. Tokyo.

TUTIN, T.G. *et al.* (eds.) 1964. *Flora Europaea*, **1**:260-346. Ed.1. Cambridge.

VAUGHAN, J.G. & GORDON, E.I. 1973. A taxonomic study of *Brassica juncea* using the techniques of electrophoresis, gas-liquid chromatography and serology. *Annals of Botany, new series*, **37**:167-183.

VAUGHAN, J.G., HEMINGWAY, J.S. & SCHOFIELD, H.J. 1963. Contributions to a study of variation in *Brassica juncea* Coss. & Czern. *Botanical Journal of the Linnean Society*, **58**:435-447.

VAUGHAN, J.G., MACLEOD, A.J. & JONES, B.M.G. (eds.) 1976. *The biology and chemistry of the Cruciferae*. London.

VERMA, S.C., MALIK, R. & DHIR, I. 1977. Genetics of the incompatibility system of the crucifer *Eruca sativa* L. *Proceedings of the Royal Society of London, series B, Biological Sciences*, **96**:131-159.

WATTS, L.E. 1976. Levels of self-incompatibility in Wallflower, *Cheiranthus cheiri* L. *Euphytica*, **25**:83-88.

WEBER, W.W. 1949. Seed production in Horseradish. *Journal of Heredity*, **40**:223-227.

WHITE, J.W. 1912. *The flora of Bristol*. Bristol.

WIGGINGTON, M.J. & GRAHAM G.G. 1981. *Guide to the identification of some of the more difficult vascular plant species*. NCC England Field Unit, Occasional Paper No. 1. Banbury.

WILLIS, S.J. 1953. *Cardaria draba* - A globe-trotting weed. *World Crops*, **5**:310-312.

WINGE, Ö. (1940). Taxonomic and evolutionary studies in *Erophila* based on cytogenetic investigations. *Comptes-rendus des Travaux du Carlsberg laboratoriet; serie physiologique. Copenhagen*, **23**:41-74.

WOODHEAD, N. 1951. Biological Flora of the British Isles. *Subularia aquatica* L. *Journal of Ecology*, **39**:465-469.

WRIGHT, F.R.E. 1936. The Lundy Brassica, with some additions. *Journal of Botany*, **74. Supplement 2**:1-8.

GLOSSARY

This glossary is primarily based on Radford *et al.* (1974) and Clapham *et al.* (1962). The terms have been applied according to the definitions below, and many may be more strictly defined than the general usage.

Acuminate: with a long, fine point, sometimes with the margins turning inwards before the tip.

Acute: sharply pointed, angle within apex less than 90 degrees.

Aggregate: a group of closely related taxa which are difficult to distinguish.

Allopolyploid: a polyploid derived by hybridization between two species followed by doubling of the chromosome number.

Angustiseptate: of a fruit with the septum across the narrowest diameter (e.g. *Capsella*).

Annual: completing its life cycle within 12 months of germination and flowering once only.

Anther: the part of the stamen containing the pollen.

Apiculate: rounded but with a short, central point.

Appressed: pressed closely to the axis upwardly with an angle of divergence of 15 degrees or less.

Ascending: held upwards at an angle of 16-45 degrees to the axis.

Auricle: small ear-like projection at the base of a leaf blade where it joins the stem (often clasping the stem).

Awn: a small projection at the apex of the sepal.

Axil: the angle above the junction between the leaf and stem.

Beak: portion of the gynoecium beyond the tip of the valve. As used here, there is a gradation from persistent style to beak to terminal segment, depending on size and structure.

Biennial: completing its life cycle between 12 and 24 months of germination and flowering once only.

Bifid: split deeply in two (used here to mean cut more than 1/4 of length).

Bipinnate: pinnate leaf in which the lobes are themselves pinnate.

Biseriate: seeds arranged side by side in 2 rows in each loculus of the fruit.

Blade: the expanded part of a leaf.

Bracteolate: with bracteoles.

Bracteole: as used here, a bracteole is a leaf-like structure which subtends an individual flower. (Bracteole is preferred to "bract" which is less precisely defined).

Bulbil: a small bulb or tuber arising is the axil of a stem leaf.

Calcicole: more frequently found upon or confined to soils with a high pH.

Capitate: with a knob-like head or tip, head-like.

Casual: an introduced plant, rarely persistent for more than a few years.

Ciliate: margin fringed with hairs.

Clavate: club-shaped.

Claw: the stalk of a petal.

Cline: a gradation of characters, usually over a geographic area.

Compressed: slightly flattened, as used here intermediate between terete and distinctly flattened.

Cordate: heart-shaped (more deeply indented than emarginate) (used here only to describe fruit or leaf base, not leaf shape).

Corymb: a flat-topped, or slightly convex, dense cluster of flowers.

Crenate: with blunt, rounded teeth.

Cuneate: wedge-shaped.

Decumbent: of stems lying on the ground but rising at the tips.

Dehiscent: opening to shed seeds or pollen (opposite = indehiscent).

Digitate: consisting of more than 3 leaflets arising from the same point.

Diploid: having 2 sets of chromosomes.

Discoid: shaped like a disk.

Ebracteolate: without bracteoles.

Ellipsoid: 3-d form of elliptic.

Elliptic: broadest at the middle with margins symmetrically curved. Elliptic is here used in preference to "oval" which has often been misapplied and is considered ambiguous.

Emarginate: shallowly notched.

Ensiform: very narrowly lanceolate.

Entire: margins even, not toothed or lobed.

Erect: held upward at angle of 0-15 degrees to the axis, upright.

Filament: the stalk of the anther, the two together forming the stamen.

Filiform: thread-like.

Flexuous: wavy, zig-zagging from side to side.

Foetid: foul-smelling.

Glabrescent: becoming glabrous with age.

Glabrous: lacking hairs, bald.

Gland: a small secretory vesicle. Glandular hairs are stalked glands.

Glaucous: with a waxy coating turning the plant bluish-grey.

Globose: spherical, round, like a ball.

Gynoecium: the female part of the flower consisting of the ovary, style and stigma.

Hastate: of a leaf or leaflet with the basal lobes directed outwards.

Horned: of sepals with a small horn at apex (best seen in buds).

Inclined: held upward at an angle of 46-80 degrees to the axis.

Indehiscent: not opening to release seeds (opposite of dehiscent).

Lamina: the expanded part of the leaf blade.

Lanceolate: shaped like the head of a lance, broadest below the middle and more than 3 times as long as wide.

Laciniate: deeply and irregularly cut.

Lateral: side (i.e. not terminal).

Latiseptate: of a fruit with the septum across the broadest diameter (e.g. *Lunaria*).

Limb: the expanded part of the petal (above the claw).

Linear: narrow with the two opposite margins ± parallel, and more than 12 times as long as wide.

Loculus: the cavity of the ovary/fruit containing the ovules/seeds (plural = loculi).

Mediseptate: of a septum in a ± terete fruit (neither angustiseptate or latiseptate).

Medifixed: of hairs attached in the middle. These are often appressed and may appear simple to a hand lens.

Node: point on the stem where a leaf or branch arises.

Ob-: inverted, e.g. oblanceolate = shaped like a lance, but broadest *above* the middle and more than 3 times as long as wide.

Oblong: roughly rectangular, less than 12 times as long as wide.

Obpyriform: like an inverted pear.

Obtuse: blunt, angle within apex more than 90 degrees.

Orbicular: round.

Ovary: the part of the gynoecium containing the ovules.

Ovate: egg-shaped, broadest below the middle and less than 3 times as long as wide.

Ovoid: 3-D version of ovate.

Ovule: the egg which develops into a seed once fertilised.

Patent: held at an angle of 80 - 100 degrees to the axis.

Pedicel: the stalk of a flower or fruit.

Pendent: hanging down.

Perennial: of a plant living for more than 24 months, or of a plant flowering in more than one year.

Perfoliate: of a leaf, the base of which completely surrounds the stem.

Persistent: not falling off.

Petiolate: of a leaf with a petiole.

Petiole: the stalk of a leaf.

Pinnate: of a leaf divided into a terminal lobe and 2 or more separate lateral lobes arranged in 2 rows along the midrib (the lobes free to the midrib).

Pinnatifid: like pinnate but only cut 1/3 - 2/3 way to the midrib (the lobes thus joined by lamina).

Pinnatisect: like pinnate but cut 2/3 - 5/6 way to the midrib (the lobes thus joined by a narrow strip of lamina).

Polymorphic: represented by many forms or variants.

Polyploid: having more than 2 sets of chromosomes.

Prostrate: lying flat along the ground.

Quadripinnate: like tripinnate but with the lobes pinnate again.

Raceme: an unbranched, elongating inflorescence with pedicelled flowers.

Receptacle: the top of the pedicel where the sepals, petals, stamens and gynoecium are inserted.

Recurved: bent backwards or downwards at an angle of 110 degrees or more to the upward axis.

Reniform: kidney-shaped.

Rhizomatous: of a plant with rhizomes.

Rhizome: an underground stem, spreading ± horizontally.

Rhombic: diamond-shaped.

Rugose: wrinkled.

Saccate: of sepals with a marked pouch or sac at their base (often associated with nectar secretion). Typically only the two larger outer sepals are saccate whilst the inner are not.

Scabrid: rough to the touch.

Self-compatible: capable of self-fertilization.

Self-incompatible: incapable of self-fertilization.

Sensu lato: in a broad sense (s.l., sens.lat.)

Sensu stricto: in a strict sense (s.s., sens.str.)

Septum: the membranous partition between the loculi (a good example is the large silvery septum of *Lunaria* used for decoration).

Serrate: sharply toothed like a saw.

Sessile: without a stalk or petiole.

Simple: unbranched, undivided (leaves may be simple but have toothed margins).

Sinuate: with a wavy outline or margin, wavy in and out.

Spathulate: paddle- or spoon-shaped.

Spreading: used loosely here to include anything directed at an angle of 45-135 degrees to the upward axis.

Stamen: one of the male reproductive organs in the flower consisting of a pollen-bearing sac (anther) on a stalk (filament).

Stellate: of hairs that are star-shaped with 3 or more rays.

Stigma: the tip of the style which receives the pollen (typically covered with minute papillae).

Stipe: a stalk between the receptacle and the bottom of the valve.

Stipitate: of a fruit with a stipe.

Stolon: a creeping, rooting lateral stem above ground.

Style: the part of the gynoecium connecting the ovary with the stigma (see also beak).

Sub-: nearly (e.g. subacute = nearly acute).

Subulate: awl-shaped.

Taxon: any taxonomic grouping, such as a species, or a genus, etc. (plural = taxa).

Terete: ± round in cross section (i.e. not strongly angled, compressed or flattened).

Terminal: end (as opposed to lateral = side).

Terminal segment: the end segment of a fruit, often large and/or fertile (small terminal segments are beaks).

Tetraploid: having 4 sets of chromosomes.

Torulose: beaded, like a string of beads.

Trifoliate: with 3 leaflets.

Tripinnate: like bipinnate but with the lobes pinnate again.

Triploid: having 3 sets of chromosomes.

Truncate: cut square across.

Undulate: wavy up and down.

Valve: the part of a dehiscent fruit covering the seeds and opening or falling off.

Vein: a strand of strengthening or conductive tissue running through an organ (here used on petals and fruits only).

Vesicle: a hollow, swollen cell on the surface of a fruit.

Wing: a flattened extension, appendage or projection of the fruit valve.

INDEX

The index lists Latin and English names widely used in British Floras. Latin names accepted in this book are given in ***bold***; Latin synonyms are given in *italics*; English names are given in light face.

Species for which there is a map are marked (**m**).